Quark Confinement and the Hadron Spectrum III

Quark Confinement and the Hadron Spectrum III

Newport News, Virginia, USA 7 – 12 June 1998

Editor

Nathan Isgur

Jefferson Laboratory, USA

World Scientific

Singapore • New Jersey • London • Hong Kong

Published by

World Scientific Publishing Co. Pte. Ltd.

P O Box 128, Farrer Road, Singapore 912805

USA office: Suite 1B, 1060 Main Street, River Edge, NJ 07661

UK office: 57 Shelton Street, Covent Garden, London WC2H 9HE

British Library Cataloguing-in-Publication Data
A catalogue record for this book is available from the British Library.

QUARK CONFINEMENT AND THE HADRON SPECTRUM III

ISBN 981-02-4236-0

Printed in Singapore by Uto-Print

ADVISORY COMMITTEE

J. Ball (Salt Lake City)
M. Cruetz (BNL)
H. G. Dosch (Heidelberg)
B. Durand (Madison)
L. Durand (Madison)
L. Fulcher (Bowling Green)
D. Gromes (Heidelberg)
H. Leutwyler (Bern)
W. Lucha (Wien)
A.Manohar (San Diego)
G. Marchesini (Milano)
A. Martin (CERN)
Y. Ne'eman (Tel Aviv)
M. Neubert (CERN)
M. G. Olsson (Madison)
E. Predazzi (Torino)
A. Radyushkin (O.D.U.)
E. Schoeberl (Wien)
M. Testa (Roma)
H. Toki (RCNP)
M. Shifman (Minnesota)
N. Törnqvist (Helsinki)
F. Yndurain (Madrid)
F. Zachariasen (Caltech)

ORGANIZING COMMITTEE

Marshall Baker (Seattle)
Nora Brambilla (Heidelberg)
Warren Buck (Hampton)
Nathan Isgur (Jefferson Lab,Chair)
José Goity (Hampton)
Khin Maung Maung (Hampton)
Giovanni Prosperi (Milan)

Preface

The National Institute for Nuclear Theory Series

The national Institute for Nuclear Theory (INT) was established by the US Department of Energy in March, 1990. The goals of the INT include:

1. Creating a productive research environment where visiting scientists can focus their energies and exchange ideas on key issues facing the field of nuclear physics, including those crucial to the success of existing and future experimental facilities;
2. Encouraging interdisciplinary research at the intersections of nuclear physics with related subfields, including particle physics, astrophysics, atomic physics, and condensed matter;
3. Furthering the development and advancement of physicists with recent Ph.D.s;
4. Contributing to scientific education through graduate student research, INT summer schools, undergraduate summer research programs, and graduate student participation in INT workshops and programs;
5. Strengthening international cooperation in physics research through exchanges and other interactions.

While the INT strives to achieve these goals in a variety of ways, its most important efforts are the three-month programs, workshops, and schools it sponsors. These typically attract 300 visitors to the INT each year. In order to make selected INT workshops and summer schools available to a wider audience, the INT and World Scientific established the series of books to which this volume belongs.

The Thomas Jefferson National Accelerator Facility and the INT inaugurated an annual workshop in 1992. The motivation was to provide an informal forum for discussions between theorists and experimentalists, with the goal of optimizing the physics coming from this new accelerator. These joint workshops have traditionally encouraged the participation of younger physicists, both as speakers and as organizers. In 1998, under the auspices of this joint agreement, a more ambitious project was undertaken, the organization of Confinement III, the Third International Conference on Quark Confinement and the Hadron Spectrum. Confinement III was hosted and sponsored by Jefferson Lab and Hampton University, with the INT participating as a co-sponsor. The organizing committee consisted of Marshall Baker, Nora Brambilla, Warren Buck, Nathan Isgur (chair), Jose Goity, Khin Maung Maung, and Giovanni Prosperi. This eighth volume of the INT series is dedicated to the proceedings of this conference.

Earlier volumes of this series include the proceedings of the 1991 and 1993 Uehling summer schools on Nucleon Resonances and Nucleon Structure and on Phenomenology and Lattice QCD; the 1994 INT workshop on Solar Modeling; the 1997 Jefferson Lab/INT workshop on Nucleon Resonance Physics; the tutorials of the spring 1997 INT program on Tunneling in Complex Systems; the 1998 Caltech/INT workshop on Nuclear Physics with Effective Field Theory; and the proceedings of the 1998 RHIC Winter

Workshop on Quarkonium Production in Relativistic Nuclear Collisions. We will continue to publish those proceedings of INT workshops and schools, which we judge to be of broad interest to the physics community.

Wick C. Haxton and Ernest Henley
Seattle, Washington, December, 1998

PROLOGUE

Confinement III was the first of this series of conferences to be held on this side of the Atlantic Ocean: the first two meetings were organized at Lake Como. The rapid progress of the field made it an exciting time to host the conference. It was also an auspicious venue since Jefferson Lab, which promises to be the focus of experimental work on confinement-dominated physics, had just come into full operation and was beginning to produce its expected flood of exciting new data.

One of the unique characteristics of the Confinement conference series is that it brings together three communities which do not always mix enough: practitioners of lattice QCD, QCD field theorists, and quark modellers. It was apparent from the strong interactions that occurred that this was a healthy mixture, with each group hearing about progress in other areas that had potentially major impacts on their own.

On behalf of myself and the participants at Confinement III, I want to thank our sponsors (Jefferson Lab, Hampton University, and the Institute for Nuclear Theory) for their generous support. We also want to thank the "Unstrung Heroes" of Confinement III (who actually never even came close to going unstrung): head Hero Sue Ewing, the Conference Secretary, along with troopers Marti Hightower, Ruth Bizot, Cynthia Sloane, and Cindy Garwood, as well as the many other behind-the-scenes individuals who contributed to the enormously successful running of this Conference. I also want to express my regret that several presentations made at the conference were not available for inclusion in these proceedings, including the talks by M. Polykarpov ("Monopoles in QCD"), A. Manohar ("Large N_c QCD"), P. Lepage ("Nonrelativistic QCD"), C. Meyer ("Experimental Status of Glueballs"), and A. Petrov ("Hybrid Charmonium Production in B Decays"), and the summaries of the parallel sessions on "Effective Hamiltonians and Quark Models" by D. Gromes and "Confinement Phenomenology and Experiments" by T. Barnes.

I was honored to be able to host Confinement III, and want to express my appreciation to Nora Brambilla and Giovanni Prosperi for entrusting this installment of the Conference to me. That it was not only an honor but an enjoyable task was in large measure because of the tremendous support I received from Marshall Baker (Seattle), Warren Buck (Hampton University), Jose Goity (Hampton University), Nora Brambilla (Heidelberg), Giovanni Prosperi (Milan), and Khin Maung Maung (Hampton University) of the Organizing Committee, the excellent suggestions made by the International Advisory Committee, and the outstanding work of the Parallel Session Convenors. I sincerely thank them all, and eagerly look forward to attending Confinement IV in Vienna in 2000.

Nathan Isgur
Jefferson Lab
December 1998

CONTENTS

Section C

Section D

PLENARY SESSION PRESENTATIONS

CONFINEMENT III: OVERVIEW OF CONFINEMENT*

Howard Georgi
Lyman Laboratory of Physics, Harvard University, Cambridge, MA 02138, USA

ABSTRACT

I discuss some recent developments in the theory of "topcolor" a hypothetical **non-confining** strong interaction that might bind a composite Higgs boson and thus be responsible for the breaking of the electroweak symmetry.

Outside University Hall at Harvard, there is a famous statue that lots of people photograph. It is labeled

John Harvard
Founder
1638

It is called the statue of the three lies because every line on the label is wrong — it is not John Harvard — he wasn't really the founder anyway — and the date is wrong. This talk is similar. It is not on overview of confinement. And the date is wrong, because I will give an overview of sorts at the conference banquet in a couple of days. I am actually Howard Georgi, so there are only two out-and-out lies in the title, but even the name is a half-truth, because in this talk, almost all of the work I will describe was done with **Sekhar Chivukula.**[1]

Just to make some attempt to make this sound like a kick-off talk, I will begin by giving my own answer to the question — Why study confinement? Some people might argue that this is no longer a terribly fundamental issue, because, after all, we already know the QCD Lagrangian, so why do we need to beat a dead horse! I sympathise with this view — I don't think that confinement *per se* is likely to lead to any deep insights. However, in my view, there are a number of good reasons to try to deepen our understanding of QCD in general —

- It's a strongly coupled quantum field theory.
- It's the only one where we can look up the answers.
- Someday, knowing how to deal with strongly coupled quantum field theory might be important for "fundamental" physics.

*Research supported in part by the National Science Foundation under grant number NSF-PHY/98-02709. Harvard preprint HUTP-98/A062.

In particular, dynamical electroweak symmetry breaking and/or SUSY breaking, one or both of which might well be a component of what happens at and beyond the eletroweak scale, are interesting precisely because they are NOT like QCD. In this talk, I will not discuss SUSY, but will focus on **Dynamical Electroweak Symmetry Breaking**

What's the problem? It is very easy to give mass to the W and Z. **TECHNICOLOR** is a beautiful and simple mechanism that does this quite nicely. But it is not so easy to get quark and lepton masses and mixings, and a large t mass, without running afoul of phenomenological. An obvious solution is to build a composite Higgs multiplet[2] out of t quarks.[3] The virtue of building a composite Higgs out of t quarks is that at least you get the Yukawa coupling of the Higgs to the t in a natural way, from graphs like

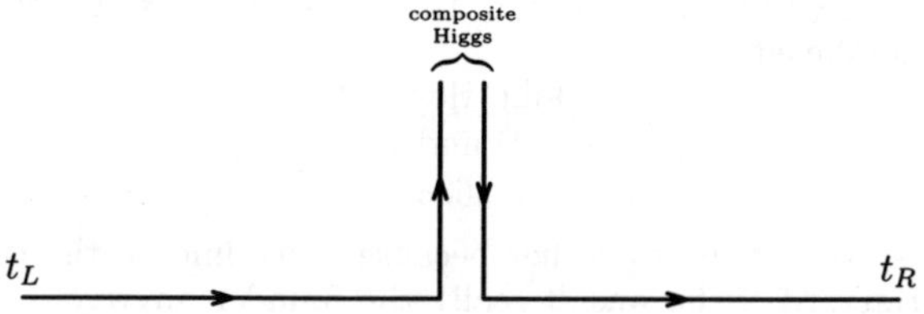

What binds the composite Higgs? The first idea was a Nambu-Jona-Lasinio (NJL) interaction — a Chiral invariant 4-fermion interaction.[3,4] Unfortunately, this is non-renormalizable — it makes sense only for scales below the cut-off. However, if you ignore this problem and analyze things in the large N_c limit with a momentum space cut-off M, the bubble sum gives a Higgs propagator:

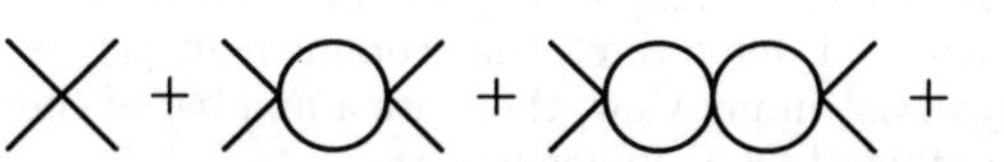

This is CRAZY PHYSICS! Bubbles are proportional to M^2 which means that the mechanism requires important contributions from the cut-off scale, where theory makes no sense.

A BETTER IDEA is TOPCOLOR.[5] at high scales color has two components t and b quarks and maybe others carry "topcolor" (TC) $SU(3)$ gauge interaction like color but stronger. The light quarks (or at least gluons) carry "protocolor" (PC). At some large scale, TC$\times$PC is spontaneously broken down to ordinary color (C) generated by sum of TC and PC, and the gauge couplings satisfy

$$\alpha_{\rm TC} \gg \alpha_{\rm PC} \approx \alpha_{\rm C}$$

In addition to the color octet of massless gluons, there is an octet of massive TC gauge bosons called "colorons".

What happens in this model at low energies? The original analysis was based on the NJL picture. No TC at low energies but a remnant of the strong TC interaction survives as a 4-fermion operator which binds a composite Higgs boson. This picture is better because now the cut-off has a physical meaning — it is just the coloron mass. But NJL approximation is still not controlled.

Another, I think better, way to think about it is to construct an **Effective Field Theory**. I will illustrate this in a toy model of just the topcolored quarks — no light quarks. I will also turn off the PC and $SU(2)_L$ gauge couplings, leaving PC and $SU(2)_L$ as global symmetries. This means that after the spontaneous breaking of topcolor, there is no gauged color interaction — just a global $SU(3)$. The matter content of the toy model is $n_f \geq 2$ flavors of quarks carrying the topcolor interation, and a set of scalars to spontaneously break topcolor:

$$\text{fermions} \begin{cases} Q_L = (3, 1, n_f, 1) \\ Q_R = (3, 1, 1, n_f) \end{cases}$$
$$\text{scalars} \;\; \xi = (\overline{3}, 3, 1, 1)$$

Then the symmetry sturcture of the model is

$$\overbrace{\begin{matrix} \text{TC} & \text{PC} \\ \text{gauge} & \text{global} \\ SU(3) & SU(3) \end{matrix}}^{\text{“color”}} \quad \overbrace{\begin{matrix} \text{F}_L & \text{F}_R \\ \text{global} & \text{global} \\ SU(n_f) & SU(n_f) \end{matrix}}^{\text{“flavor”}}$$

Q comprises t and b and possibly other quarks

$$Q = \begin{pmatrix} t \\ b \\ \vdots \end{pmatrix}$$

If ξ develops a VEV

$$\langle\xi\rangle = \begin{pmatrix} v & 0 & 0 \\ 0 & v & 0 \\ 0 & 0 & v \end{pmatrix}$$

then topcolor breaks spontaneously:

$$\text{TC} \times \text{PC} \to \text{C}\,,$$

and the mass of the coloron (for small g_{TC}) is

$$M = g_{\text{TC}} v\,.$$

Nothing except perhaps gauge anomalies prevents us from replacing ξ by some dynamical mechanism for doing the symmetry breaking.

In this toy model, there are two dimensional parameters, m_ξ^2 (which determines v) and $\Lambda_{\rm TC}$, the scale where TC would get strong if it were not broken. Let's consider how physics depend on m_ξ^2 parameter.

For m_ξ^2 very negative, $m_\xi^2 \ll -\Lambda_{\rm TC}^2$, TC symmetry is badly broken. This is the Higgs phase. In this region, the TC interactions never get strong. The quarks remain massless and they are triplets under C, because the were triplets under TC and TC×PC→C; and the quarks carry unbroken chiral flavor symmetry.

For m_ξ^2 very positive, $\Lambda_{\rm TC}^2 \ll m_\xi^2$, TC symmetry is not broken. This is the confining phase. In this region, the chiral flavor symmetry is spontaneously broken when TC gets strong at the scale $\Lambda_{\rm TC}$. There are massless Goldstone bosons, and the usual sort of massive mesons. But there are also heavy "quarks" — Q-ξ bound states, which transform like triplets under C, this time because ξ is a triplet under C.

What happens in between the two limits? Clearly there is a transition between the two phases. What is the nature of the transition? The large N NJL arguments suggest that it is a second order transition where all that happens is that the composite Higgs mass term in the effective Lagrangian changes sign. If so, then close to the transition, the light bound states really look like a Higgs multiplet because their masses are much smaller than the energy scale at which compositeness effect appear. But does this really work?

There is another possible hint. The Higgs phase with $\xi = (\overline{3}, 3, 1, 1)$ satisfies complementarity[6] — there is no boundary between confining and Higgs phases. There is a smooth mapping between states in Higgs and confining pictures

Higgs	confining
Q	$Q\,\xi$
colorons	$\xi\,\overline{\xi}$
$Q\,\overline{Q}$	$Q\,\overline{Q}$

This means that "confinement" is a fuzzy concept here. Now there is a weak argument for second orderness from considering how the picture depends on N and n_f. For large N, # of flavors, n_f, should not matter. For $m = 1$, there should be no transitions at all! There is no deconfining transition because of complementarity. There is no chiral symmetry breaking transition because the only chiral symmetry is the $U(1)$ chiral symmetry is broken by instanton effects. Since everything is smooth for $n_f = 1$ and n_f is not very important for large N, the transition should be smooth for any small n_f. This is not an ironclad argument — but it is suggestive — we will assume it works and go on to discuss the tilting of the vacuum.

Look at a topcolor theory with only the t and the b, for simplicity. The composite Higgs now is now a 2×2 matrix that couples both the the t and to the b:

$$\Phi = \begin{pmatrix} t \\ b \end{pmatrix} \begin{pmatrix} \bar{t} & \bar{b} \end{pmatrix} = \begin{pmatrix} \phi_t & \phi_b \end{pmatrix}$$

To get large t mass but avoid large b mass, Hill[5] wants $m_{\phi_t}^2 < 0$ and $m_{\phi_b}^2 > 0$. But topcolor gives $m_{\phi_t}^2 = m_{\phi_b}^2$. We need additional physics to "tilt" vacuum. One possibility is a strong $U(1)$s. In the remainder of this talk, I will show how the NJL approximation give misleading results for such tilting, and how the effective field theory picture resolves these difficulties.

Typically,[7] such a tilting interaction is also a spontaneously broken gauge interaction, and in the NJL approximation is again described by a four-fermion operator. The simplest possibility is to have the spinless field or condensate that produces the topcolor breaking transform nontrivially under the strong $U(1)$ as well. The details simplify if we imagine that this tilting $U(1)$ has a much larger coupling than any other $U(1)$ in the theory. Then the low-energy four-fermion operator due to $U(1)$ boson exchange looks like

$$\frac{1}{2}\,\frac{g_1^2\,j_1^\mu\,j_{1\mu}}{m^2}=\frac{1}{2}\,\frac{j_1^\mu\,j_{1\mu}}{Q^2\,v^2}\,, \tag{1}$$

where Q is the $U(1)$ charge of the field that breaks the symmetry, j_1^μ is the $U(1)$ current, and m is the $U(1)$ gauge boson mass, satisfying

$$m^2 \equiv g_1^2\,Q^2\,v^2\,. \tag{2}$$

There are two peculiar things about (1):

1. It does not depend on g_1.
2. It does depend on Q, and the interaction can be made very large by taking Q small (while leaving the couplings to the t- and b-quarks fixed).

Both of these features are physically unreasonable. Do we really believe that the $U(1)$ interaction tilts the vacuum even in the limit when its coupling g_1 is very tiny? Should we believe that for moderate g_1 we can still get very large tilting just by taking Q and thus m very small? Fortunately, these puzzling features do not persist in an effective field theory analysis that goes beyond the NJL approximation.

We consider the effect of a spontaneously broken $U(1)$ on the effective field theory describing the composite Higgs in a topcolor model near the critical coupling. We will work in a large N expansion for the topcolor interactions, but will not assume that the NJL approximation is valid. We will assume instead that the low energy theory is consistently defined in a mass independent renormalization scheme such as dimensional regularization with modified minimal subtraction ($DR\bar{M}S$). We assume that the topcolor theory can be matched at the scale M (the mass of the colorons) onto a theory of massless colored quarks coupled to light scalars (Φ), and study the effect of the massive $U(1)$ gauge boson. While the matching at the scale M is nonperturbative, below the scale M, without the strongly coupled topcolor gauge bosons, the theory can be analyzed perturbatively (until we get down to the QCD scale where ordinary color gets strong). After reviewing some simple consequences of the $1/N$ expansion and effective field theory, we will discuss some general properties of the matching that we can establish simply by requiring that the physics be continuous as we vary the mass of the $U(1)$ gauge boson.

It is convenient to begin with a calculation that is trivial. Consider the matching of a current,

$$j^\mu=\bar{\psi}_L\,\gamma^\mu\,Q_L\,\psi_L+\bar{\psi}_R\,\gamma^\mu\,Q_R\,\psi_R \tag{3}$$

in the high energy theory onto the low energy theory. This is the conserved Noether current associated with a $U(1)$ symmetry in the high energy theory. It is therefore mapped onto the corresponding Noether current in the low energy theory, which has the form:

$$\tilde{j}^\mu = \bar{\psi}_L \gamma^\mu Q_L \psi_L + \bar{\psi}_R \gamma^\mu Q_R \psi_R + i \,\mathrm{Tr}\left[\Phi^\dagger \overleftrightarrow{\partial}^\mu (Q_L \Phi - \Phi Q_R)\right]. \tag{4}$$

Note that the current $\tilde{j}^\mu$ in the low energy theory contains contributions from both the fermions and the composite Higgs, as it must, because both transform under the $U(1)$ symmetry associated with the current.

Let us now consider the only slightly less trivial calculation of integrating out a $U(1)$ gauge boson, which couples to the L and R topcolored fermions with charges Q_L and Q_R, respectively. If the $U(1)$ gauge boson's mass is greater than M, we integrate out the gauge boson before matching onto the low energy theory to obtain the 4-fermion operator

$$\frac{g_1^2}{2\,m^2}\, j^\mu\, j_\mu \tag{5}$$

where g_1 is the coupling, m is the $U(1)$ gauge boson mass and j^m is the current in (3). In leading order in N, the matching is trivial. Because the currents are invariant under the topcolor gauge transformations, the leading contributions factor and the result is*

$$\frac{g_1^2}{2\,m^2}\, \tilde{j}^\mu\, \tilde{j}_\mu \tag{6}$$

This simple result contains an important lesson. In general, the results of matching onto the low energy theory will involve both the fermions and the scalars. We will come back to this momentarily.

If instead, the $U(1)$ gauge boson's mass is less than M, the gauge boson survives into the low energy theory, where it now couples to the low energy current $\tilde{j}^\mu$. Now when the gauge boson is integrated out, we again obtain (6). In this case, continuity of the physics across the boundary $m = M$ is automatically satisfied and does not give us any more information.

Now consider the matching of a mass operator

$$\Sigma \equiv \bar{\psi}_R\, \psi_L\,. \tag{7}$$

In this case, we cannot evaluate the matching exactly, but we can use symmetry arguments, N and loop counting, and dimensional analysis to write the dominant contribution in terms of a small number of parameters of order one.† The matching

*Note that here, renormalization group running plays no role because both operators are renormalized at the the same scale, M.

†Another difference between this case and the case of the Noether current is that the mass operator has an anomalous dimension. Such anomalous dimensions are very important for the structure of the low energy theory at scales much smaller than M, because they give rise to large logarithms from renormalization that allow us to calculate some quantities reliably in spite of the nonperturbative nature of the matching.[4] Here, however, we will be discussing matchings at or near the scale M, so there are no large logs.

produces an operator in the low energy theory of the form

$$\tilde{\Sigma} = A\,\bar{\psi}_R\,\psi_L + B\,\frac{\sqrt{N}}{4\pi}\,M^2\,\Phi + C\,\frac{4\pi}{\sqrt{N}}\,\Phi\,\Phi^\dagger\,\Phi + \cdots\,, \tag{8}$$

where the $\cdots$ are higher dimension terms suppressed by powers of $1/M$ in the low energy theory. The parameters A, B, and C are of order one (at the scale M), but their precise value depend on the details of the strong topcolor physics. The factors of $\frac{\sqrt{N}}{4\pi}$ come from N and loop counting. One can think about them as follows. The coupling of Φ to the fermions must be of order $4\pi/\sqrt{N}$, in order that the Φ kinetic energy term which comes from planar diagrams like Fig. 1 have the standard normalization.

Fig. 1. Typical Feynman graph contributing to the Φ kinetic energy term. Such graphs, with arbitrary planar coloron dressing, are proportional to $\frac{N}{16\pi^2}$, and thus the Φ coupling should be of order $\frac{4\pi}{\sqrt{N}}$.

With the result, (8), we can immediately write down the result for matching of a 4-fermion operator of the form $\mathrm{Tr}(\Sigma\,\Sigma^\dagger)$. In leading order in the $1/N$ expansion, this just goes into $\mathrm{Tr}(\tilde{\Sigma}\,\tilde{\Sigma}^\dagger)$. This follows immediately from factorization for large N. This result and (8) might seem slightly strange to a reader steeped in the NJL approximation. In the NJL approximation, (8) smacks of double counting, because the mass operator Σ plays much the same role as the scalar field Φ. In particular, one might worry that the $\mathrm{Tr}(\Sigma\,\Sigma^\dagger)$ term in $\mathrm{Tr}(\tilde{\Sigma}\,\tilde{\Sigma}^\dagger)$ would produce additional contributions to the Φ dependence from loop graphs like those shown in Fig. 2. But in the low energy theory, these graphs vanish identically. They are relevant in the NJL approximation because of the use of a momentum space cut-off of order M. In some way, the use of the momentum space cut-off is supposed to mock up the nonperturbative physics of the strong coloron exchange near the critical coupling. But in our language, these contributions (for zero momentum transfer through the loops) are simply eliminated by $DR\bar{M}S$

Dimensional Analysis

As in any composite Higgs theory we can systematically estimate the size of any term by dimensional analysis.[8,9] The intrinsic scale of the compositeness interactions is the

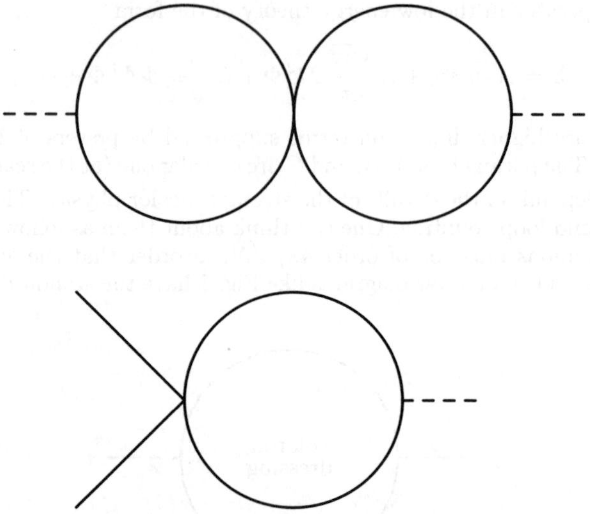

Fig. 2. These graphs vanish in the low energy theory at zero momentum transfer.

topcolor scale, M. Hence we expect derivatives to be suppressed by a factor of M. If the amplitude for the creation of a composite Higgs field is proportional to f (the analog of f_π in chiral Lagrangian dimensional analysis in QCD, each field Φ comes suppressed by a factor of f. In a strongly interacting theory, we expect there are no additional relevant parameters[8] in the low-energy theory. Requiring that the kinetic energy terms of the fields Φ and ψ have the canonical normalization then determines that the overall size of each term in the Lagrangian is of order $M^2 f^2$.

A priori, the dimensionless ratio M/f is undetermined. The effective Lagrangian contains an infinite series of terms of arbitrary dimension. As in the QCD chiral Lagrangian loop diagrams involving lower dimension operators will require counterterms of higher dimension. Since the renormalization point associated with the logarithms in the results of the loops is undetermined to $\mathcal{O}(1)$, the *smallest* size of the coefficients of these higher dimension operators is of order the typical size of the coefficient of the chiral log for which the operator is a counterterm. This consistency condition implies

$$\frac{M}{f} \sim \frac{4\pi}{\sqrt{N}}, \tag{9}$$

and the results given in the previous section immediately follow. In general, the N counting implicit in (9) will give the dependence of the *leading* contribution in large-N and should only be taken as a guide. As we will see below, there are cases in which the leading contribution vanishes and we will need to adjust the N-dependence to correspond to the diagrams which would be evaluated in matching the full and effective theories.

Using these rules to estimate the generic size of the composite Higgs mass term in the effective theory, we find $m_\Phi^2 \propto M^2$. This is the hierarchy problem. In the absence of some other symmetry not accounted for in these rules, some adjustment is required to obtain $m_\Phi^2 \ll M^2$.

$U(1)$ gauge boson exchange — beyond leading order in N

Now finally we are ready to discuss the effect of $U(1)$ gauge boson exchange beyond the leading order in N. This is important, because it is just such a nonleading effect that produces the "tilting" of the vacuum alignment that is crucial to the phenomenology of topcolor models.[5,7,10] We begin by discussing the situation for $m > M$ where we first integrate out the $U(1)$ gauge boson to obtain the 4-fermion operator (5). In particular, the interesting effect involves both the left-handed and right-handed parts of the currents:

$$\frac{g_1^2}{m^2}\left[\bar{\psi_L}\,\gamma^\mu\,Q_L\,\psi_L\right]\left[\bar{\psi_R}\,\gamma_\mu\,Q_R\,\psi_R\right] \tag{10}$$

This gives rise to symmetry-breaking matching contributions to the Φ mass and Yukawa coupling from arbitrary planar coloron dressings of the diagrams in Figs. 3 and 4, where the dotted line connects the left-handed and right-handed gauge invariant parts of the 4-fermion operator in (10) (note that the 4-fermion contribution is already included in (6)). These give following contributions to the Φ mass and Yukawa coupling

$$-D\,\frac{g_1^2\,M^4}{16\pi^2 m^2}\,\mathrm{Tr}\left[Q_L\,\Phi\,Q_R\,\Phi^\dagger\right]+E\,\frac{g_1^2\,M^2}{16\pi^2 m^2}\,\frac{4\pi}{\sqrt{N}}\,\bar{\psi}_L\,Q_L\,\Phi\,Q_R\,\psi_R+\text{h.c.}+\cdots \tag{11}$$

where D and E are of order 1.[‡] Naive application of dimensional analysis would yield an additional factor of N. However, these contributions (see Fig. 3 and 4) are formally nonleading in N because the fermion lines connect the two gauge invariant parts of the operator (10). This can also be seen directly from the diagrams: the factors of $1/\sqrt{N}$ in the Φ coupling cancel against the N from the single fermion loop. Note that "planar" in this case includes graphs in which the coloron lines go through the dotted line in Fig. 3 and 4. Thus this contribution does not factorize, and we cannot relate it to (8). Note also that the $U(1)$ gauge coupling goes like $1/\sqrt{N}$ for large N[10] so the contribution of (11) to the Φ mass goes to zero as $N \to \infty$.

Note that the diagrams in Figs. 3 and 4 are really the same graphs as those in Fig. 2, except that they contain coloron dressing. It is the coloron effects that produce the matching contribution, and the scale M that is required to give a nonzero result.

Now what happens if $m < M$? Now there are a number of contributions that arise from the diagrams in Figs. 5 and 6, both with and without coloron dressing.

‡In the NJL approximation, D is positive.

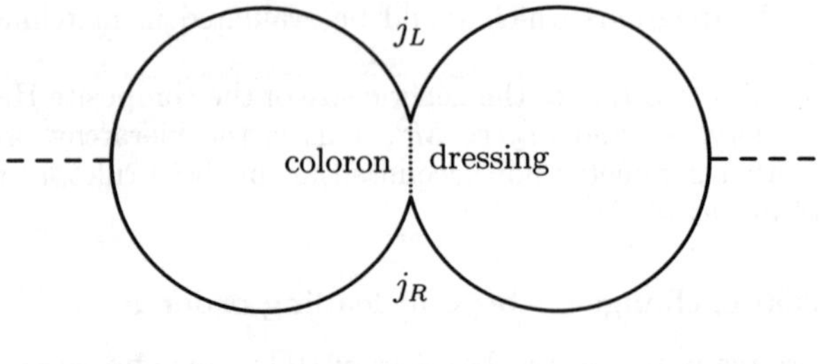

Fig. 3. Symmetry breaking matching contribution to the Φ mass. The dotted line connects two gauge invariant parts of the 4-fermion operator. The matching contribution incorporates the effect of the coloron dressing which is not present in the low energy theory.

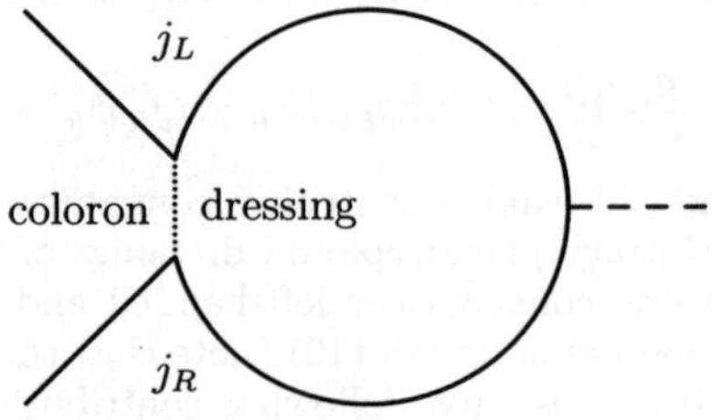

Fig. 4. Symmetry breaking matching contribution to the Φ Yukawa coupling. The dotted line connects two gauge invariant parts of the 4-fermion operator. The matching contribution incorporates the effect of the coloron dressing which is not present in the low energy theory.

There are effects from integrating out the colorons at the scale M; these contributions to the Φ mass and Yukawa coupling have the form

$$-D' \frac{g_1^2}{16\pi^2} M^2 \operatorname{Tr}\left[Q_L \Phi Q_R \Phi^\dagger\right] + E' \frac{g_1^2}{16\pi^2} \frac{4\pi}{\sqrt{N}} \bar{\psi}_L Q_L \Phi Q_R \psi_R + \text{h.c.} + \cdots . \quad (12)$$

There also are effects from from integrating out the massive $U(1)$ gauge boson at the scale m in the low energy theory; these contributions to the Φ mass and Yukawa coupling have the form

$$-D'' \frac{g_1^2}{16\pi^2} m^2 \operatorname{Tr}\left[Q_L \Phi Q_R \Phi^\dagger\right] + E'' \frac{g_1^2}{16\pi^2} \frac{4\pi}{\sqrt{N}} \bar{\psi}_L Q_L \Phi Q_R \psi_R + \text{h.c.} + \cdots . \quad (13)$$

where the parameters D', E', D'' and E'' are of order 1. In addition, there are calculable effects from the renormalization group running in the low energy theory that causes these coupling to evolve with the energy scale. The interesting difference between (11) in the high energy theory and (12) and (13) in the low energy theory is the $1/m^2$ dependence in (11). There is no way that either the coloron effect at the scale M, or the running and matching below M can produce the $1/m^2$ dependence. At M, the typical momentum in the matching calculation is of order M, so the $U(1)$ gauge boson propagator does not produce a $1/m^2$. Below M, we can never get an m^2 in the denominator of the mass term or Yukawa coupling (or the Φ^4 coupling, which we have not written), because it would have to be compensated by factors of M^2 in the numerator, and there is no way that such factors can arise in the low energy theory. They can originate in the matching at the scale M from the nonperturbative physics of the coloron exchange, but new factors of M cannot occur in the low energy theory.

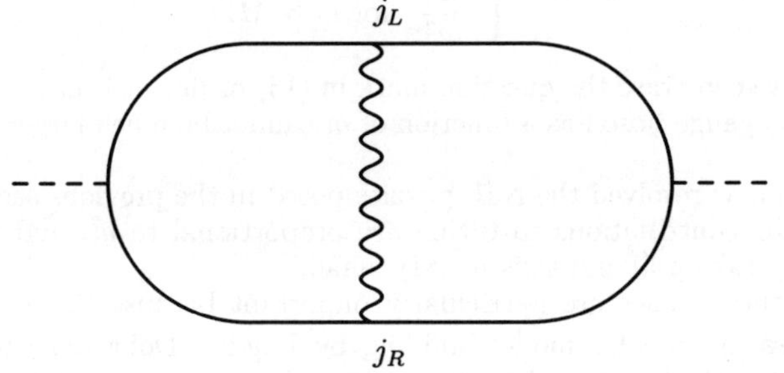

Fig. 5. Symmetry breaking contribution to the Φ mass in the low energy theory.

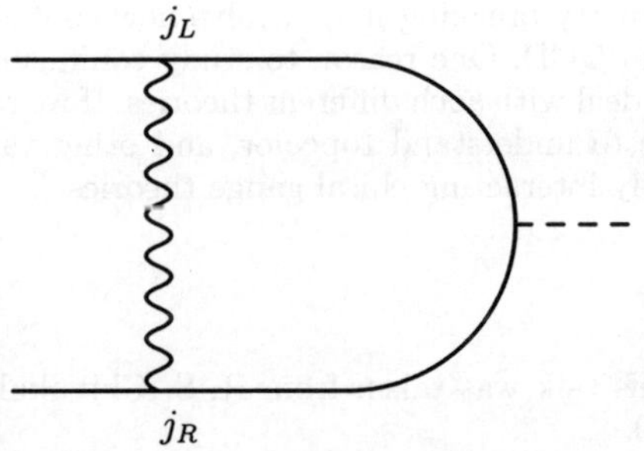

Fig. 6. Symmetry breaking contribution to the Φ Yukawa coupling in the low energy theory.

Now we can impose the constraint that the physics should be continuous at the boundary $m = M$. Clearly, this implies

$$D = D' + D'', \qquad E = E' + E''. \tag{14}$$

This determines the Yukawa coupling below the scale m because it is only the sum $E' + E''$ that contributes.

In fact, though we cannot prove it using these methods, we think it likely that the most important symmetry breaking contribution to the Φ mass comes from integrating out the $U(1)$ gauge boson, in which case we would conclude

$$D'' \overset{?}{\gg} D'. \qquad (15)$$

This would have the interesting consequence that the $U(1)$ interaction is most efficient at tilting the vacuum when $m = M$. Then the contribution to the symmetry breaking mass of the Φ would be

$$\Delta m_\Phi^2 \propto \begin{cases} m^2 & \text{for } m < M, \text{ and} \\ \dfrac{M^4}{m^2} & \text{for } m > M. \end{cases} \qquad (16)$$

But whether we can erase the question mark in (15) or not, it is clear that the tilting effect of a $U(1)$ gauge boson as a function of m cannot be much larger that the effect at $m = M$.

We have now resolved the NJL puzzles posed in the previous section. When g_1^2 is small, all the contributions to tilting are proportional to g_1^2, and we cannot get large tilting by taking Q and thus m very small.

I think these issues are particularly important because there has been some recent progress in topcolor model building by Bogdan Dobrescu and Chris Hill.[?] Their idea is that the basic condensate in a topcolor model is actually does not break the electroweak symmetry and occurs at a higher scale than the t quark mass. I do not have time to discuss their model, but I believe that you will soon see a flurry of model building based on these ideas.

Electroweak symmetry breaking may involve new confining physics very different from what we see in QCD. One reason to study confinement in QCD is to hone our tools to allow us to deal with such different theories. If we really understand QCD, we should also be able to understand topcolor, and other variants such as walking technicolor, and strongly interacting chiral gauge theories.

References

1. Much of this talk was taken from R. S. Chivukula and H. Georgi, hep-ph/9806289.
2. D. B. Kaplan and H. Georgi, Phys. Lett. **136B** (1984) 183.
3. Y. Nambu, "BCS Mechansim, Quasi-Supersymmetry, and Fermion Mass Matrix," Talk presented at the Kasimirz Conference, EFI 88-39 (July 1988); "Quasi-Supersymmetry, Bootstrap Symmetry Breaking, and Fermion Masses," EFI 88-62 (August 1988); a version of this work appears in "*1988 International Workshop on New Trends in Strong Coupling Gauge Theories*," Nagoya, Japan, ed. Bando, Muta and Yamawaki; report EFI-89-08 (1989); V. A. Miransky, M. Tanabashi and K. Yamawaki, Mod. Phys. Lett. **A4** (1989) 1043; Phys. Lett. **B 221** (1989) 177; W. J. Marciano, Phys. Rev. Lett. **62** (1989) 2793.

4. W.A. Bardeen, C.T. Hill and M. Lindner, Phys. Rev. **D 41** (1990) 1647.
5. C. T. Hill, Phys. Lett. **B 266** (1991) 419; for a recent review see G. Cvetic, hep-ph/9702381.
6. E. Fradkin and S. H. Shenker, Phys. Rev. **D19** (1979) 3682.
7. C. T. Hill, Phys. Lett. **B345** (1995) 483, hep-ph/9411426.
8. H. Georgi, Phys. Lett. **B298** (1993) 187, hep-ph/9207278
9. R. S. Chivukula and E. H. Simmons, Phys. Lett. **B388** (1996) 788, hep-ph/9608320.
10. R. S. Chivukula and H. Georgi, Large-N and Vacuum Alignment in Topcolor Models, hep-ph/9805478.
11. B. A. Dobrescu and C. T. Hill, "Electroweak Symmetry Breaking vie a Top Condensation Seesaw," FERMILAB-PUB-97-409-T (Dec 1997), hep-ph/9712319, to appear in *Phys. Rev. Lett.*

THE MECHANISM OF QUARK CONFINEMENT

Gunnar S. Bali
Institut für Physik, Humboldt Universität, Invalidenstraße 110, 10115 Berlin, Germany

ABSTRACT

I summarise recent lattice results on the QCD confinement mechanism in the maximally Abelian projection.

1. Introduction

The phenomenology of strong interactions contains two fundamental ingredients: asymptotic freedom and the confinement of colour charges. The former requirement led to the invention of QCD. A mathematically rigorous proof that QCD as the *microscopic* theory of strong interactions indeed gives rise to the *macroscopic* property of linear quark confinement as indicated by Reggé trajectories and quarkonia spectra is, after a quarter of a century, still lacking. Meanwhile, lattice gauge theory simulations have provided convincing numerical evidence for this conjecture.

The difficulty in deriving infra red properties of QCD illustrates that something qualitatively new is happening: unlike in previously existing elementary physical theories, it is not possible to reduce everything down to two-body interactions but collective excitations of quark and gluon states have to be accounted for. For the first time, excitations of the vacuum that are considered to be fundamental do not occur as initial or final states anymore. Therefore, *understanding* confinement, in my opinion, is one of the most exciting challenges of modern physics. New physical and mathematical techniques that can successfully be applied to non-perturbative QCD might be required for dealing with other strongly interacting theories or theories with a non-trivial vacuum structure, in general. Vice versa techniques developed in a different context might help to prove QCD confinement and in solving QCD. A recent example is the proof of confinement in SUSY Yang-Mills theories.[1] Of course QCD as we know it does not obey super-symmetry but nonetheless such activities point into a promising direction. Just as QCD can serve as a guinea pig and development centre for non-perturbative techniques, lattice simulations help probing the validity range of effective low-energy models or in verifying certain conjectures.

QCD in itself is sufficiently complicated to keep many physicists busy. Solving QCD is still important and even more so since (almost) everyone believes in it. It is widely accepted that perturbative QCD (pQCD) successfully describes high energy scattering processes. However, without understanding non-perturbative aspects of QCD, even in this case, it is not clear why pQCD works at all. Only hadrons, leptons and photons but no quarks or gluons are observed in experiment. In order to fill the gap, the formation of colour-neutral hadronic jets from quarks and gluons has to be modelled. Furthermore, it is commonly conjectured that, once the fundamental scattering on the quark and gluon level has taken place, further interactions can be neglected. It is demanding to derive the ingredients of fragmentation models directly from truly non-perturbative QCD, as well as to verify the factorisation hypothesis.

Many phenomenologically important questions are posed in low-energy QCD that eagerly await an answer: is the same set of fundamental parameters (QCD coupling and quark masses) that describes for instance the hadron spectrum consistent with high energy QCD or is there place for new physics? Are all hadronic states correctly classified by the naïve quark model or do glueballs, hybrid states and molecules play a rôle? At what temperatures/densities does the transition to a quark-gluon plasma occur? What are the experimental signatures of quark-gluon matter? Can we solve nuclear physics on the quark and gluon level? Clearly, complex systems like iron nuclei are unlikely ever to be solved from *first principles* alone but *modelling* and certain *approximations* will always be required. Therefore, it is desirable to test model assumptions, to gain control over approximations and, eventually, to derive low-energy effective Lagrangians from QCD. Lattice simulations are a very promising tool for this purpose.

In the past decades, many explanations of the confinement mechanism have been proposed, most of which share the feature that topological excitations of the vacuum play a major rôle. A list of these theories includes the dual superconductor picture of confinement,[2,3] the centre vortex model,[4] the instanton liquid model,[5] and the anti-ferromagnetic vacuum.[6] All these interpretations have been explored in lattice studies. The situation with respect to an anti-ferromagnetic vacuum is still somewhat inconclusive.[7] Instantons seem to be more vital for chiral symmetry related properties than for confinement.[8] Depending on the picture, the excitations giving rise to confinement are thought to be magnetic monopoles, instantons, dyons, centre vortices, etc.. I would like to stress that the above ideas are not completely disjoint and do not necessarily exclude each other. For instance, all the above mentioned topological excitations are found to be correlated with each other in numerical as well as analytical studies.

I will restrict myself to the, at present, most popular superconductor picture which is based on the concept of electro-magnetic duality after an Abelian gauge projection. Recently, the centre vortex model has been resuscitated too. In the latter picture, the centre group that is directly related to the traditional order parameter of the deconfinement phase transition in finite temperature pure gauge theories, the Polyakov line, plays an essential rôle. Another striking feature is that — unlike monopole currents — centre vortices form two-dimensional objects, such that in *four* space-time dimensions, a linking number between a Wegner-Wilson loop and centre vortices can unambiguously be defined, providing a geometric interpretation of the confinement mechanism.[9] Unfortunately, I have not enough space to review these exciting developments. Therefore, I refer the interested reader to a review by Greensite.[10]

Another obvious omission are questions related to chiral properties of QCD. Real world QCD is not only a theory of gluons but includes quarks which means that the deconfinement phase transition will eventually be replaced by a chiral phase transition. In the phase with broken chiral symmetry colour charges are still being anti-screened and linear confinement approximately holds. However, if the binding energy within a hadron exceeds a critical value, the hadron will break up into two or more colour-neutral parts (string breaking).

Based on different confinement pictures various effective models of infra red QCD have been proposed in the past. Three such examples are the Abelian Higgs model,[11,9] the stochastic vacuum model[12] and dual QCD.[13] As well as *understanding* the mechanism of confinement, it is desirable to verify aspects of these models and eventually to derive certain model parameters directly from QCD.

This article is organised as follows. I will start with a brief motivation and introduction into lattice methods in Sec. 2, before explaining the concept of the Abelian projection in Sec. 3. Subsequently, I will introduce some notations and review lattice results in Sec. 4 and conclude with a summary of answered and open questions.

2. Why Lattice?

Lattice methods allow for a somewhat ***brute force*** but ***first principles*** numerical evaluation of expectation values of a given quantum field theory that is defined by an action S. The lattice volume and spacing are limited due to finite computer speed and memory. Simulations are performed in Euclidean space and an analytic continuation to Minkowski space of numerical results that have been obtained on discrete points with finite precision is virtually impossible. Therefore, not every physically meaningful quantity can be calculated in a straight forward manner. The obvious strength of lattice methods are hadron mass determinations. Even if one is unimpressed by post-dictions of experimental values that have been known with high precision for decades, within an accuracy of only 5 %, such simulations allow one to fix fundamental standard model parameters like quark masses and QCD running coupling in the low-energy domain. Of course, plenty of other applications of phenomenological importance exist.

Unfortunately, only the lowest lying one or two radial excitations of a hadronic state are accessible in practice. Lattice predictions are restricted to rather simple systems too. Even the deuterium is beyond the reach of present day super-computers. Therefore, it is desirable to supplement lattice simulations by analytical methods. The computer alone acts as a ***black box.*** In order to understand and interprete the output values and to predict their dependence on the input parameters, some modelling is required. Vice versa, the lattice itself is a strong tool to validate models and approximations. Unlike in the *"real"* world, we can vary the quark masses m_q^i, the number of colours N_c, the number of flavours n_f, the temperature, the volume, the space-time dimension of our *femto*-universe and even the boundary conditions in order to expose models to thorough tests in many situations. Instead of indirectly and in a somewhat uncontrolled fashion deriving parameter values from experiment we can compute *custom designed* observables. What is thought to work in the real world ought to work on the lattice too! Moreover, many models rely on certain approximations. Experimentalists cannot switch off quark flavours but we can!

In a lattice simulation, hyper-cubic Euclidean space-time is discretised on a box with, say, L^4 lattice points or sites, x. Two adjacent points are connected by an oriented bond or link (x, μ). The product of four links, enclosing an elementary square, is a plaquette. Quarks are living on sites, gauge fields on links and the plaquette determines the curvature within the $SU(N_c)$ group manifold and corresponds to the field strength tensor. For simplicity, we assume an isotropic lattice with equal lattice spacing a in all directions. The lattice spacing provides an ultra violet cut-off on the gluon momenta $q < \pi/a$ and regulates the theory.

We simulate an action $S(\beta, m_q^i)$ on the lattice, which contains the quark masses as well as an inverse QCD coupling $\beta = 2N_c/g^2$ as free parameters. By varying β (and m_q^i) the lattice spacing a is changed. Asymptotic freedom tells us that the ultra violet cut-off is removed as $\beta \to \infty$. The physical dimension of a is determined by calculating a dimensionful quantity on the lattice and associating it to its experimental value. If the right theory is being simulated we should be able to reproduce all experimental mass ratios in the continuum limit $a \to 0$, such that it becomes irrelevant what experimental quantity we have chosen to set the scale*. In practice one does not get all the quark masses right, such that there are always systematic uncertainties in a, due to the ambiguity of the choice of this experimental input quantity.

The discretised lattice action is formulated in a manifestly gauge-invariant way and approaches the continuum action with $a \to 0$. One ideally extrapolates to this limit at fixed physical lattice extent, $La = \text{const.}$. Later on, the thermodynamic limit $La \to \infty$ should be investigated. Due to the finiteness of a computer these two extrapolations, as well as extrapolations and interpolations between results obtained at different quark masses, are subject to systematic uncertainties that have to be carefully estimated. For most hadronic processes, a lattice spacing $a < 0.1$ fm is considered to be *close* to the continuum limit while an extent $La \approx 2$ fm is comfortably large to accommodate ground state mesons and baryons.

Expectation values of operators O are determined by computation of the path integral,

$$\langle O \rangle = \frac{1}{Z} \int [dU][d\psi][d\bar{\psi}]\, O[U] e^{-S[U,\psi,\bar{\psi}]}. \qquad (1)$$

The normalisation factor Z is such that $\langle 1 \rangle = 1$. $U_{x,\mu} \in SU(N_c)$ denotes a gauge field and the $4 \times n_f \times N_c$ tuple ψ_x is a fermion field. The high-dimensional integral is evaluated by means of a (stochastic) Monte-Carlo method as an average over an ensemble of n *representative* gauge configurations, $C_i = \{U_{x,\mu}^{(i)}\}, i = 1, \ldots, n$. Therefore, the result on the expectation value is subject to a statistical error that will basically decrease like $1/\sqrt{n}$: the longer we *measure* the more precise the prediction becomes. Therefore, we might speak of *lattice measurements* and *lattice experiments* in analogy to *"real" experiments*. Ideally, the sample size n is such that the statistical precision is similar or smaller than the systematic uncertainty of the extrapolations.

In Fig. 1 I illustrate that confinement is a numerical fact by displaying the potential between a static quark and an anti-quark, separated by a distance r. The data have been obtained in $SU(3)$ gauge theory with and without two light (mass-degenerate) quark flavours at lattice spacings somewhat smaller than 0.1 fm. The scale r_0 corresponds to 0.5 fm. The (partially) unquenched potential, that has been obtained at a quark mass $m_q \approx 50$ MeV, exhibits a more pronounced singularity at small r, due to the

*In the results presented here, I set the scale (somewhat arbitrarily) by the value $\sqrt{\sigma} = 440$ MeV for the string tension in case of $SU(2)$ and by the value $r_0 = 0.5$ fm in case of $SU(3)$ chromodynamics. r_0 is the distance at which $r_0^2\, dV(r)/dr|_{r_0} = 1.65$ where $V(r)$ denotes the potential between two static colour sources, separated by a distance r.

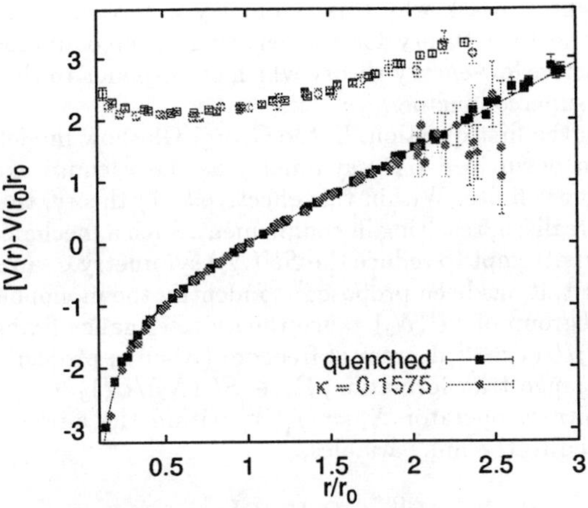

Fig. 1. Ground state (full symbols) and E_u hybrid potential (open symbols) at $\beta = 6.2$ (quenched) and $\beta = 5.6$, $\kappa = 0.1575$.[14]

slower running of the QCD coupling towards *zero* as the momentum scale is increased (or r is decreased). Furthermore, from $r \approx 1.2$ fm onwards, the (partially) unquenched potential eventually flattens: the QCD string "breaks". Indeed, many interesting features of the theory can be studied on the lattice that are not at all accessible to *real* experiment. As an example, I have included a so-called E_u (or continuum Π_u) hybrid potential[15,16] into the plot that corresponds to the interaction energy of two static quark sources for the case that gluons contribute one unit of angular momentum along the interquark axis.

3. (Abelian) photons and monopoles

Soon after the advent of QCD, 't Hooft and Mandelstam[2] proposed the dual superconductor scenario of confinement; the QCD vacuum is thought to behave analogously to an electrodynamic superconductor but with the rôles of electric and magnetic fields being interchanged: a condensate of magnetic monopoles expels electric fields from the vacuum. If one now puts electric charge and anti-charge into this medium, the electric flux that forms between them will be squeezed into a thin, eventually string-like, Abrikosov-Nielsen-Oleson (ANO) vortex which results in linear confinement.

In all quantum field theories in which confinement has been proven, namely in compact $U(1)$ gauge theory in the Villain formulation,[17] the Georgi-Glashow model[18] and SUSY Yang-Mills theories,[1] this scenario is indeed realised. Before one can apply this picture to QCD or $SU(N_c)$ chromodynamics one has to identify the relevant dynamical variables: it is not straight forward to generalise the electro-magnetic duality of a $U(1)$ gauge theory to $SU(N_c)$ since gluons carry colour charges. How can one define electric fields in a gauge invariant way? What fields are dual to the electric fields? How can one identify a monopole current? The next question is that for the order parameter of the confinement-deconfinement phase transition: what symmetry is broken? One might also ask why this symmetry is broken — a question that has been answered by the BCS theory for conventional superconductors. Finally, we are interested in an effective low-energy theory which corresponds to the Ginzburg-Landau theory of standard superconductors.

Let us address the first question. In the Georgi-Glashow model, the $SO(3)$ gauge symmetry is broken down to a $U(1)$ symmetry as the vacuum expectation value of the Higgs field becomes finite. Within this effective $U(1)$ theory, the standard electromagnetic duality is realised, resulting in confinement. Such a mechanism is not provided by QCD but one can attempt to reduce the $SU(N_c)$ symmetry to an Abelian symmetry *by hand.* In this spirit, it has been proposed[3] to identify the monopoles in the $U(1)^{N_c-1}$ diagonal Cartan subgroup of $SU(N_c)$ gauge theory after gauge fixing in respect to the off-diagonal $SU(N_c)/U(1)^{N_c-1}$ degrees of freedom (Abelian projection).

In general, a gauge transformation, $\Omega_x \in SU(N_c)/U(1)^{N_c-1}$, can be found that diagonalises an arbitrary operator $X_x = f[U^\Omega]$ within the adjoint representation of $SU(N_c)$ when applied to the link variables,

$$U^\Omega_{x,\mu} = \Omega_x U_{x,\mu} \Omega^\dagger_{x+a\hat{\mu}}. \tag{2}$$

Subsequently, a coset decomposition of the gauge transformed links is performed,

$$U^\Omega_{x,\mu} = C_{x,\mu} u_{x,\mu} \quad \text{with} \quad C_{x,\mu} \in SU(N_c)/U(1)^{N_c-1}, \quad u_{x,\mu} \in U(1)^{N_c-1}. \tag{3}$$

If we now apply a residual gauge transformation, $\omega_x \in U(1)^{N_c-1}$, we find,

$$u_{x,\mu} \longrightarrow \omega_x u_{x,\mu} \omega^\dagger_{x+a\hat{\mu}}, \quad C_{x,\mu} \longrightarrow \omega_x u_{x,\mu} \omega^\dagger_x, \tag{4}$$

i.e. $u_{x,\mu}$ transforms like a gauge field while $C_{x,\mu}$ transform like matter fields. Therefore, we will refer to the diagonal gluon field $u_{x,\mu}$ as "photon" field. While the gauge fields of electrodynamics in the non-compact continuum formulation are free of singularities and, therefore, cannot form magnetic monopole solutions, the gauge transformation Ω_x is singular wherever two eigenvalues of X_x coincide. Therefore, the gauge field $u_{x,\mu}$ will in general contain such monopoles.

After Abelian gauge fixing QCD can be regarded as a theory of interacting photons, monopoles and matter fields (i.e. off-diagonal gluons and quarks). One might assume that the off-diagonal gluons do not affect long range interactions. This conjecture is known as *Abelian dominance.*[19] In addition, *monopole dominance* of non-perturbative physics has been proposed.[20]

The identification of photon fields and monopoles is a gauge invariant process. However, the choice of the operator X_x, that defines the Abelian projection, is ambiguous. In his original work, 't Hooft suggested that the dual superconductor scenario would be realised in any Abelian projection. Indeed, the expectation value of a monopole creation operator has been found to be an order parameter of the deconfinement phase transition in quite a few different projections of $SU(2)$ chromodynamics.[21] As long as any $U(1)^{N_c-1}$ projection of the theory yields similar results, one does not have to specify the mechanism that is thought to break the $SU(N_c)$ gauge symmetry of QCD. However, numerical simulations suggest that Abelian and monopole dominance are not at all universal. The most popular gauge projection applied and discussed is the maximally Abelian projection (MAP).[3] One feature that is not shared by almost all other projections that have been investigated on the lattice so far is that the (local) MAP gauge condition gives rise to non-propagating ghost fields only, guaranteeing renormalisability of the Abelian projected theory. Another — and possibly related — fact is that both, Abelian and monopole dominance have qualitatively been verified in this projection.

One should mention that the analogy to an ordinary superconductor after Abelian projection is not complete. The electrons that form Cooper pairs in BCS theory are all negatively charged. However, the monopoles that are thought to condense in QCD can carry both, negative and positive magnetic charges. Therefore, the composition of the condensate is very different. The origin of the interaction that results in an attractive force between monopoles and anti-monopoles also differs from the periodic background potentials of the BCS theory.

I will briefly explain how MAP is performed on the lattice for the case of $SU(2)$ gauge theory. In the first step one maximises the functional,[22]

$$F(\Omega) = \sum_{x,\mu} \mathrm{tr}\left(\tau_3 U^{\Omega}_{x,\mu} \tau_3 U^{\Omega\dagger}_{x,\mu}\right), \tag{5}$$

by means of a gauge transformation Ω_x. τ_3 denotes a Pauli-matrix†. After the maximisation, all link variables are *as diagonal as possible*. The resulting gauge fields $U^{\Omega}_{x,\mu} = \exp\left(i\sum_c A^c_{x,\mu}\tau_c/2\right)$ satisfy 't Hooft's differential MA gauge fixing condition,[3]

$$\left(\partial_\mu \pm iA^3_{x,\mu}\right) A^{\pm}_{x,\mu} = 0, \quad A^{\pm}_{x,\mu} = \frac{1}{\sqrt{2}}\left(A^1_{x,\mu} \pm A^2_{x,\mu}\right). \tag{6}$$

After gauge fixing, the projection is performed: observables are calculated on Abelian configurations $\{\theta_{x,\mu}\}$ rather than $\{U_{x,\mu}\}$. We refer to the field theory, defined in this way, as Abelian projected $SU(2)$ gauge theory [APSU(2)]. For convenience the Abelian links are represented in the Lie algebra rather than the group itself, $u_{x,\mu} = \exp(i\theta_{x,\mu}\tau_3)$. Note that the normalisation has been chosen such that the periodicity of θ_μ is 2π as opposed to the periodicity 4π of A^3_μ. With this convention we find magnetic charges to be multiples of $2\pi/e$ just like in electrodynamics rather than $2N_c\pi/e$ as one might have expected in $SU(N_c)$ gauge theory. We also use the convention $e = 1$ for the fundamental electric charge.

Recently, some articles have appeared whose authors claim to have "proven" Abelian dominance, either in general or specifically in the MAP. In one case the argument is based on the fact that the off-diagonal gluon fields acquire mass and, therefore, are thought to affect ultra violet physics only, a statement that does not necessarily apply to confining field theories. Other arguments are either based on misinterpretations

†In principle, we can fix the gauge along any direction within the $SU(2)$ group space. The resulting gauges will only differ from each other by a global gauge transformation that will not affect expectation values — as long as we perform the subsequent coset decomposition with respect to the same $U(1)$ subgroup. Also note that if we forced an adjoint Higgs field with a δ-like potential into the 3-direction, the form of the interaction term with the gauge fields would be identical to F.

of the transfer matrix formalism or on rewriting the original theory in terms of Abelian projected variables plus perturbations (the off-diagonal fields) that are proportional to a *small* parameter ϵ. In order to avoid going too much into details I only mention that both, the static potential and Wilson loops in MAP are very different from their counter parts in the original theory, even for large distances. In case of the potential, the asymptotic slope (string tension) comes out to be similar. However, APSU(2) and $SU(2)$ potentials are shifted with respect to each other by a substantial constant due to different self-energy contributions. Any *"proof"* that fails to account for the latter fact is necessarily wrong to some extent.

4. Results

4.1. Some Definitions

The language of differential forms[23] turns out to be very convenient for the present purpose and this is even more so on the lattice. Let us start with a $U(1)$ field theory in $D = 4$ dimensions. The anti-symmetric field strength tensor F (a 2-form) can be decomposed as follows,

$$F = dA + \delta * C, \tag{7}$$

where A denotes the standard magnetic four-potential (1-form) and C denotes an electric four-potential which is a 1-form on the dual lattice. A and C are not uniquely determined but subject to a $U_{el}(1) \times U_{mag}(1)$ gauge invariance since gradients of scalar fields can be added. "d" denotes the exterior derivative that, applied to an n-form, results in an $n + 1$-form. "$*$" is the pull-back operator that connects an n-form to a $(D - n)$-form on the dual lattice while the dual derivative $\delta = *d*$ turns n-forms into $(n - 1)$-forms. We have $*^2 = 1$ and $d^2 = \delta^2 = 0$. We define a Laplacian $\Box = (-)^D(d\delta + \delta d)$.

In this notation, we obtain the generalised Maxwell equations,

$$\delta F = j, \quad dF = *k. \tag{8}$$

The electric current j is a 1-form on the original lattice while the magnetic (monopole) current k is a 1-form on the dual lattice. In Landau gauge ($\delta A = 0$ or $\delta C = 0$, respectively) this means,

$$j = \Box A, \quad k = \Box C. \tag{9}$$

Note that, unlike the continuum four-potential A, the link angles $\theta \in (-\pi, \pi]$ are subject to a 2π shift-periodicity. Therefore, the identification,

$$F = \frac{1}{a^2} \sin(d\theta) \left[1 + \mathcal{O}(a^2)\right], \tag{10}$$

is natural for electro-magnetic fields. We can factorise the plaquette $d\theta \in (-4\pi, 4\pi]$ into a regular part $\overline{\theta}_\Box \in (-\pi, \pi]$ and a singular part $m \in \{-2, -1, 0, 1, 2\}$:

$$d\theta = \overline{\theta}_\Box + 2\pi m. \tag{11}$$

While $dd\theta = 0$, $d\overline{\theta}_\Box = -2\pi dm$ does not necessarily vanish. Let us consider the magnetic flux through a (spatial) plaquette, $\Phi_{mag} = a^2 d\theta \left[1 + \mathcal{O}(a^2)\right]$, which, in the absence of monopole strings, would be identical to a contour integral of the vector potential A around the plaquette. However, in presence of monopoles, only $\exp(i\int_\Box d^2f\,d\theta) = \exp(i\int_{\partial\Box} dx\,A)$ holds: the argument can be shifted by multiples of 2π which, with the normalisation $e = 1$, is the elementary monopole charge. Thus, $-2\pi m$ counts the monopole contribution to the magnetic flux through the plaquette and $-2\pi dm$ corresponds to the "flux" out of a 3-cube, such that[24]

$$k = -2\pi *dm = *d\overline{\theta}_\Box \tag{12}$$

is a magnetic monopole current‡. k is conserved since $\delta k = -2\pi\delta^2 *m = 0$, i.e. monopole currents form closed loops on the dual lattice. The individual components of k can take values $2\pi n$ with $n = -4, -3, \ldots, 4$.

Monopole currents can alternatively be defined through,[25] $\tilde{k} = *dF = *d\sin(d\theta)$, the advantage being that the second Maxwell equation [Eq. (8)] is automatically fulfilled. The current $\tilde{k}$ is obviously conserved too. The quantisation of magnetic charges, however, is obscured by lattice artefacts. Magnetic monopoles become extended objects and a geometric interpretation is not as straight forward as for the definition presented before. For all results I am going to review, the first definition has been used. Note that the locations of monopoles in either of the definitions are not necessarily identical to positions of singularities of the adjoint operator X that has been diagonalised by the gauge fixing, as originally proposed by 't Hooft.

Each link variable can be factorised[26–28] into a *singular* part θ^{sing} that is induced by magnetic monopoles and a *regular* (or photon) part $\theta^{reg} = \theta - \theta^{sing}$ that obeys the homogenous Maxwell equation $dF = d\sin(d\theta^{reg}) = 0$. If we take the divergence of Eq. (11) in Landau gauge ($\delta\theta = 0$) we obtain,

$$2\pi\delta m = \delta d\theta = \Box\theta^{sing}. \tag{13}$$

One solution of this equation is obviously,

$$\theta^{sing}_x = -2\pi \sum_y D_{xy}\delta m_y, \tag{14}$$

where D_{xy} denotes the lattice Coulomb propagator in position space.

Instead of using the Abelian links $\{\theta_{x,\mu}\}$ one can evaluate observables from the *monopole* and *photon* parts $\{\theta^{sing}_{x,\mu}\}$ and $\{\theta^{reg}_{x,\mu}\}$, separately. We will refer to such expectation values as *monopole* and *photon contributions*, respectively.

4.2. Successes

I will present some facts in support of the superconductor confinement scenario in the MAP: approximate Abelian dominance of the static potential has been verified.[20] In a recent study on a large lattice,[26] the APSU(2) string tension has been confirmed to account for $(92 \pm 4)\%$ of the $SU(2)$ string tension. It is an open question whether the agreement will improve as the continuum limit is approached.

‡I assume the quantities to be given in lattice units a. k has dimension a^{-3}, F has dimension a^{-2}.

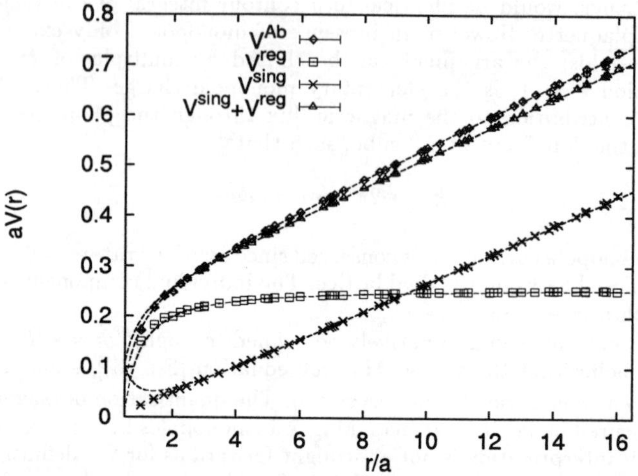

Fig. 2. Photon and monopole contributions V^{reg} and V^{sing} to the APSU(2) potential[26] V^{Ab} in units $a \approx 0.081$ fm.

The photon part of the potential does not contain a linear contribution. Monopole dominance and the factorisation $V^{Ab}(r) = V^{sing}(r) + V^{reg}(r)$ approximately hold. In Fig. 2, we display the result of a recent lattice study.[26] Everything is plotted in lattice units $a \approx 0.081$ fm. The monopole contribution amounts to $(95 \pm 1)\%$ of the Abelian string tension.

Approximate Abelian and monopole dominance has been confirmed for the light hadron spectrum[29] of $SU(3)$ gauge theory. In Fig. 3, I display recent results on the nucleon and ρ masses that have been obtained on a lattice with extent $La \approx 2.8$ fm and spacing $a \approx 0.175$ fm. On the horizontal axis, the squared pion mass (in units of the string tension) is displayed, which changes as the quark mass is varied. Results for $SU(3)$, APSU(3) and Abelian monopole contributions as well as for $SU(3)$ with the photon part removed lie on the same curve. In the same study, π and ρ masses are found to become degenerate, as soon as the monopole contribution to the gauge fields is subtracted, both in $SU(3)$ chromodynamics and the Abelian projected theory, i.e. Abelian monopoles are required for chiral symmetry breaking. Consistent results have been found by Lee *et al.*[27] and Bielefeld *et al.*.[30] Of course, one can also argue that since instantons are always accompanied by Abelian monopoles,[31] removing these monopoles naturally results in a trivial vacuum topology.

The one-loop β-functions in APSU(N_c) and $SU(N_c)$ gauge theories agree,[32] i.e. at weak coupling, masses obtained in both theories should be proportional to each other. A recent analytical investigation has also confirmed anti-screening of colour fields in APSU(2).[33] These results are in agreement with numerical data.[26,34]

For the next few observations, I refer to M. Polikarpov's talk at this conference: monopoles have been found to be condensed in the confined phase of $SU(2)$ gauge theory.[21,35] Furthermore, APSU(2) monopoles are spatially correlated with regions of

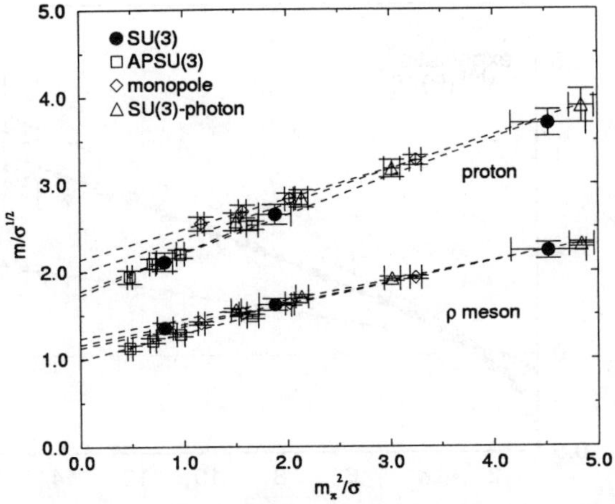

Fig. 3. $SU(3)$ ρ and nucleon masses: Abelian, monopole and non-photon contributions.[29]

access in the $SU(2)$ action density[36] which rules out that the monopoles are mere gauge fixing artefacts since their presence is reflected in gauge invariant observables. An effective monopole Lagrangian has been constructed which can (approximately) be mapped onto the Abelian Higgs model.[9]

4.3. Puzzles

Of course, we do not expect $SU(N_c)$ gauge theories to be *identical* to an Abelian Higgs model. So, we should become suspicious if the picture did not fail to describe the QCD reality at some point. Obviously, APSU(2) becomes very different from $SU(2)$ gauge theory in the ultra violet. I will restrict myself to two points where something in the infra red might go wrong that I consider as being serious.

The static potential between sources within the adjoint representation of $SU(2)$ (adjoint potential) will saturate at a distance at which the binding energy exceeds the gluelump-gluelump threshold.[15] Therefore, we expect for the adjoint string tension σ_{adj} the asymptotic value,

$$\sigma_{adj} = \lim_{r\to\infty} \frac{dV_{adj}(r)}{dr} = 0. \tag{15}$$

In APSU(2) we obtain for adjoint Wilson loops around a rectangle S,

$$W^{Ab}_{adj}(S) = \frac{1}{3}\left[1 + 2W^{Ab,2}(S)\right], \quad W^{Ab,2} = \cos\left(2 \sum_{(x,\mu)\in\partial S} \theta_{x,\mu}\right), \tag{16}$$

where we call $W^{Ab,2}$ the *charge two* Wilson loop; the adjoint source corresponds to a *neutral* and two *charge two* Abelian components. Obviously, as the area S goes to

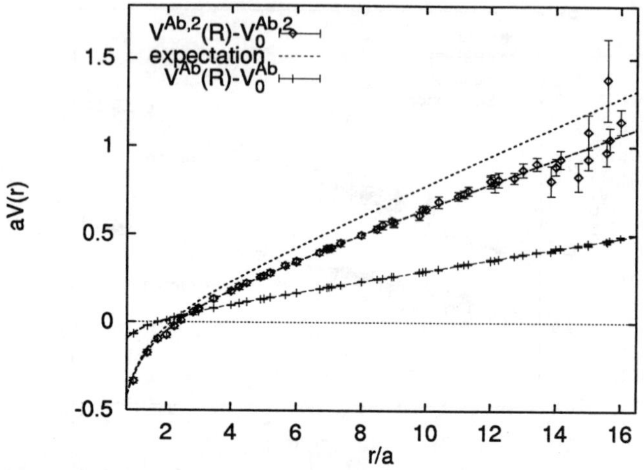

Fig. 4. The charge two potential in APSU(2) in lattice units $a \approx 0.081$ fm.[26]

infinity, $W_{adj}^{Ab}(S)$ approaches the constant value 1/3. Therefore,

$$V_{adj}^{Ab} = \frac{1}{a} \lim_{t\to\infty} \log \frac{W(r,t)}{W(r,t+a)} = 0 \quad \forall \quad r, \tag{17}$$

i.e. the correct asymptotic value $\sigma_{adj}^{Ab} = 0$ is produced.

However, lattice results indicate that the adjoint string will only break deep in the infra red at a distance slightly smaller than 1.5 fm,[15] a region in which we would expect Abelian dominance to hold if we consider the MAP theory to have relevance for hadronic physics. Simple models predict a linear rise of the potential in the intermediate region, with a slope that is proportional to the Casimir charge of the representation, i.e. $\sigma_{adj} = 8/3\,\sigma$ for $SU(2)$ gauge theory. Indeed, lattice simulations yield a linear increase, however with slightly smaller slope: $2\sigma \le \sigma_{adj} < 8/3\,\sigma$. In Fig. 4 the fundamental potential $V^{Ab}(r)$ is displayed, together with the *charge two* potential $V^{Ab,2}(r)$.[26] The curves are fits in accord to the parametrisation,

$$V^{Ab(,2)}(r) = -\frac{e^{Ab(,2)}}{r} + \sigma^{Ab(,2)} r. \tag{18}$$

The upmost curve corresponds to $e^{Ab,2} = 4e^{Ab}$, $\sigma^{Ab,2} = 8/3\,\sigma^{Ab}$. As in $SU(2)$, we find the slope in APSU(2) to be somewhat smaller than expected.

In conclusion, it seems that the *charge two* potential resembles the features of the *adjoint* potential within $SU(2)$. However, it is not clear how we can get rid of the *neutral* contribution to the adjoint Wilson loop. For this purpose, interactions with the off-diagonal gluons have to be considered.

A second puzzle that has recently been noticed[34] is related to the energy distribution within the ANO vortex between static sources. In Fig. 5, I show the $1/e$

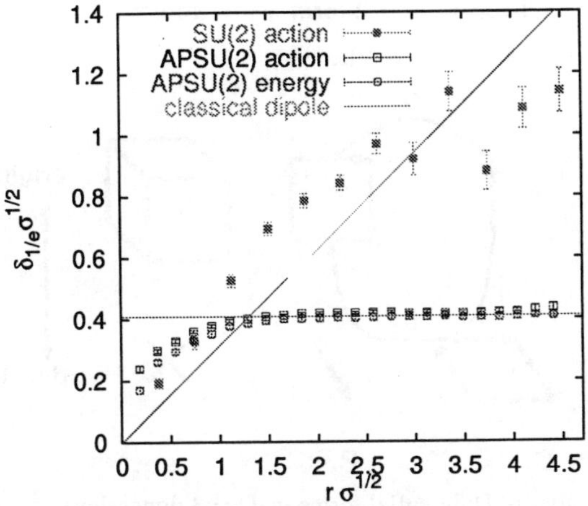

Fig. 5. $1/e$-radius of the flux tube in $SU(2)$ and APSU(2).[34]

radius of the flux tube as a function of the source separation. The scale is set by $\sigma^{-1/2} \approx 0.45$ fm. In $SU(2)$ gauge theory only for the action density distribution a statistically reasonable signal has been obtained while for APSU(2) we are able to display both, action and energy flux tube radii. While the width of the action density distribution in $SU(2)$ increases before it eventually saturates around a separation $r \approx 1.2$ fm at a value $2\delta_{1/e} \approx 0.9$ fm, both APSU(2) widths become consistent with a constant $2\delta_{1/e} \approx 0.36$ fm as soon as $r > 0.5$ fm.

Obviously, the matter fields change the structure of the flux tube and, therefore, affect an infra red observable. However, only the shape differs while the cross section, i.e. the string tension, remains (almost) invariant. The tremendous width of the $SU(2)$ flux tube is somewhat counter-intuitive in view of the linear potential and the phenomenological success of string-like descriptions of hadrons; obviously, the $SU(2)$ flux tube is not one-dimensional and transverse degrees of freedom should be considered. However, the projected theory yields what we would have expected: a thin, string-like object. We might speculate that the Abelian projection reveals the physically relevant core of the vortex that becomes obscured by the matter fields of the full theory. However, we are left with a puzzle: why does the amplitude of the underlying transverse string fluctuations remain that constant? One should be able to reproduce such a vortex in terms of an effective string theory. The simplest such theories are not renormalisable in *four* dimensions. Therefore, the width should increase logarithmically with the rate being controlled by an ultra violet cut-off, i.e. the scale at which longitudinal fluctuations become important and the string description breaks down. This is obviously not the case for our APSU(2) "string". What is the physical mechanism that prevents the string from widening? Is the effective string theory supersymmetric[37]? What correction terms have to be added in order to produce such a string?

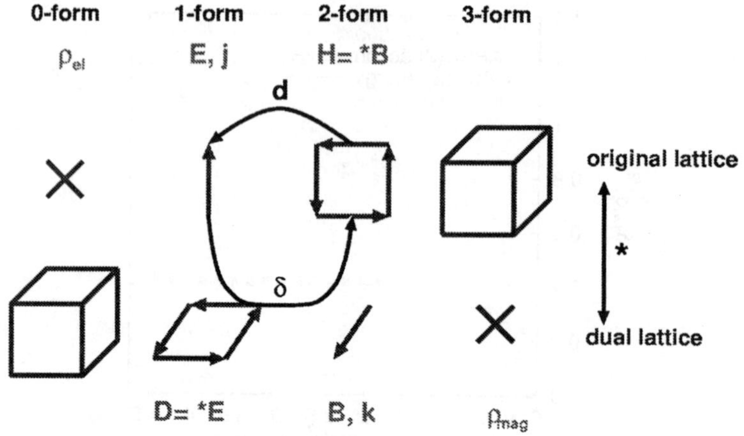

Fig. 6. Differential forms in $D = 3$ dimensions.

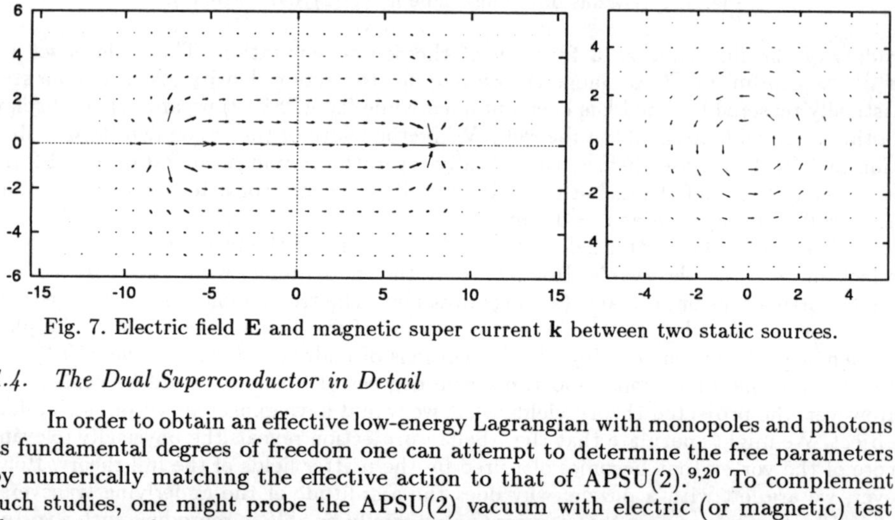

Fig. 7. Electric field **E** and magnetic super current **k** between two static sources.

4.4. The Dual Superconductor in Detail

In order to obtain an effective low-energy Lagrangian with monopoles and photons as fundamental degrees of freedom one can attempt to determine the free parameters by numerically matching the effective action to that of APSU(2).[9,20] To complement such studies, one might probe the APSU(2) vacuum with electric (or magnetic) test charges to verify predictions of the effective theory and *measure* the values of the model parameters, which is the line I am going to follow here. Investigations of field distributions in presence of charges have been performed previously.[38] I will concentrate on the results from a more recent study.[39]

We are probing the vacuum with *static* electric sources. For this purpose we consider three dimensional spatial cross sections (time slices) of the lattice. In Fig. 6, I have visualised where on the lattice different objects are "living". The advantage in working with differential forms is that the Stokes theorem is guaranteed to be exact and not subject to lattice artefacts: if the differential Maxwell equations are fulfilled, the integrated versions automatically hold too. Of course, when finally relating lattice results to continuum physics, lattice artefacts re-enter the game. The (generalised) Maxwell equations [Eq. (8)] read,

$$\delta \mathbf{E} = \rho_{el}, \quad d\mathbf{E} = *\mathbf{k} - *\dot{\mathbf{B}}, \tag{19}$$

$$\delta \mathbf{B} = \rho_{mag}, \quad d\mathbf{B} = *\mathbf{j} + *\dot{\mathbf{E}}. \tag{20}$$

The charge densities ρ_{el} and ρ_{mag} are the 4-components of j and k, respectively. The "*dot*"-symbol denotes a temporal derivative. For a static problem, electric and magnetic fields decouple. In the absence of magnetic test charges, this implies $\rho_{mag} = 0$, $\mathbf{j} = \mathbf{0}$ and, therefore, $\mathbf{B} = \mathbf{0}$ such that we are left with two Maxwell equations only, $\delta\mathbf{E} = \rho_{el}$ and $d\mathbf{E} = *\mathbf{k}$.

As mentioned above, one can define the monopole current via the latter equation (Ampère's law) which is then trivially fulfilled. However, with our definition of $\mathbf{k}$ [Eq. (12)] neither of the equations necessarily holds. Strong fields may give rise to non-linear quantum corrections. Moreover, APSU(2) is not electrodynamics; in general, the action will be non-local and the equations of motion might differ. Our numerical study, however, verifies div $\mathbf{E}$ to disappear outside of the vicinity of the sources while the curl of the electric field is identical to the magnetic current.[39] I display the electric field for a distance $r = 15a \approx 1.2$ fm in Fig. 7. The vortex is stabilised by the surrounding super current $\mathbf{k}$.

The starting point of our investigation is the London limit. The electric vector potential $\mathbf{C}$ on the dual lattice is defined through,

$$\mathbf{E} = *d\mathbf{C}. \tag{21}$$

One representation of the (dual) London equations reads,

$$\mathbf{C} + \lambda^2 \mathbf{k} = 0, \tag{22}$$

with λ being the inverse mass of the dual photon. By building the curl of this relation, we obtain $\mathbf{E} = -\lambda^2 * d\mathbf{k}$ which, together with the dual Ampère law results in,

$$\mathbf{E} = -\lambda^2 \delta d\mathbf{E} = \lambda^2 \left(\nabla^2 \mathbf{E} + d\rho_{el}\right). \tag{23}$$

Without an electric current $\mathbf{j}$, the Maxwell equations imply, $d\left(\mathbf{B} + \lambda^2 d\dot{\mathbf{k}}\right) = 0$, i.e. even in the absence of a magnetic field, the super current $\mathbf{k}$ remains constant and in general non-*zero*. From Eq. (23) it is also obvious why λ is called the *penetration length*: if we expose a superconducting probe to a homogeneous electric field, $\mathbf{E} = E_z \mathbf{e}_x$, the field strength will decay with the distance x towards the centre of the medium: $E_z(x) = E_z(0)\exp(-x/\lambda)$.

We create a charge-anticharge pair at a separation r parallel to the z-axis of our lattice, $\rho_{el} = \delta^3(-r/2\,\mathbf{e}_z) - \delta^3(r/2\,\mathbf{e}_z)$. x denotes the transverse distance from the core of the ANO vortex. For the electric flux through any surface enclosing the charge, we expect $\Phi_{el} = \int d^2x\, E_z(x,0) = 1$. For our geometry Eq. (23) reads in the centre plane perpendicular to the vortex,

$$E_z(x) = \lambda^2 \nabla_2^2 E_z(x) + \Phi \delta^2(x), \quad E_x(x) = 0. \tag{24}$$

A solution can be expressed in terms of the modified Bessel function K_0,

$$E_z(x) = \frac{\Phi_{el}}{2\pi\lambda^2} K_0(x/\lambda). \tag{25}$$

Within statistical accuracy, we find our data to be compatible with such a functional form for $x > x_{\min} = 4.2a \approx 0.35$ fm[§] with parameter values,

$$\lambda = (1.82 \pm 0.07)a \approx (1.3\,\text{GeV})^{-1}, \quad \Phi_{el} = 1.44 \pm 0.08. \tag{26}$$

For small x the data are overestimated by the fit since $K_0(x)$ diverges as $x \to 0$ which explains why the electric flux comes out to be significantly larger than *one*. A dual photon mass of 1.3 GeV is compatible with the mass of the lightest glueball in $SU(2)$ gauge theory. However, the quantum numbers of a dual photon are $J^{PC} = 1^{+-}$ as opposed to 0^+ for this glueball.

How can we refine our description such that the electric field not only in the surface region but also closer to the centre of the vortex is correctly reproduced? Obviously, a second scale ξ that is smaller than the 0.035 fm, above which the London limit seems to apply, has to be introduced. Such a scale appears in the Ginzburg-Landau (GL) equations as the *coherence length* of the GL wave function $\psi(\mathbf{x})$ that describes the spatial density of superconducting monopole charges, $n(\mathbf{x}) = |\psi(\mathbf{x})|^2$. We decompose ψ into a phase and an amplitude f,

$$\psi(\mathbf{x}) = \psi_\infty f(\mathbf{x}) e^{i\theta(\mathbf{x})}, \quad f(\mathbf{x}) \stackrel{x\to\infty}{\longrightarrow} 1. \tag{27}$$

ξ characterises the decay of the monopole density towards the centre of the vortex, where f will vanish as superconductivity breaks down, while λ controls the penetration of the vortex field into the surrounding vacuum. The case $\xi = 0$ corresponds to the London limit. If we increase the width of the flux tube, the dia-electric energy of the vortex is reduced by an amount roughly in proportion to λE^2 while the amount of energy we have to pay for pushing the monopoles further into the vacuum increases like ξE^2. Values $\xi < \sqrt{2}\lambda$ correspond to a negative surface energy in accord to the Abrikosov criterium for a classical system. This tendency to maximise the surface results in retardation of flux tubes: we obtain a type II superconductor while values $\kappa = \lambda/\xi < 1/\sqrt{2}$ correspond to a type I superconductor. From our experience we are prejudiced to expect a type II scenario since in electrodynamics type I flux tubes cannot be realised due to the absence of isolated magnetic charges. In the present situation, however, the presence of two isolated electric sources forces the field lines through the surrounding vacuum, regardless of the type of the superconductor.

If we restrict ourselves to the perpendicular centre plane, the equation $d\mathbf{C} = *\mathbf{E}$ implies for the azimuthal component of $\mathbf{C}$ (up to gauge transformations),

$$C_\theta(x) = \frac{1}{2\pi x}\int_{x'<x} d^2x'\, E_z(x') \stackrel{x\to\infty}{\longrightarrow} \frac{\Phi_{el}}{2\pi x}, \tag{28}$$

[§] Along off-axis lattice directions, we obtain data for non-integer multiples of the lattice spacing.

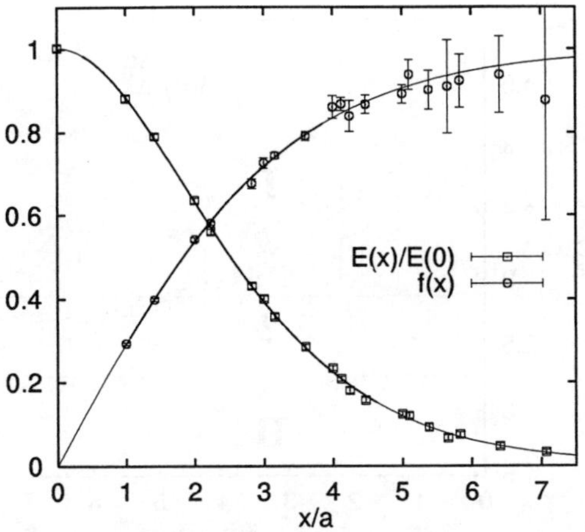

Fig. 8. The electric field and the amplitude of the Ginzburg-Landau wave function against the distance from the centre of the ANO vortex.[39]

while the other components vanish. For non-constant density of magnetic charges, the London equation Eq. (22) is modified and becomes the second GL equation,

$$\left(C_\theta(x) - \frac{\Phi_{el}}{2\pi x}\right) + \frac{\lambda^2}{f^2(x)} k_\theta(x) = 0. \tag{29}$$

We can solve this equation with respect to $F(x) = f(x)/\lambda$ after having reconstructed $C_\theta(x)$ via Eq. (28).

The result is displayed in Fig. 8, together with $E_z(x)$. Data obtained at $x < 2.2a$ has to be treated with care since the difference between lattice and continuum versions of "curl" turns out to be bigger than our statistical uncertainty. For $x > 4.2a$ the errors on f explode: here, no contradiction to the London limit has been found. We fit $F(x)$ with the ansatz,

$$F(x) = \frac{f(x)}{\lambda} = \frac{1}{\lambda} \tanh(x/\alpha), \tag{30}$$

which conforms to the right boundary conditions. The fit is included into the figure as well as the result of a fit of E_z to a more involved four parameter ansatz that also respects the boundary conditions on f.[39] From the fit Eq. (30) we obtain $\lambda = 1.62(2)a$. The fit to E_z yields $\lambda = 1.84(8)a$ while a simultaneous fit to E_z and k_θ yields $\lambda = 1.99(5)a$. This has to be compared with the value $\lambda = 1.82(7)a$ from the London limit fit of Eq. (25). We end up with the conservative estimate,

$$\lambda = 1.84^{+20}_{-24}a = (0.15 \pm 0.02)\,\text{fm}, \quad \Phi_{el} = 1.08 \pm 0.02. \tag{31}$$

One should settle in a scaling study whether the deviation from $\Phi_{el} = 1$ can be attributed to a non-trivial vacuum dielectricity constant due to anti-screening.

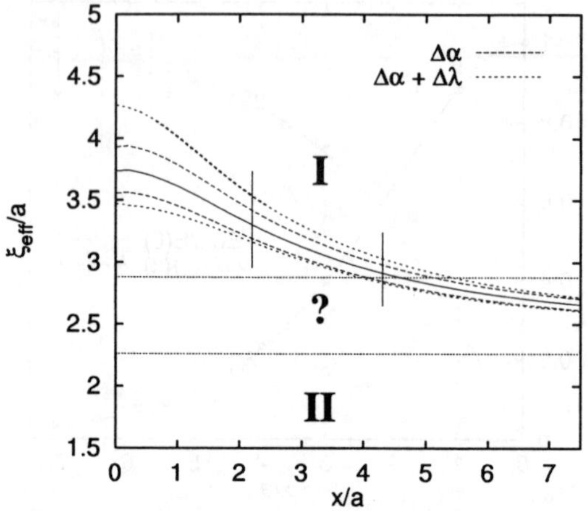

Fig. 9. Effective coherence length versus distance from the centre of the ANO vertex.[39]

Since the first GL equation is non-linear, we cannot consistently formulate it in terms of differential forms on a lattice, i.e. — unlike the Maxwell equations — we have to verify it in the *continuum* and discard data obtained at small x/a values where the lattice structure is still apparent. For our geometry the first GL equation reads,

$$f(x) = f^3(x) + \xi^2 h(x) \left\{ \left[\frac{1}{x} - \frac{2\pi C_\theta(x)}{\Phi_{el}} \right]^2 - \frac{1}{x}\frac{d}{dx}\left(x\frac{d}{dx} \right) \right\} f(x). \tag{32}$$

Our strategy is to solve this equation with respect to an *effective* coherence length $\xi_{\text{eff}}(x)$, employing the above mentioned parametrisations of $E_z(x)$ and $f(x)$ to interpolate the data. The result is visualised in Fig. 9. Results outside of the window of observation $2.2 < x/a < 4.2$ are unreliable, for small x due to lattice artefacts and for large x due to lacking precision data on f. The figure contains error bands that are related to the uncertainties in α as well as in λ. Within the window, ξ_{eff} varies by only 10 %, i.e. the GL equation is qualitatively satisfied. Taking this variation of ξ with x into account, we obtain the result,

$$\xi = 3.10^{+43}_{-35} a = (0.251 \pm 0.032)\,\text{fm}. \tag{33}$$

The central value corresponds to $\xi = \xi_{\text{eff}}(\xi)$.

For the ratio of penetration and coherence lengths we determine,

$$\kappa = 0.59^{+13}_{-14} < \frac{1}{\sqrt{2}}, \tag{34}$$

i.e. we have evidence of a *weak* type I superconductor. However, we are rather close to the Abrikosov limit whose position might differ from $1/\sqrt{2}$ due to quantum corrections.

In order to finally settle the question of the type of the superconductor, interactions between two flux tubes should be investigated. This has been done in recent studies of the confined phase of $U(1)$ gauge theory[40] as well as in three-dimensional Z_2 gauge theory,[41] the result being attraction in both cases, i.e. type I superconductivity.

5. Summary and Open Questions

The lattice is an ideal tool to test ideas on the confinement mechanism. Many infra red aspects of QCD are reproduced in the maximally Abelian projection. After the projection only the monopole contribution to the original gauge fields seems to be relevant for most low-energy properties. The dual Maxwell equations have been verified in APSU(2) and the fields are adequately described by the dual Ginzburg-Landau equations with the values $\lambda = 0.15(2)$ fm and $\xi = 0.25(3)$ fm for penetration and coherence length, respectively. These values correspond to a (dual) photon mass $m_\gamma \approx 1.3$ GeV$\approx 3\sigma$ and a Higgs mass of $m_H \approx 0.8$ GeV$\approx 2\sigma$, the ratio of which, $\kappa = \lambda/\xi = 0.59(13)$, indicates the vacuum of $SU(2)$ gauge theory to be a (weak) type I superconductor. It is demanding to clarify whether flux tubes in $SU(N_c)$ gauge theory as well as in the Abelian projected theory attract or repel each other.

Electric flux tubes are found to be significantly thinner after the Abelian projection. Contrary to $SU(2)$ gauge theory, their width seems to saturate at a separation $r \approx 0.5$ fm at a value, $2\delta_{1/e} \approx 0.36$ fm. The Abelian projection also seems to suffer under problems with charges in non-fundamental representations. These observations require further thought. In the end we would like to clarify the rôle that charged gluons play. Ideally, one would like to completely circumvent the Abelian projection and arrive at a genuinely non-Abelian description of the superconductor scenario. In view of the existence of other reasonable proposals for the confinement mechanism more thought should be spend on relations between different such pictures.

An extension of the detailed superconductor study presented from $SU(2)$ to $SU(3)$ gauge theory should be attempted. It is interesting to simulate the 4D Abelian Higgs model with the m_γ and m_H parameters that have been predicted above and compare the resulting flux distributions with the result from APSU(2). This approach has recently been pursued by Chernodub and collaborators.[9]

Acknowledgements

This work has been supported by DFG grant Ba 1564/3. I thank my collaborators, in particular V. Bornyakov, M. Müller-Preußker, K. Schilling, and C. Schlichter.

References

1. N. Seiberg and E. Witten, Nucl. Phys. **B426**, 19 (1994).
2. G. 't Hooft, in *High Energy Physics*, ed. A. Zichici (Editrice Compositori, Bologna, 1976); S. Mandelstam, Phys. Rept. C **23**, 245 (1976).
3. G. 't Hooft, Nucl. Phys. **B190**, 455 (1981).
4. J. Ambjorn and P. Oleson, Nucl. Phys. **B170**, 265 (1980).
5. D.I. Diakonov and V.Yu. Petrov, Nucl. Phys. **B245**, 259 (1984).
6. G.K. Savvidy, Phys. Lett. **71B**, 133 (1977).

7. P. Cea and L. Cosmai, Phys. Rev. D **43,** 620 (1991); H.D. Trottier and R.M. Woloshyn, Phys. Rev. Lett. **70,** 2053 (1993).
8. See e.g., J.W. Negele, hep-lat/9804017.
9. M. Polikarpov, *Proc. Confinement III*, TJ Lab., 1998.
10. L. Del Debbio, M. Faber, J. Giedt, J. Greensite, and S. Olejnik, hep-lat/9801027 and references therein.
11. T. Suzuki, Prog. Theor. Phys. **80,** 929 (1988).
12. H.G. Dosch, *Proc. Confinement III*, TJ Lab., 1998.
13. M. Baker, *Proc. Confinement III*, TJ Lab., 1998.
14. $T\chi L$ collaboration: G.S. Bali *et al.*, in preparation.
15. C. Michael, *Proc. Confinement III*, TJ Lab., 1998, hep-lat/9809211.
16. C. Morningstar, *Proc. Confinement III*, TJ Lab., 1998, hep-lat/9809305; J.E. Paton, *ibid.*; T.J Allen, *ibid.*.
17. T. Banks, R. Myerson, and J. Kogut, Nucl. Phys. **B129,** 493 (1977).
18. A.M. Polyakov, JETP Lett. **20,** 194 (1974); G. 't Hooft, Nucl. Phys. **B79,** 276 (1974).
19. Z.F. Ezawa and A. Iwazaki, Phys. Rev. D **25,** 2681 (1982).
20. See e.g., T. Suzuki, Prog. Theor. Phys. Suppl. **122,** 75 (1996).
21. See e.g., A. Di Giacomo, hep-th/9809047.
22. A.S. Kronfeld, G. Schierholz, U.-J. Wiese, Nucl. Phys. **B293,** 461 (1987).
23. See e.g., P. Becher and H. Joos, Zeit. Phys. **C15,** 343 (1982).
24. T.A. DeGrand and D. Toussaint, Phys. Rev. D **22,** 2478 (1980).
25. M. Zach, M. Faber, W. Kainz and P. Skala, Phys. Lett. **B358,** 325 (1995).
26. G.S. Bali, V. Bornyakov, M. Müller-Preußker, and K. Schilling, Phys. Rev. D **54,** 2863 (1996).
27. F.X. Lee, R.M. Woloshyn, and H.D. Trottier, Phys. Rev. D **53,** 1532 (1996).
28. J. Smit and A.J. van der Sijs, Nucl. Phys. **B355,** 603 (1991).
29. S. Kitahara *et al.*, hep-lat/9803020.
30. T. Bielefeld, S. Hands, J. Stack, J. Wensley, Phys. Lett. **B415,** 150 (1998).
31. V. Bornyakov and G. Schierholz, Phys. Lett. **B384,** 190 (1996); A. Hart and M. Teper, Phys. Lett. **B371,** 261 (1996); H. Suganuma, K. Itakura, H. Toki, and O. Miyamura, hep-ph/9512347.
32. M. Quandt and H. Reinhardt, Phys. Lett. **B424,** 115 (1998); K.-I. Kendo, *Proc. Confinement III*, TJ Lab., 1998, hep-th/9808186.
33. G. Di Cecio, A. Hart, and R.W. Haymaker, hep-lat/9807001.
34. G.S. Bali, K. Schilling, and C. Schlichter, in preparation.
35. M. Chernodub, M. Polikarpov, A. Veselov, Phys. Lett. **B399,** 267 (1997).
36. B.L.G. Bakker, M.N. Chernodub, and M.I. Polikarpov, Phys. Rev. Lett. **80,** 30 (1998); E.-M. Ilgenfritz, H. Markum, M. Müller-Preußker, and S. Thurner, hep-lat/9801040.
37. J. Greensite and P. Oleson, hep-th/9806235.
38. P. Cea and L. Cosmai, Phys. Rev. D **52,** 5152 (1995); R.W. Haymaker, hep-lat/9510035.
39. G.S. Bali, K. Schilling, and C. Schlichter, hep-lat/9802005.
40. M. Zach, M. Faber, and P. Skala, *Proc. Confinement III*, TJ Lab., 1998, hep-lat/9808055; hep-lat/9709017.
41. F. Gliozzi, in *Proc. Confinement II*, eds. N. Brambilla, G. Prosperi, (World Scientific, Singapore, 1997), hep-lat/9609040.

QUARK DYNAMICS and FLUX TUBE STRUCTURE in $q\bar{q}$ BOUND STATES

Nora Brambilla
Institut für Theoretische Physik, Boltzmanngasse 5, A-1090, Vienna, Austria

ABSTRACT

I discuss the bound state quark dynamics focusing on the nonperturbative interaction and the flux tube structure.

1. Introduction

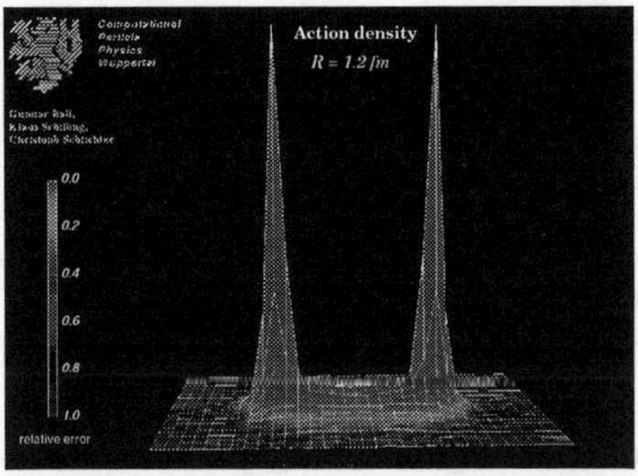

Fig. 1. Lattice evidence of the interquark flux tube. Figure provided by G. Bali, see.[1,2]

Confinement and chiral symmetry breaking are the two main facts in low-energy QCD and show up explicitly in the spectrum of quark bound states. Confinement implies that the nature of the quark dynamics has to change strongly in dependence on the scale of energy or distance associated with the interaction. Let us consider the interaction between a quark and an antiquark. At short quark-antiquark distance r, the interaction is dominated by the one gluon exchange (dipole configuration) and is obtained by a perturbative expansion in α_s. With increasing distance, the interaction becomes linear in r (flux tube configuration). The formation of a flux tube of constant energy density $\sigma \simeq 1\text{GeV/fm}$ (the so called string tension) confining the quark and the antiquark to remain bound, is the most striking feature of confinement and is inherently nonperturbative. We have hints about this flux tube from the phenomenological spectrum and the Regge trajectories. Phenomenological potential models include interactions rising with the quark-antiquark distance. Dual models explain this feature as a chromoelectric flux tube formation due to a dual Meissner effect, the QCD vacuum behaving like a t'Hooft-Mandelstam superconductor. Confinement is then due to monopoles condensation.

Lattice studies of the electric and magnetic fields strength distributions around a static quark-antiquark pair yield direct evidence and a detailed picture of the string formation (see Fig.1). Lattice data provide us with: a quark-antiquark static potential linear for $r > 0.2\,\mathrm{fm}$; an interquark flux tube structure with a transversal rms width ranging between 0.5 fm and 0.75 fm; positive tests of monopole condensations; Abelian projected electric fields and currents distributions in agreement with the dual Ginzburg Landau predictions.[2,3]

The lattice remains the most suitable tool to obtain nonperturbative information. However, most of these appear as independent and uncorrelated measurements. On the other side we have many phenomenological models implementing confinement. Finally, high order perturbative calculations are also available. In order to gain some insight in the mechanism of confinement and flux tube formation we need a model-independent, exact (in the sense that it contains perturbative and nonperturbative physics and is valid at any scale), gauge-invariant and *systematic* approach to the quark interaction. It helps to have a formulation which is closely related to the lattice, allows analytic calculations and is physically transparent.

In Secs. 2 and 3 we show that this program is realizable in the heavy quark case since a small expansion parameter (the quark velocity v) still exists. The result obtained is model independent and suitable for lattice evaluation. An analytic evaluation is possible once a QCD vacuum model is considered. Therefore it is possible to test models of confinement in an unambiguous and direct relation with the phenomenological and lattice data.

We know that in the low energy region new degrees of freedom become relevant. The flux tube formation hints to the fact that dynamical gluon effects are vital when attempting to incorporate nonperturbative interaction, indeed the gluonic degrees of freedom are condensed into stringlike flux tube. Hence, the flux tube should determine the nonperturbative bound state quark dynamics. On one side, this means that also excited states of the interquark string has to be considered. This leads to the consideration of hybrids states and was discussed by Michael[4] at this Conference. On the other side, this means that even when considering the ground state of the interquark glue, the flux tube energy should be taken into account properly. Moreover, to understand the transition region, we should be able to control precisely how the collective behavior of gluons modifies and alters single gluon exchange.

In Secs. 4, 5 and 6 we discuss the heavy quark dynamics in QCD vacuum models that reproduce the flux tube structure, in comparison with the lattice data and with the "$1/Q^4$" models extensively used in the literature. We find that all the 'flux tube models' predict the same form for the heavy quark interaction also in the intermediate region. We show that the Gaussian dominance of the Wilson loop average has to be related with the Abelian dominance of infrared QCD establishing a connection with the dual superconductor mechanism.

In Secs. 7 and 8 we extend our discussion to bound systems involving at least one light quark. Here the spontaneous chiral symmetry symmetry breaking emerges in an interplay with confinement.

2. Gauge-Invariant approach to quark bound states

We consider a gauge invariant quark-antiquark singlet state*

$$|\phi^{lj}_{\alpha\beta}\rangle \equiv \frac{\delta_{lj}}{\sqrt{3}}\bar{\psi}^i_\alpha(x)U^{ik}(x,y,C)\psi^k_\beta(y)|0\rangle \tag{1}$$

where $i,j,k,l = 1,\ldots 3$ are color indices and the Schwinger string line U has been inserted to ensure gauge-invariance

$$U(x,y;C) = P\exp\{ig\int_{y,C}^{x} A_\mu(z)dz^\mu\}. \tag{2}$$

At a time $t=0$, a quark and an antiquark are created, interact while propagating for a time time $t=T$ at which they are annihilated. Then, the quark-antiquark interaction is contained in the gauge-invariant four-point Green function ($x_j = (\mathbf{x}_j, T)$, $y_j = (\mathbf{y}_j, 0)$)

$$G(x_1,x_2,y_1,y_2) = \frac{1}{Z}\int \mathcal{D}A\, Tr S(x_2,y_2;A)U(y_2,y_1)S(y_1,x_1;A)U(x_1,x_2)e^{iS_{\rm YM}} \tag{3}$$

where we neglected the fermionic determinant and the annihilation terms†, Tr is the color trace and $S(y,x;A)$ is the quark propagator in the external field A. To find a closed expression for the Green function in Eq.(3) is quite a formidable task. However, the situation becomes simpler if we consider the case of heavy quarks. In the static limit $m\to\infty$ the equation for the static quark propagator

$$(i\gamma_0 D_0 - m)S_0(x,y;A) = \delta^4(x-y) \tag{4}$$

is solvable in closed form[6,7] and we find[8,6,9]

$$G(T) \stackrel{m_j\to\infty}{\longrightarrow} \delta^3(\mathbf{x}_1-\mathbf{y}_1)\delta^3(\mathbf{x}_2-\mathbf{y}_2)e^{-i(m_1+m_2)T}\langle W(\Gamma_0)\rangle \tag{5}$$

where

$$\langle W(\Gamma_0)\rangle = \frac{1}{Z}\int \mathcal{D}A\, Tr P\exp\{ig\oint_{\Gamma_0} dz^\mu A_\mu(z)\}\exp\{igS_{\rm YM}\} \tag{6}$$

is the static Wilson loop average, the loop Γ_0 being a rectangle. From the Feynman–Kac formula and Eq. (5) we get

$$V_0(r) = \lim_{T\to\infty}\frac{i}{T}\ln\langle W(\Gamma_0)\rangle. \tag{7}$$

---

*Actually this state is not an eigenstate of the Hamiltonian. It serves as a trial state to extract the energy of the lowest eigenstate having a non-vanishing projection on $|\phi\rangle$. This is equivalent to say that the contribution of the string is negligible in the limit $T\to\infty$, which will become more apparent when dealing with the Wilson loop.

†Since we neglect the fermionic determinant we work in the quenched approximation. In the case of heavy quarks however we can introduce into the effective action the expansion of the determinant.[5]

The static quark-antiquark potential, *if exists*, is given in terms of the static Wilson loop average. Notice that this is an exact expression that contains both the perturbative and the nonperturbative dynamics. Indeed, no expansion in the coupling constant has been performed. Here and in the following sections we implicitly assume the existence of the potential and we postpone this issue to Sec 6. The spin-dependent $1/m^2$ corrections to the potential were calculated in[7,10] considering the first correction to Eq. (4) and evaluating it on the zero order exact solution. The result is given in terms of vacuum expectation value of (color) electric and magnetic field insertions in the static Wilson loop and controls the fine and hyperfine separation in quarkonia.

However there were some residual difficulties:

- The kinetic energy terms was not considered‡ lacking in consistency with the virial theorem.
- The spin-dependent potentials appeared as an expansion in $1/m^2$ completely missing the terms in $\ln m$ obtained at one loop in the S matrix perturbative calculation of the interaction.[11]
- The calculation of the nonperturbative contributions was based on several assumptions and there was no clear and unambiguous procedure to calculate the nonperturbative behavior of the v.e.v. of the field strength insertions in the static Wilson loop.
- It was not clear when a potential description was holding.

The solution is *to disentangle the different scales of the bound system* (m, $p = mv$, $E = mv^2$, v being the heavy quark velocity) and to establish a *systematic* , *unambiguous* and *exact* expansion procedure. This will be reported in the next section.

3. Analytic closed expression for the heavy quark interaction

In the heavy quark systems, the existence of an expansion parameter (the inverse of the mass m in the Lagrangian and the velocity v of the quark as a dynamical defined power counting parameter) makes possible to establish a systematic expansion procedure. The tool is provided by NRQCD.[12] This is an effective theory equivalent to QCD and obtained from QCD by integrating out the hard energy scale m. The Lagrangian comes from the original QCD Lagrangian via a Foldy–Wouthuysen transformation. The ultraviolet regime of QCD (at energy scale m) is perturbatively encoded order by order in the coupling constant α_s in the matching coefficients which appear in front of the new operators of the effective theory. This ensures the equivalence between the effective theory and the original one at a given order in $1/m$ and α_s. At order $1/m^2$ the NRQCD Lagrangian describing a bound state between a quark of mass m_1 and an antiquark of mass m_2 is[13,14]

$$L = \psi_1^\dagger \Bigg(iD_0 + c_2^{(1)} \frac{\mathbf{D}^2}{2m_1} + c_4^{(1)} \frac{\mathbf{D}^4}{8m_1^3} + c_F^{(1)} g \frac{\sigma \cdot \mathbf{B}}{2m_1} \, c_D^{(1)} g \frac{\mathbf{D}\cdot\mathbf{E} - \mathbf{E}\cdot\mathbf{D}}{8m_1^2}$$

‡The static quark propagator is the zero order solution. This is similar to the HQET approach.

$$+ ic_S^{(1)} g \frac{\sigma\cdot(\mathbf{D}\times\mathbf{E} - \mathbf{E}\times\mathbf{D})}{8m_1^2}\Bigg) \psi_1 + \text{ antiquark terms } (1 \leftrightarrow 2) + \frac{d_1}{m_1 m_2}\psi_1^\dagger\psi_2\psi_2^\dagger\psi_1$$

$$+\frac{d_2}{m_1m_2}\psi_1^\dagger\sigma\psi_2\psi_2^\dagger\sigma\psi_1 + \frac{d_3}{m_1m_2}\psi_1^\dagger T^a\psi_2 Q_2^\dagger T^a\psi_1 + \frac{d_4}{m_1m_2}\psi_1^\dagger T^a\sigma\psi_2\psi_2^\dagger T^a\sigma\psi_1. \quad (8)$$

This is the relevant Lagrangian in order to calculate the bound state energies up to order $O(v^4)$. The coefficients $c_2^{(j)}$, $c_4^{(j)}$, ... are evaluated at a matching scale μ for a particle of mass m_j. The matching coefficients are 1 or 0 at the tree level. At one loop they contain terms in $\ln m_j$ which establish[15] the agreement of the present calculation with the result for the perturbative interaction obtained at one loop in the S matrix formalism.[11] For simplicity in the following we consider the tree level values [§].

With Eq. (8) we have reduced the heavy quark dynamics to a nonrelativistic Schrödinger theory for the heavy quark (antiquark) with Pauli propagator K coupled to the usual relativistic field theory for light quark (omitted here, quenched appr.) and gluons. This result is especially useful for calculation on coarse lattice.[12] However, here we are looking for analytic results. The strategy is the following.[16,17] At the $O(v^2)$ the Pauli propagator satisfies the equation

$$i\frac{\partial}{\partial x^0}K_j = \Bigg[m_j + \frac{1}{2m_j}(\mathbf{p}_j - g\mathbf{A})^2 - \frac{1}{8m_j^3}(\mathbf{p}_j - g\mathbf{A})^4 + gA^0 \quad (9)$$

$$-\frac{g}{m_j}\mathbf{S}_j\cdot\mathbf{B} - \frac{g}{8m_j^2}(\partial_i E^i - ig[A^i, E^i]) + \frac{g}{4m_j^2}\varepsilon^{ihk}S_j^k\{(p_j - gA)^i, E^h\}\Bigg] K_j$$

and therefore admits a path integral representation

$$K_j(x,y|A) = \int \mathcal{D}[\mathbf{z}_j, \mathbf{p}_j]\, \mathrm{T}_s\, e^{i\int dt\,[\mathbf{p}_j\cdot\dot{\mathbf{z}}_j - H]} \quad (10)$$

where H is the Hamiltonian appearing in (9). Starting, as in Sec. 2, from the $q\bar{q}$ Green function (3), we can substitute the quark Dirac propagator with the Pauli propagator and use Eq. (10) to obtain the closed path integral representation

$$G(T) \longrightarrow K(T) = \int \mathcal{D}[\mathbf{z}_1,\mathbf{p}_1]\int\mathcal{D}[\mathbf{z}_2,\mathbf{p}_2]\exp\left\{i\int dt\sum\left[\mathbf{p}_j\cdot\dot{\mathbf{z}}_j - m_j - \frac{\mathbf{p}_j^2}{2m_j} + \frac{\mathbf{p}_j^4}{8m_j^3}\right]\right\}$$

$$\times\Bigg\langle\frac{1}{3}\mathrm{Tr}\,\mathrm{T_s}\,\mathrm{P}\exp\Bigg\{ig\oint_\Gamma dz^\mu A_\mu(z) + \quad (11)$$

$$\sum_{j=1}^{2}\frac{ig}{m_j}\int_{\Gamma_j} dz^\mu(S_j^l\hat{F}_{l\mu}(z) - \frac{1}{2m_j}S_j^l\varepsilon^{lkr}p_j^kF_{\mu r}(z) - \frac{1}{8m_j}D^\nu F_{\nu\mu}(z))\Bigg\}\Bigg\rangle$$

with $\langle f[A]\rangle \equiv \frac{1}{Z}\int\mathcal{D}A\, f[A]\exp\{iS_{\rm YM}\}$, T_s spin ordering and P path ordering. Notice that now the kinetic energy of the quark is explicitly considered and the contour integral in A_μ and $F_{\mu\nu}$ is extended to the distorted Wilson loop in Fig. 2. Indeed, due to the presence of the path integral sum, we are considering any possible trajectory for the quark (antiquark) at variance with the static path of the previous section.

[§] For a discussion of the matching coefficients, the 4-fermions interaction and the power in v of the operators in (8) see.[5]

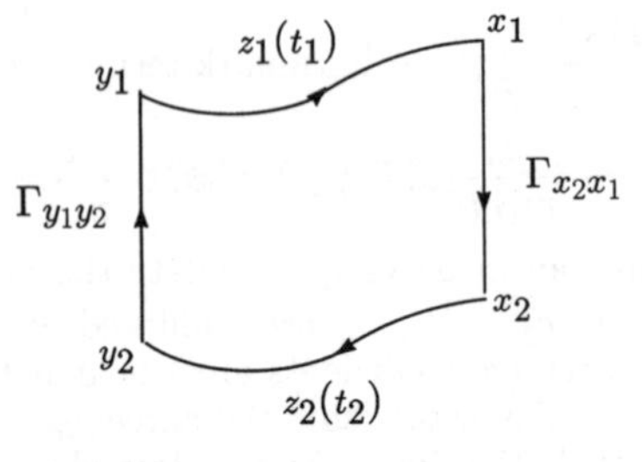

Fig. 2. Distorted Wilson loop.

Then, the potential exists if we can make the following identification in (11)

$$\left\langle \frac{1}{3}\mathrm{Tr}\cdots\right\rangle = \mathrm{T_s}\exp\left[-i\int dt\, V(\mathbf{z},\mathbf{p},\mathbf{S})\right]. \tag{12}$$

From Eqs.(11) and (12) we find[16]

$$\int_{t_i}^{t_f} dt V_{\mathrm{Q\bar{Q}}} = i\log\langle W(\Gamma)\rangle \tag{13}$$

$$-\sum_{j=1}^{2}\frac{g}{m_j}\int_{\Gamma_j} dz^\mu \left(S_j^l \langle\!\langle \hat{F}_{l\mu}(z)\rangle\!\rangle - \frac{1}{2m_j}S_j^l\varepsilon^{lkr}p_j^k \langle\!\langle F_{\mu r}(z)\rangle\!\rangle - \frac{1}{8m_j}\langle\!\langle D^\nu F_{\nu\mu}(z)\rangle\!\rangle \right)$$

$$-\sum_{j,j'=1}^{2}\frac{ig^2}{2m_j m_{j'}}\mathrm{T_s}\int_{\Gamma_j} dz^\mu \int_{\Gamma_{j'}} dz'^\sigma\, S_j^l S_{j'}^k \left(\langle\!\langle \hat{F}_{l\mu}(z)\hat{F}_{k\sigma}(z')\rangle\!\rangle - \langle\!\langle \hat{F}_{l\mu}(x)\rangle\!\rangle \langle\!\langle \hat{F}_{k\sigma}(x')\rangle\!\rangle\right)$$

where $\langle\!\langle f(A)\rangle\!\rangle \equiv \langle f(A)W(\Gamma)\rangle/\langle W(\Gamma)\rangle$ is the vacuum expectation value in presence of a quark–antiquark pair (i.e. the Wilson loop). Eq. (13) is the quark-antiquark potential at order v^2 of the systematic expansion in v. No expansion in the coupling constant has been performed. The result is physically transparent. In Eq. (13) the first term contains the static and the velocity-dependent potential, the second is the magnetic interaction, the third is the Thomas precession, the fourth is the Darwin term and the last one is the spin-spin interaction. All the dynamics is contained in the Wilson loop and in the v.e.v. of field strength in presence of the Wilson loop, hence is pure gluodynamics. Had we worked in QED, we would obtain the same result as in (13). Notice that the coupling underlying (13) is the vectorial coupling of the Dirac equation. The difference between QED and QCD is contained in the behavior of the v.e.v. of field strengths and Wilson loop. The non linear dynamics of QCD determines the nonperturbative behavior of these v.e.v and accounts for the difference with QED.

Varying the quark path $z_j^\mu(t) \to z_j^\mu(t) + \delta z_j^\mu(t)$, the Mandelstam formula can be obtained

$$g\langle\!\langle F_{\mu\nu}(z_j)\rangle\!\rangle = (-1)^{j+1}\frac{\delta i\log\langle W(\Gamma)\rangle}{\delta S^{\mu\nu}(z_j)}$$

$$g^2(\langle\!\langle F_{\mu\nu}(z_1)F_{\lambda\rho}(z_2)\rangle\!\rangle - \langle\!\langle F_{\mu\nu}(z_1)\rangle\!\rangle\langle\!\langle F_{\lambda\rho}(z_2)\rangle\!\rangle) = -ig\frac{\delta}{\delta S^{\lambda\rho}(z_2)}\langle\!\langle F_{\mu\nu}(z_1)\rangle\!\rangle$$

with $\delta S^{\mu\nu}(z_j) = dz_j^\mu \delta z_j^\nu - dz_j^\nu \delta z_j^\mu$. We conclude that to obtain the complete quark-antiquark order $O(v^2)$ interaction (quenched) no other assumptions are needed than the behavior of $\langle W(\Gamma)\rangle$: *given* $\langle W(\Gamma)\rangle$ *everything is analytically calculable.* On the other side, expanding the average on the distorted Wilson loop Γ in terms of the static Wilson loop Γ_0 we get expressions for the potentials suitable for lattice evaluation.[18] Then, *the way to validation of analytic models of the QCD vacuum via lattice data and phenomenological data is open.* The aim is to obtain information as much as possible model-independent on the nonperturbative dynamics.

3.1. General form of the potential

At the order $O(v^2)$ and with tree level matching coefficients the quark-antiquark interaction reads

$$V = V_0 + V_{\rm SD} + V_{\rm VD} \tag{14}$$

with the spin-dependent interaction

$$V_{\rm SD} = \frac{1}{8}\left(\frac{1}{m_1^2}+\frac{1}{m_2^2}\right) + \Delta\left[V_0(r)+V_{\rm a}(r)\right]\left(\frac{\mathbf{L}_1\cdot\mathbf{S}_1}{2m_1^2}-\frac{\mathbf{L}_2\cdot\mathbf{S}_2}{2m_2^2}\right)\frac{1}{r}\left[V_0'(r)+2V_1'(r)\right] \tag{15}$$
$$+\frac{1}{m_1m_2}(\mathbf{L}_1\cdot\mathbf{S}_2-\mathbf{L}_2\cdot\mathbf{S}_1)\frac{V_2'(r)}{r}+\frac{1}{m_1m_2}\left(\frac{\mathbf{S}_1\cdot\mathbf{r}\,\mathbf{S}_2\cdot\mathbf{r}}{r^2}-\frac{\mathbf{S}_1\cdot\mathbf{S}_2}{3}\right)V_3(r)+\frac{\mathbf{S}_1\cdot\mathbf{S}_2}{3m_1m_2}V_4(r)$$

where $\mathbf{L}_j = \mathbf{r}\times\mathbf{p}_j$, and the velocity-dependent interaction is given by ($\{\ \}$ = Weyl ordering),

$$V_{\rm VD} = \frac{1}{m_1m_2}\{\mathbf{p}_1\cdot\mathbf{p}_2V_{\rm b}(r)\}+\frac{1}{m_1m_2}\left\{\left(\frac{\mathbf{p}_1\cdot\mathbf{p}_2}{3}-\frac{\mathbf{p}_1\cdot\mathbf{r}\,\mathbf{p}_2\cdot\mathbf{r}}{r^2}\right)V_{\rm c}(r)\right\}$$
$$+\sum\frac{1}{m_j^2}\{p_j^2V_{\rm d}(r)\}+\sum\frac{1}{m_j^2}\left\{\left(\frac{p_j^2}{3}-\frac{\mathbf{p}_j\cdot\mathbf{r}\,\mathbf{p}_j\cdot\mathbf{r}}{r^2}\right)V_{\rm e}(r)\right\}. \tag{16}$$

The dynamics is contained in the $V_i(r)$ functions. These can be evaluated analytically via functional derivatives or expansion of the Wilson loop (once its nonperturbative behavior is assigned in a vacuum model), see[19,16] or can be evaluated numerically on the lattice from v.e.v. of field strength insertions in the static Wilson loop, see.[18,20] The spin-dependent potentials agree with the Eichten-Feinberg result.[7] The velocity dependent potentials were obtained in[16,18] and *have to be taken into account* to be consistent at the order v^2 of the systematic expansion. The lattice results are presented in Sec. 5 while in the next Section we summarize the analytic results obtained in various models of the QCD vacuum. The common thread of these models is reproducing the interquark flux tube structure.

4. Infrared dynamics and flux tube structure

The expression (13) for the heavy quark interaction is exact at the order of the expansion in v considered. It is gauge-invariant and this allows us to insert approximations, i.e. vacuum models to get the nonperturbative behavior of the Wilson loop average. We retain the relevant configuration in the nonperturbative regime, which turns out to be a flux-tube like configuration.

4.1. A simple model calculation

The simplest model consist in assuming that the short and the long range contributions to the Wilson loop factorize and evaluating the last one using an area law (as suggested by the strong coupling expansion)

$$\log\langle W(\Gamma)\rangle = \log\left[\langle W(\Gamma)\rangle^{\rm SR}\cdot\langle W(\Gamma)\rangle^{\rm LR}\right] = -\frac{4}{3}g^2\oint_\Gamma dx^\mu \oint_\Gamma dy^\nu D_{\mu\nu}(x-y)+\sigma S_{\rm min}+\frac{C}{2}\mathcal{P}. \tag{17}$$

$D_{\mu\nu}$ is the free gluon propagator, σ is the string tension, $S_{\rm min}$ is the minimal area enclosed in the *distorted* Wilson loop contour (cf. Fig. 2), C is a constant and $\mathcal{P}$ is the perimeter of the loop. The general expression for the minimal surface with contour Γ is the Nambu-Goto action. In ref.,[16] it is shown that, at the order v^2 of the interaction, the minimal area is equivalent to the area spanned by straight lines connecting points at equal time on the quark and antiquark trajectories. For any details we refer to.[16,19] Here, we only present the final result of this minimal area law model (MAL)

$$V_0(r) = -\frac{4\alpha_s}{3r}+\sigma r + C \quad V_b(r) = \frac{8\alpha_s}{9r}-\frac{\sigma}{9}r \quad V_c(r) = -\frac{2\alpha_s}{3r}-\frac{\sigma}{6}r \quad V_d(r) = -\frac{\sigma}{9}r-\frac{C}{4}$$
$$V_e(r) = -\frac{\sigma}{6}r \quad V_1'(r) = -\sigma \quad V_2'(r) = \frac{4\alpha_s}{3r^2} \quad V_3(r) = 4\frac{\alpha_s}{r^3} \quad V_4(r) = \frac{32}{3}\pi\alpha_s\delta^3(\mathbf{r}) \tag{18}$$

and $\Delta V_a = 0$. The one gluon exchange term (proportional to α_s) coincides with the usual Breit-Fermi potential. Let us focus on the nonperturbative part (in σ and C). The $V_1 - V_4$ spin potentials (sd) agree with the electric confinement calculation of Eichten and Feinberg[7] (taking into account the Gromes correction[10]). The $V_a - V_e$ velocity potentials (vd) were first calculated in MAL.[16,18] To make clear the physical content of the result, let us write the nonperturbative contributions in the center of mass system for equal masses. The spin-dependent part is

$$V_{\rm SD}^{NP} = -\frac{\sigma}{2m^2 r}\mathbf{L}\cdot\mathbf{S} \tag{19}$$

and corresponds to *pure Thomas precession*, i.e. the magnetic contribution is zero. The velocity dependent part is

$$V_{\rm VD}^{NP} = -\frac{\sigma}{6}\frac{\mathbf{L}^2}{m^2 r} \tag{20}$$

and it is proportional to the flux tube angular momentum. Indeed, the same vd correction is obtained in the relativistic flux tube model[21] where the energy of the interquark flux tube is explicitly added to the Hamiltonian¶. Then, it is clear that our gauge-invariant Wilson loop approach has automatically included the energy of the flux tube‖.

¶In the relativistic flux tube model the Lagrangian is composed by the relativistic energy of the quarks plus the relativistic energy of the flux tube considered as a mechanical tube with constant energy density and velocity transversal to the interquark line, see.[21]

‖However, we are not using any 'mechanical model.'

The result is even more interesting since *it cannot be obtained* within the semirelativistic reduction of a pure convolution (i.e. depending only on the momentum transfer $Q = p_1 - p_1'$) Bethe-Salpeter kernel, with any mixture of vector and scalar components. In general, a scalar $1/Q^4$ BS kernel is used in order to reproduce (19), however the velocity dependent correction differ from (20), see Sec. 5. On the other hand, the result (19)-(20) is physically transparent. Imagine a quark-antiquark pair connected by a chromoelectric flux tube. The magnetic field is zero in the comoving frame and then the spin-interaction is purely Thomas precession; the electric tube is moving with a transversal velocity and then its energy originates the vd term. This is the Buchmüller picture. We emphasize that the 'scalar-like' character of the spin splitting we obtained is *dynamically* generated through *the collective nature of the gluonic degrees of freedom.* It is interesting that in[22] this same conclusion has been obtained working in the framework of the Coulomb gauge Hamiltonian and diagonalizing a sector of the Fock space. Again, the scalar interaction is effectively generated. However, the Wilson loop approach appears to be simpler and more powerful (at least for heavy quarks); the intermediates states that are explicitly considered in the Fock Hamiltonian approach here are summed up into the functional formalism. Expressing the potential in terms of the Wilson loop we keep the relevant degrees of freedom. Our special 'trial function' constructed with the string interquark operator has selected out the flux tube configuration.

The MAL model is in many respects too rough (e.g. the $\langle\langle F_{\mu\nu} F_{\rho\sigma} \rangle\rangle$ v.e.v. turns out to be zero and the flux tube turns out to be infinitely thin) and it may play some role only in the limit of very large interquark distances where an effective relativistic string model description is expected to hold. In the next Sections we discuss more sophisticated assumptions.

4.2. Gaussian dominance in gluodynamics versus Abelian dominance in infrared QCD

We need models of the QCD vacuum which provide us with the nonperturbative behavior of the Wilson loop average. To this aim we want to exploit all the available lattice information on the mechanism of confinement and the measurements of Wilson loop and field strength v.e. v.. About the mechanism of confinement, at this Conference, evidence[3,2] was presented that the QCD infrared dynamics is well approximated by a (dual) Abelian Higgs model. Indeed, in the Abelian projection the QCD gluodynamics is reduced to Abelian fields, Abelian monopoles and charged matter fields degrees of freedom. Lattice simulations show Abelian dominance and monopole dominance in the long range features of QCD and condensation of monopoles in the confined phase. For details we refer to.[3,2] The measured electric fields and magnetic currents in the presence of static quark sources are consistent with dual Ginzburg–Landau type of equations. The penetration length $\lambda = 1/M$ (M = dual gluon mass) and the correlation length $\xi = 1/M_\phi$ (M_ϕ = Higgs mass) are measured. It is found $M \simeq M_\phi$ and therefore the QCD vacuum behaves as a dual superconductor on the border between type I and type II. Flux tube solutions (of the type of Abrikosov-Nielsen-Olesen vortices) exist and the structure of these solutions is controlled by the penetration and the correlation length. However these are lattice measurements and it is not clear e.g. the relation of λ and ξ with the parameters of QCD.

On the other side, let us consider the v.e.v. of the Wilson loop. It pays to expand this average in terms of fields strength expectation values, by using the non-Abelian Stokes theorem[23]

$$\langle W(\Gamma)\rangle \equiv \langle \exp\{ig\oint_\Gamma dz_\mu A_\mu(z)\}\rangle = \langle P\exp ig\int_{S(\Gamma)} dS_{\mu\nu}(1)U(0,1)F_{\mu\nu}(1)U(1,0)\rangle = \quad (21)$$

$$\exp\{\sum_{n=0}^{\infty}\frac{(ig)^n}{n!}\int_{S(\Gamma)} dS_{\mu_1\nu_1}(1)\cdots dS_{\mu_n\nu_n}(n)\langle U(0,1)F_{\mu_1\nu_1}(1)U(1,0)\cdots F_{\mu_n\nu_n}(n)U(n,0)\rangle_{\text{cum}}\}$$

where $i \equiv x_i$, $S(\Gamma)$ denotes a surface with contour Γ and $\langle\ldots\rangle_{\text{cum}}$ stands for the cumulant average.[23] In principle, the v.e.v. of the Wilson loop is given by the sum of all the cumulants. However, in a recent lattice investigation,[24] evidence of the Gaussian dominance in the cumulant expansion of quasi-static Wilson loop average was found. Therefore, it follows that

$$\langle W(\Gamma)\rangle \simeq \exp\{-\frac{1}{2}\int_{S(\Gamma)} dS_{\mu\nu}(0)\int_{S(\Gamma)} dS_{\rho\sigma}(x)\langle g^2U(0,x)F_{\mu\nu}(x)U(x,0)F_{\rho\sigma}(0)\rangle\} \quad (22)$$

is a good approximation. This is the basic assumption in the (Gaussian) stochastic vacuum model[23] and it was phenomenologically confirmed by calculation in high energy scattering[25] and quarkonia. Then, the heavy quark interaction is determined by the two-point field strength correlator

$$g^2\langle U(0,x)F_{\mu\nu}(x)U(x,0)F_{\lambda\rho}(0)\rangle = \frac{g^2\langle F^2(0)\rangle}{24N_c}\Big\{(\delta_{\mu\lambda}\delta_{\nu\rho}-\delta_{\mu\rho}\delta_{\nu\lambda})(D(x^2)+D_1(x^2))$$
$$+(x_\mu x_\lambda\delta_{\nu\rho}-x_\mu x_\rho\delta_{\nu\lambda}+x_\nu x_\rho\delta_{\mu\lambda}-x_\nu x_\lambda\delta_{\mu\rho})\frac{d}{dx^2}D_1(x^2)\Big\}. \quad (23)$$

In (23) the Lorentz decomposition is general and the dynamics is contained in the form factors D and D_1. The function D is responsible for area law and confinement (indeed in QED, due to the Bianchi identity, we have $D=0$). For D and D_1 the lattice calculations[24,26] gives an exponential long-range decreasing behavior $\simeq G_2\exp\{-|x|/T_g\}$, where $G_2 \equiv \langle\alpha_s F^2(0)\rangle/\pi$ is the gluon condensate and $T_g \simeq 0.2\,\text{fm}$ ** is the gluon correlation length.

In ref.,[27] the QCD two–point field strength correlator (23) has been related to the dual field propagator of the effective Abelian Higgs model describing infrared QCD. In this way the Gaussian dominance in the Wilson loop average is understood as following from the classical approximation†† in the dual theory. Moreover, it is possible to relate the QCD parameter T_g and G_2 to the dual parameters. In the London limit T_g is identified with the dual gluon mass M, without the London limit the relation is more involved but still T_g is expressed in terms of the dual theory parameters.

The conclusion is that we need two parameters T_g and G_2 to describe the heavy quark dynamics and indeed they are necessary to control the structure of the flux tube.

**Phenomenological calculation in high energy scattering indicates $T_g \equiv 0.3 \div 0.35\,\text{fm}$.[25]

††Surely valid in the dual description.

Had we only one parameter, like the string tension σ, we could encode the information of a constant energy density in the flux tube. However, the whole structure is important, and also the information about the width of the flux tube has to be considered. In the limit of very large interquark distances and in particular dynamical regimes, we can store the relevant information in one parameter, the string tension.

4.3. Vacuum models for the Wilson loop

As shown above, one obtains the $O(v^2)$ quenched quark dynamics in a given vacuum model, simply evaluating the Wilson loop in that model. In this way the phenomenological data are put in direct relation with the assumptions on the QCD vacuum. Here we consider three models: Stochastic vacuum model (SVM),[25] dual QCD (DQCD)[28] and the flux tube model of Isgur and Paton.[29] We emphasize that these are models of the QCD vacuum, valid at the confinement scale.

•*Stochastic vacuum model.* The stochastic vacuum model[25] is based on the idea that the infrared part of the QCD functional integral can be approximated by a stochastic process with a converging cluster expansion and a finite correlation length T_g. This assumption is well confirmed by the lattice data. Then, the Wilson loop is given by Eqs.(22)-(23) with parameters T_g and G_2. The behavior of D and D_1 and the value of T_g are taken from the lattice measurements, G_2 from the phenomenological data. The static potential is

$$V_0(r) \simeq G_2 \int_0^\infty d\tau \int_0^r d\lambda (r-\lambda) D(\tau^2+\lambda^2) \tag{24}$$

and the string tension σ emerges as an integral on the D function $\sigma \simeq G_2 \int_0^\infty d\tau \int_0^\infty d\lambda\, D(\tau^2+\lambda^2)$ in the limit $T_g/r \to 0$. The field distribution between the quark and the antiquark is a flux tube[30] with $r_{MS} \sim 1.8\, T_g$. Similarly the sd and vd potential are obtained. I refer to[23,19] for the details. Here we restrict the discussion to the long range limits of the potentials in the three vacuum models, (see the final subsection).

•*Dual QCD.* DQCD[28] is a concrete realization of the Mandelstam t'Hooft dual superconductor mechanism of confinement. It describes the QCD vacuum as a dual superconductor on the border between type I and II. The Wilson loop approach supplies with a simple method to connect averaged local quantities in QCD and in dual QCD (=dual gluodynamics). For large loops we assume[28,31]

$$\langle W(\Gamma)\rangle \simeq \langle W(\Gamma)\rangle_{Dual} = \frac{\int \mathcal{D}C\,\mathcal{D}\mathcal{B}\, \exp\left[i\int dx L(G^{\rm S}_{\mu\nu})\right]}{\int \mathcal{D}C\,\mathcal{D}\mathcal{B}\, \exp\left[i\int dx L(0)\right]} \tag{25}$$

where C_μ are the dual potentials, $G_{\mu\nu}$ the dual field strengths, $\mathcal{B}_i$ the monopole fields, L the effective dual Lagrangian in the presence of quarks $L(G^{\rm S}_{\mu\nu}) = 2\ {\rm Tr}\{-\frac{1}{4}\mathcal{G}^{\mu\nu}\mathcal{G}_{\mu\nu} + \frac{1}{2}(\mathcal{D}_\mu\mathcal{B}_i)^2\} - U(\mathcal{B}_i)$, and $U(\mathcal{B}_i)$ the Higgs potential. The quark sources moving around the Wilson loop are inserted via the Dirac string tensor $G^{\rm S}_{\mu\nu}(x) = g\,\epsilon_{\mu\nu\alpha\beta}\int ds \int dt\, y'^\alpha \dot{y}^\beta\, \delta(x-y)$. From (25) it follows

$$g\langle\langle F_{\mu\nu}(z_j)\rangle\rangle = \frac{2}{3} g\,\varepsilon_{\mu\nu\rho\sigma}\langle\langle G^{\rho\sigma}(z_j)\rangle\rangle_{Dual} \tag{26}$$

and it is possible to relate the averaged values of local quantities in QCD and in the dual theory. Then Eqs.(15)-(16) produce the $O(v^2)$ interaction in the dual formalism, for details cf..[31] We discuss only the long range limits in the final subsection. The nonperturbative parameters are the v.e.v. of the Higgs field, $\mathcal{B}_0$, and κ, the coupling constant of the Higgs potential (from these the penetration length and the correlation length can be constructed). The flux tube configurations ($\simeq$ Abrikosov-Nielsen-Olesen vortex) with finite $r_{MS} \sim 1/M$, M = mass of the dual gluon, arise[32] from the numerical solution of the classical dual Ginzburg-Landau type of equations obtained from L. It is the presence of the Higgs field which confines transversally the energy distribution in a flux tube. Again σ comes from the integration on a function exponentially decreasing approximately with M.

•*Isgur-Paton flux tube model.* This model is extracted from the strong coupling limit of the QCD lattice Hamiltonian. A N-body discrete string-like model Hamiltonian describes the gluonic degrees of freedom. The limit $N \to \infty$ corresponds to a localized string with an infinite number of degrees of freedom; the radius of the flux tube is proportional to $1/N$. For any detail we refer to the talk of Paton in these Proceedings. In a recent paper,[22] the sd potential has been obtained in this model.

4.4. Potentials and flux tube structure

It is interesting to consider the potentials obtained using these different vacuum models in the limit of large interquark distance r.

•*SVM*

$$V_0(r) = \sigma r + \frac{1}{2}c^{(1)}\sigma T_g - c\sigma T_g, \quad \frac{d}{dr}V_1(r) = -\sigma + \frac{c\sigma T_g}{r}, \quad \frac{d}{dr}V_2(r) = \frac{c\sigma T_g}{r}, \tag{27}$$

V_3 and V_4 fall off exponentially as $\exp\{-r/T_g\}$ and

$$V_b(r) = -\frac{\sigma}{9}r - \frac{2}{3}\frac{d\sigma T_g^2}{r} + \frac{8}{3}\frac{e\sigma T_g^3}{r^2}, \qquad V_c(r) = -\frac{\sigma}{6}r - \frac{d\sigma T_g^2}{r} + \frac{2}{3}\frac{e\sigma T_g^3}{r^2}, \tag{28}$$

$$V_d(r) = -\frac{\sigma}{9}r + \frac{c}{4}\sigma T_g - \frac{c^{(1)}}{8}\sigma T_g + \frac{d\sigma T_g^2}{3\ r} - \frac{2}{9}\frac{e\sigma T_g^3}{r^2} \quad V_e(r) = -\frac{\sigma}{6}r + \frac{d\sigma T_g^2}{2\ r} - \frac{e\sigma T_g^3}{3\ r^2}.$$

•*DQCD*

$$V_0(r) = \sigma r - (C + C^{(1)})\frac{\sigma}{M} \qquad \frac{d}{dr}V_1(r) = -\sigma + C\frac{\sigma}{M}\frac{1}{r} \qquad \frac{d}{dr}V_2(r) = C\frac{\sigma}{M}\frac{1}{r} \tag{29}$$

and V_3 and V_4 fall off exponentially as $\exp\{-Mr\}$.

•*I-P flux tube model*

$$\frac{d}{dr}V_1(r) = -\sigma \quad \frac{d}{dr}V_2(r) = \lim_{N\to\infty}\frac{\sigma}{N} \tag{30}$$

and V_3 and V_4 are $\simeq 1/N$. The $c, c^{(1)}, d, e, C, C^{(1)}$ above are numerical constants known in terms of the parameters of the models. We learn that

1) The gluon correlation length T_g in SVM has the same role as the inverse of the dual gluon mass M in DQCD and T_g/r as the inverse of the number of degrees of freedom N in the I-P flux tube model. All these quantities act like a correlation length. From now on, we use T_g to refer to any of them.

2) The MAL results are completely contained in all these models[‡‡]. The MAL describes the limit of large interquark distances or better the limit $T_g/r \to 0$ which is also the limit in which a potential exists, cf. Sec. 6, or the string limit. Indeed, all the corrections to the MAL are proportional to the gluon correlation length or equivalently to the width of the flux tube: the form of these corrections is *the same* in all these flux tube-like models (models that predict the flux tube structure).

3) The magnetic interaction is zero only in the limit $T_g/r \to 0$. In the intermediate distance region is different from zero as one can see from the value of V_2. Then a purely scalar effective interaction emerges only in the long range limit.

4) T_g controls the width and the shape of the flux tube as well as the validity of the potential description. Physically it should correspond to the size of the color domain as well as the fluctuation of the color fields.

We conclude that *the results for the heavy quark interaction are essentially the same for any flux-tube model of the QCD vacuum.*

5. Lattice calculations, flux tube models and $1/Q^4$ models

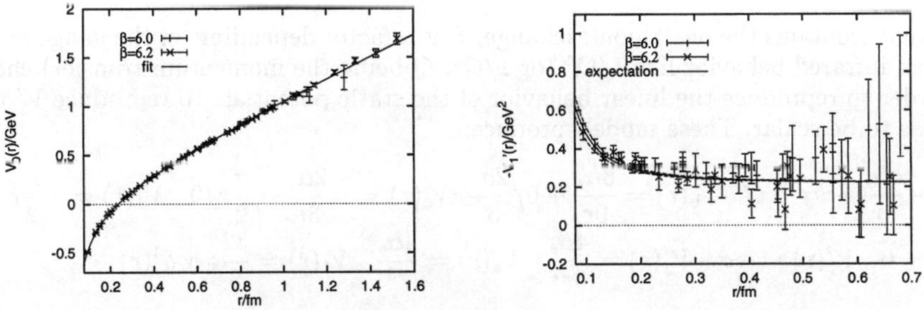

Fig. 3. V_0 and V_1' potentials from.[20]

An entire class of confining models for the heavy quark dynamics comes from the semirelativistic reduction of a pure convolution Bethe-Salpeter kernel (see[33] and refs. therein)

$$I(Q^2) = \gamma_1^\mu \gamma_2^\nu P_{\mu\nu} J_v(Q) + J_s(Q) \tag{31}$$

‡‡For the vd potential in DQCD see.[31]

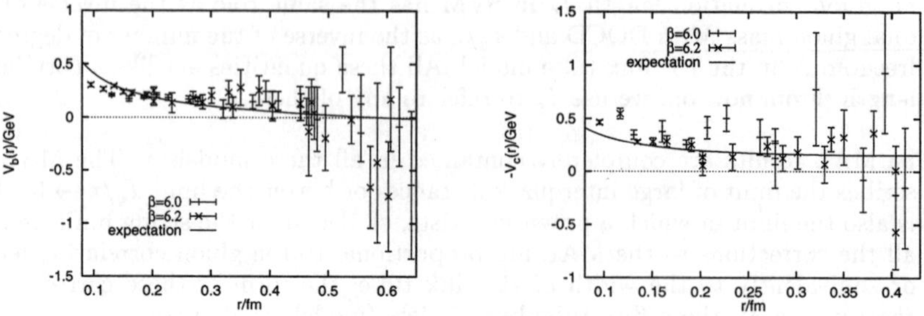

Fig. 4. $V_{\rm b}$ and $V_{\rm c}$ potentials from.[20]

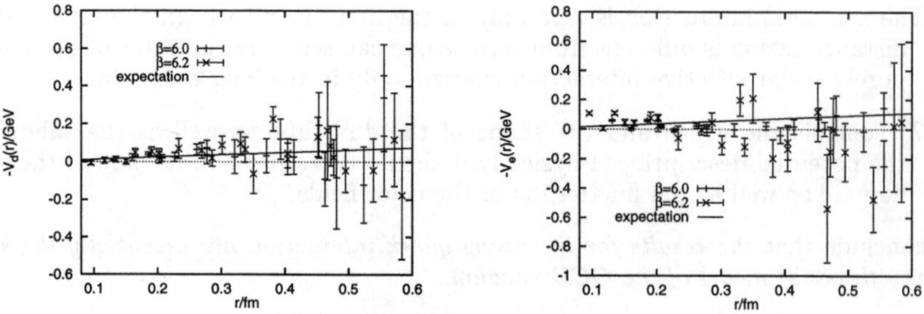

Fig. 5. $V_{\rm d}$ and $V_{\rm e}$ potentials from.[20]

where J_v contains the one gluon exchange, P is a factor depending on the gauge and J_s has an infrared behavior like $1/\mathbf{Q}^4$ (or $1/Q^4$, Q being the momentum transfer) chosen in order to reproduce the linear behavior of the static potential. To reproduce V_1' then J_s has to be scalar. These models produce:

$$V_0 = -\frac{4\alpha_{\rm s}}{3r} + \sigma r + C \quad V_{\rm b}(r) = \frac{8\alpha_{\rm s}}{9r} + 0/\frac{2\sigma}{3} \quad V_{\rm c}(r) = -\frac{2\alpha_{\rm s}}{3r} - \frac{\sigma}{2}r/0 \quad V_{\rm d}(r) = -\frac{\sigma}{2}r - \frac{C}{4}$$

$$V_{\rm e}(r) = 0 \quad V_1'(r) = -\sigma \quad V_2'(r) = \frac{4\alpha_{\rm s}}{3r^2} \quad V_3(r) = \frac{4\alpha_{\rm s}}{r^3} \quad V_4(r) = \frac{32}{3}\pi\alpha_{\rm s}\delta^3(\mathbf{r}) \qquad (32)$$

The two results in Eq. (32) correspond respectively to the instantaneous form of the kernel $1/\mathbf{Q}^4$ or to the inclusion of retardation corrections as in.[34] Notice that: 1) These models are constructed in order to reproduce V_1 as given in Eq. (18). 2) They cannot reproduce the behavior of the $V_{\rm VD}^{NP}$ (20) (even adding retardation corrections or allowing a mixture of vector and scalar confining kernels[34]). 3) The $1/Q^4$ propagator *does not avoid the spreading of the flux tube* between the quarks *even if the integrated energy increases linearly with the distance* between the quarks.

We conclude that a linear confining behavior with the correct spin-orbit *is not sufficient to characterize* the heavy quark dynamics at the order $O(v^2)$. On the other hand the v^2 corrections to the potential are connected to v.e.v. of field strengths insertion in the Wilson loop and therefore to the energy density and to the interquark flux tube structure. *The $1/Q^4$ models fail to take into account the dynamical nature of the glue.* In Figs. 3-5 we present the latest lattice results for the potentials.[20] The lines are the MAL predictions (18). The static potential is clearly linear for $r > 0.2\,\mathrm{fm}$ (recent results indicate that the string breaking happens for $r > 0.2\,\mathrm{fm}^2$). The nonperturbative behavior of V_1' is consistent with the result (18) and (32). Unfortunately, the lattice data for the vd potential are not yet sufficiently accurate to discriminate between the flux tube model and the $1/Q^4$ predictions. Notice that an accurate measurement of V_e provides a good test since it is predicted to be zero in $1/Q^4$ models.

Phenomenological applications to the bottomonium and charmonium spectrum show that the vd corrections of the type (18) do not induce the deviations from the data originated by the vd corrections (32).[35]

6. On the validity of the potential description

In the previous sections we assumed that a potential description holds. This is not always the case. We know from the Lamb-shift in QED that at some order of the perturbative expansion the ultrasoft photons start giving a contribution to the energy not of a potential type. This is certainly true also in QCD but here we cannot rely only on the perturbative calculation. Then, the situation becomes more interesting and again the flux tube configuration plays a role. We can extract a potential as far as the adiabatic approximation holds, i.e., as far as the gluon time scale is short compared to the orbital period of the quarks $T_g \ll T_q$. In this case we are describing a $q\bar{q}$ pair moving in the adiabatic potential corresponding to the ground state energy of the gluon field, i.e. the ground state flux tube. The next adiabatic surface describes an excited state of the glue: *an hybrid.*[4] If these adiabatic surfaces are well separated, still each of them can be described with a potential. The situation is controlled by the gluon correlation length T_g which is proportional to the this difference. The quark dynamics is simply given in terms of the two-point correlator (23) whose nonperturbative exponential decreasing behavior is controlled by T_g. In the case in which T_g is small with respect to the other physical scales of the problem we recover the potential description, in the opposite limit we recover the sum rules description with local condensates à la Leutwyler-Voloshin.[36] Therefore there is no discrepancy between these two descriptions. However, the $c\bar{c}$ and $b\bar{b}$ systems lie in the situation in which T_g is small compared to the other scales. Just for the ground state of the $b\bar{b}$ both descriptions can be undertaken.

Recently a new effective theory has been obtained from NRQCD integrating out the relative momentum scale. It is called potential non relativistic QCD (pNRQCD) and was presented at this Conference by J. Soto.[37] The theory describes ultrasoft gluons and the matching coefficients are the potentials. pNRQCD could be a powerful tool to understand the relation between ultrasoft degrees of freedom and nonpotential nonperturbative glue.

7. Confinement, flux tube and chiral symmetry

As soon as the light quarks enter the game we are left without any expansion parameter to develop any well–founded calculation of the quark dynamics. Even in the simplest case, a meson formed by a light and a heavy quark, we are far away from the intuitive description developed for the heavy–heavy bound states. Indeed, in this case the heavy quark is surrounded by the fuzzy cloud of the light degrees of freedom. This picture is naively completely different from a flux tube picture. Moreover chiral symmetry breaking should appear. Here, we address the issue whether the Wilson loop formalism still helps to understand the quark dynamics. If this is the case we are in the position to study the interplay between confinement and chiral symmetry breaking.

Let us consider the simplest system, the heavy-light bound state. Heavy-light systems are understood in terms of heavy quark effective theory (HQET), but HQET cannot predict the spectrum of the 'brown muck' since this is determined by the dynamics of strong QCD. In the limit $m_Q \gg \Lambda_{QCD}$, HQET organizes a systematic expansion of the physical quantities in terms of Λ_{QCD}/m_Q and $\alpha_s(m_Q)$. The meson mass is given as

$$m_M = m_Q + \bar{\Lambda} + O\left(\frac{1}{m_Q}\right) \text{ corrections.} \tag{33}$$

The parameter $\bar{\Lambda}$ can be fixed on the data but its actual calculation needs a dynamical input. In the no-recoil limit $m_Q \to \infty$, the usual choice is to take a scalar Dirac equation to get the $\bar{\Lambda}$ parameter as an eigenvalue. The motivation is reproducing the spin-orbit splitting and the fact that this kind of equation is mathematically well behaving. There are a number of reasons against this choice:

- It breaks explicitly chiral symmetry.
- We have shown that scalar confinement arises effectively for heavy quark interaction due to the collective nature of the nonperturbative gluonic degrees of freedom.

In the next section we address the problem in the Wilson loop formalism.

7.1. *Gauge-invariant approach to heavy-light quark systems*

We start from the gauge-invariant quark-antiquark Green function in the Feynman-Schwinger representation[38,39]:

$$\begin{aligned}
G(x,u,y,v) = \frac{1}{4}\Big\langle \mathrm{Tr}\,\mathrm{P}\,(i\,\not{D}_y^{(1)} + m_1) \int_0^\infty dT_1 \int_x^y \mathcal{D}z_1 e^{-i\int_0^{T_1} dt_1 \frac{m^2+\dot{z}_1^2}{2}} \\
\times \int_0^\infty dT_2 \int_v^u \mathcal{D}z_2 e^{-i\int_0^{T_2} dt_2 \frac{m^2+\dot{z}_2^2}{2}} e^{ig\oint_\Gamma dz^\mu A_\mu(z)} \\
\times e^{i\int_0^{T_1} dt_1 \frac{g}{4}\sigma_{\mu\nu}^{(1)} F^{\mu\nu}(z_1)} e^{i\int_0^{T_2} dt_2 \frac{g}{4}\sigma_{\mu\nu}^{(2)} F^{\mu\nu}(z_2)} (-i\,\overleftarrow{\not{D}}_v^{(2)} + m_2)\Big\rangle.
\end{aligned} \tag{34}$$

Again the dynamics is contained in the Wilson loop, that now looks like Fig. 6. We can exploit the symmetry of the situation, taking the modified coordinate gauge $A_\mu(x_0, \mathbf{0}) = 0$, $x^j A_j(x_0, \mathbf{x}) = 0$ ($A_0(x) = \int_0^1 d\alpha x^k F_{k0}(x_0, \alpha\mathbf{x})$, $A_j(x) = \int_0^1 d\alpha\alpha x^k F_{kj}(x_0, \alpha\mathbf{x})$) in which

$$W(\Gamma) = Tr\,P\,\exp\left\{ig\int_x^y dz^\mu A_\mu(z)\right\}. \tag{35}$$

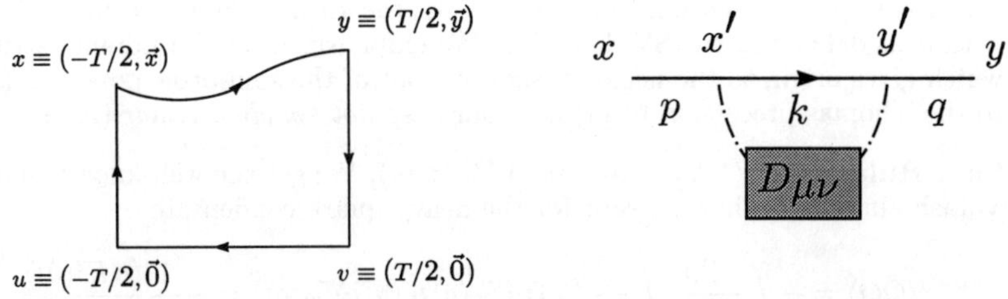

Fig. 6. The Wilson loop in the static limit of the heavy quark and the interaction kernel K.

At this point, at variance from the heavy quark case, we have to make a model dependent assumption: we consider still valid the dominance of the bilocal correlator with a finite correlation length. Under this assumption we obtain a Dirac like equation for the light quark in presence of the heavy quark. Approximating the Wilson loop as

$$\begin{aligned} \langle W(\Gamma)\rangle &= \exp\left\{-\frac{g^2}{2}\int_x^y dx'^{\mu}\int_x^y dy'^{\nu} D_{\mu\nu}(x',y')\right\}, \quad (36)\\ D_{\mu\nu}(x,y) &\equiv x^k y^l \int_0^1 d\alpha\, \alpha^{n(\mu)}\int_0^1 d\beta\, \beta^{n(\nu)}\langle F_{k\mu}(x^0,\alpha\mathbf{x})F_{l\nu}(y^0,-\beta\mathbf{y})\rangle \end{aligned}$$

where $n(0) = 0$ and $n(i) = 1$, we obtain the Dirac–like equation for the light quark propagator

$$S_D = S_0 + S_0 K S_D \quad (37)$$

with a kernel $K(y',x') \equiv \gamma^\nu S_0(y',x')\gamma^\mu D_{\mu\nu}(x',y')$, given diagrammatically in Fig. 6. Notice that[38]: 1) K *is not a translational invariant quantity*: the coordinate gauge breaks explicitly this symmetry in the propagator. Physically this is due to the presence of the heavy quark. Indeed Eq. (37) is an integral equation for the light quark propagator in the field of the heavy quark. 2) The kernel depends on $D_{\mu\nu}$ which in turns is given in terms of the two-point correlator (23). Then, the heavy light dynamics is controlled by the same two parameters controlling the heavy-heavy dynamics, T_g and G_2. 3) The problem has many relevant scales: the light mass m, the correlation length $T_g \sim \Lambda_{QCD}$, the characteristic energy and momentum of the bound states. We have different dynamical regimes in dependences on the relative values of these scales. Let us study the various situations. In the following we consider only the nonperturbative dynamics.[38]

- **Potential Case:** $m > 1/T_g > p_0 - m, \mathbf{p}, \mathbf{p}-\mathbf{q}$. We neglect the negative energy states and expand the kernel K in m. We obtain:

$$V(r) \sim G_2\{\int_{-\infty}^{+\infty} d\tau \int_0^r d\lambda (r-\lambda) D(\tau^2+\lambda^2) + \frac{\sigma\cdot\mathbf{L}}{4m^2 r}\int_{-\infty}^{+\infty} d\tau \int_0^r d\lambda \left(\frac{2\lambda}{r}-1\right) D(\tau^2+\lambda^2)\} \quad (38)$$

which coincides in the limit of large r with the Eichten-Feinberg potential (18) with σ as defined in the SVM section. We emphasize that the Lorentz structure which gives origin to the negative sign in front of the spin-orbit potential (hence to the Thomas precession term) is in our case *not simply a scalar* ($K \simeq \sigma r$).

- **Sum Rules case:** ($1/T_g < p_0 - m$, $1/T_g < m$). We get the well-known Shifman, Vainshtein and Zakharov result for the heavy quark condensate

$$\langle \bar{Q}Q \rangle = -\int \frac{d^4p}{(2\pi)^4} \int \frac{d^4q}{(2\pi)^4} \mathrm{Tr}\,\{S_0(q)K(q,p)S_0(p)\} = -\frac{1}{12}\frac{\langle \alpha F^2(0) \rangle}{\pi m}.$$

- **D_s and B_s case:** $1/T_g > m$. For D_s and B_s one can still assume that the propagator inside the kernel is free and solve the equation to get the spectrum.

- **D and B case:** $m \ll 1/T_g$. The nonlinear equation[40,41]

$$S_D = S_0 + S_0 K(S_D) S_D \tag{39}$$

has to be solved with Schwinger–Dyson like techniques. Notice that in the limit $m \to 0$ the form of the kernel K is such that the interaction turns out to be *chiral symmetric*. In Ref.[41] a solution of the gap equation corresponding to Eq. (39) (with a simplified form of the interaction) was found. In that case it was possible to disentangle the translational invariant part of the interaction (the self-energy) from the non-invariant part (the bound state interaction). In this way one obtains the light quark condensate

$$\langle 0|\bar{q}q|0\rangle \sim -T_g G_2 \tag{40}$$

and the heavy-light bound state energy spectrum. The result (40) looks appealing: it establishes a connection between the gluon condensate and the light quark condensate. The connection is possible since the non-local gluon condensate has introduced into the game a finite correlation length T_g. The same quantity that controls the width of the flux tube as well as the validity of the potential description in the case of heavy quarks.

In conclusion we have reached some insight in the interplay between confinement and chiral symmetry breaking, obtaining for the first time an unified framework for sum rules, potential models and chiral symmetry breaking studies.

8. Gauge-invariant approach to light-light quark systems

The starting point is again Eq. (34). The dynamics is contained in the distorted Wilson loop of Fig. 2. In this situation it does not exist a choice of the gauge like the modified coordinate gauge in the heavy-light case, i.e. it is not possible to get rid of both the final Schwinger strings. A result for the nonperturbative BS kernel can still be obtained in the form of effective diagrams[39] but gauge-invariance is lost. Even recovering the potential from such a kernel appears to be problematic.

9. Conclusions

We have shown that the $O(v^2)$ heavy quark interaction is controlled only by the Wilson loop behavior: heavy quarks are a nice laboratory to test QCD vacuum models with respect to lattice and phenomenological data. On the other hand, the Wilson loop is a good approach to the quark dynamics: it keeps automatically the relevant degrees of freedom and includes naturally the flux tube configurations. The description of the heavy quark interaction needs two nonperturbative parameters T_g and G_2: these give origin to σ only in particular dynamic situations. T_g controls the width and the shape of the flux tube, as well as the validity of the potential description. Any model of the QCD vacuum reproducing the interquark flux tube structure gives the same prediction for the nonperturbative heavy quark interaction. On the other hand, the class of $1/Q^4$ models give definite different predictions.

Moreover, the Wilson loop formalism appears to be useful in selecting the relevant configurations also in situations where an intuitive flux tube picture does not exist: e.g. the heavy-light system. In this way the heavy-heavy and heavy-light systems are understood in terms of the same parameters (T_g, G_2). These parameters are measured on the lattice and are also connected to the parameters of the dual superconductor mechanism.

Among the open problems/work in progress we do list: extension of the analytic calculations to the higher order corrections in NRQCD for the heavy quark interaction; inclusion of the non-potential contribution (pNRQCD?); treatment in this formalism of systems involving more than one light quark; unquenching.

10. Acknowledgements

The author acknowledges the support of the European Community, Marie Curie fellowship, TMR Contract n. ERBFMBICT961714.

References

1. G. S. Bali, C. Schlichter and K. Schilling, Phys. Rev. **D51**, 5165 (1995).
2. G. S. Bali, in these Proceedings.
3. M. Polikarpov, in these Proceedings.
4. C. Michael, in these Proceedings.
5. A. Vairo, in these Proceedings, hep-ph/9809229; N. Brambilla, J. Soto and A. Vairo in preparation; N. Brambilla and A. Vairo, *in Proceedings of QCD'98*, hep-ph/9809230.
6. L. S. Brown and Weisberger, Phys. Rev. **D20**, 3239 (1979).
7. E. Eichten and F. Feinberg, Phys. Rev. **D23**, 2724 (1981).
8. K. Wilson, Phys. Rev. **D10**, 2445 (1974).
9. M. A. Peskin, in *Proc. 11th SLAC Summer Institute*, SLAC Report No. 207, 151, edited by P. Mc. Donough (1983).
10. D. Gromes, Z. Phys. **C26**, 401 (1984).
11. Y. J. Ng, J. Pantaleone and S. H. Tye, Phys. Rev. Lett. **55**, 916 (1985).
12. B. Thacker and G. Lepage, Phys. Rev. **D43**, 196 (1991); G. Lepage, in these Proceedings.

13. A. V. Manohar, Phys. Rev. **D56**, 230 (1997).
14. A. Pineda and J. Soto, hep-ph/9802365.
15. Yu-Qi Chen, Yu-Ping Kuang and R. J. Oakes, Phys. Rev. **D52**, 264 (1995).
16. N. Brambilla, P. Consoli, G. M. Prosperi, Phys. Rev. **D50**, 5878 (1994).
17. N. Brambilla and A. Vairo, "Quark confinement and the hadron spectrum", *Lectures given at the HUGS 13th school at Cebaf*, UWThPh-1998-33.
18. A. Barchielli, N. Brambilla and G. M. Prosperi, Nuovo Cimento **103A**, 59 (1990); A. Barchielli, E. Montaldi and G. M. Prosperi, Nucl. Phys. **B296**, 625 (1988).
19. N. Brambilla and A. Vairo, Phys. Rev. **D55**, 3974 (1997); M. Baker, J. S. Ball, N. Brambilla, A. Vairo, Phys. Lett. **B389**, 577 (1996).
20. G. S. Bali, A. Wachter and K. Schilling, Phys. Rev. **D56**, 2566 (1997).
21. M. G. Olsson and K. Williams, Phys. Rev. **D48**, 417 (1993).
22. E. Swanson and P. Szczepaniak, Phys. Rev. **D55**, 3987 (1997).
23. H. G. Dosch, Phys. Lett. **B190**, 177 (1987). H. G. Dosch and Yu. A. Simonov, Phys. Lett. **205**, 339 (1998); Yu. A. Simonov, Nucl. Phys. **B324**, 67 (1989).
24. G. S. Bali, N. Brambilla and A. Vairo, Phys. Lett. **B421**, 265 (1996).
25. H. G. Dosch, in these Proceedings.
26. A. Di Giacomo, E. Meggiolaro and H. Panagopoulos, Nucl. Phys. Proc. Suppl. **54A**, 343 (1997).
27. M. Baker, N. Brambilla, H. G. Dosch and A. Vairo, Phys. Rev. **D58**, 034010 (1998).
28. M. Baker, in these Proceedings.
29. N. Isgur and J. Paton, Phys. Rev. **D31**, 2910 (1985); Phys. Lett. **B124**, 247 (1983); J. Paton in these Proceedings.
30. M. Rüter and H. G. Dosch, Z. Phys. **C66**, 245 (1995).
31. M. Baker, J. S. Ball, N. Brambilla, G.M. Prosperi and F. Zachariasen, Phys. Rev. **D54**, 2829 (1996); M. Baker, J. Ball and F. Zachariasen, Phys. Rev. **D51**, 1968 (1995).
32. M. Baker, J. S. Ball and F. Zachariasen, Int. Jour. Mod. Phys. **A11**, 343 (1996).
33. W. Lucha, F.F. Schöberl and D. Gromes, Phys. Rep. **200**, 127 (1991).
34. D. Ebert, R. N. Faustov, V. O. Galkin, hep-ph/9804335.
35. N. Brambilla and G. Prosperi, Phys. Lett. **B236**, 69 (1990).
36. U. Marquard and H. G. Dosch, Phys. Rev. **D35**, 2238 (1987).
37. J. Soto, in these Proceedings.
38. N. Brambilla and A. Vairo, Phys. Lett. **B407**, 167 (1997); N. Brambilla and A. Vairo, Nucl. Phys. B (Proc. Suppl.) **64**, 423 (1998).
39. N. Brambilla and A. Vairo, Phys. Rev. **D56**, 1445 (1997).
40. Yu. A. Simonov, Phys. Atom. Nucl. **60**, 2069 (1997); A. Nefediev, in these Proceedings.
41. P. Bicudo, N. Brambilla, E. Ribeiro and A. Vairo, hep-ph/9807460.

QCD Vacuum Models

Hans Günter Dosch
Inst. f. Theoretische Physik der Universität, Philosophenweg 16,
D69120 Heidelberg, Fed. Rep. Germany
h.g.dosch@thphys.uni-heidelberg.de

ABSTRACT

I shall present some general properties of the groundstate of QCD and two specific models for it: the model of the stochastic vacuum and dual QCD. I also point out close relations between confinement physics and diffractive high energy scattering.

1. Introduction

It is a pleasure to have the opportunity to speak here about the NOTHING, but I am afraid that though the subject seems rather empty 35 minutes are by far not enough to treat the field adequately and I therefore apologize to all those whose work is not adequately represented in this contribution.

My talk is organized as follows: First I shall shortly present what we know more or less certainly about the physical vacuum of QCD; then after some general classifying remarks I shall shortly present two vacuum models: One is the model of the stochastic vacuum, in some sense a child of sum rules, the other the model of dual QCD. On the latter field perhaps the most promising progress has been made towards an understanding of confinement in super-symmetric QCD, but this will be subject of a separate contribution of Yuval Neeman. In the last part I shall speak on a special concern of mine, namely the relation between confinement physics and high energy scattering, which adds a facet to the string dynamics presented by Nora Brambilla and Jack Paton in this conference. In order to avoid misunderstandings let me make clear that if I speak of the vacuum or physical vacuum I mean the ground-state of full QCD and condensates are expectation values of operators where all *perturbative* vacuum fluctuations are taken out, at least to the best of our faculty. So the vacuum I speak off is the *non-perturbative* vacuum of QCD.

2. Vacuum Condensates

I am so old fashioned as to give knowledge based on experiments some preference to results obtained from computer simulations on the lattice. Therefore I think best experimental evidence for a highly non-trivial structure of the QCD vacuum comes from QCD sum rules.

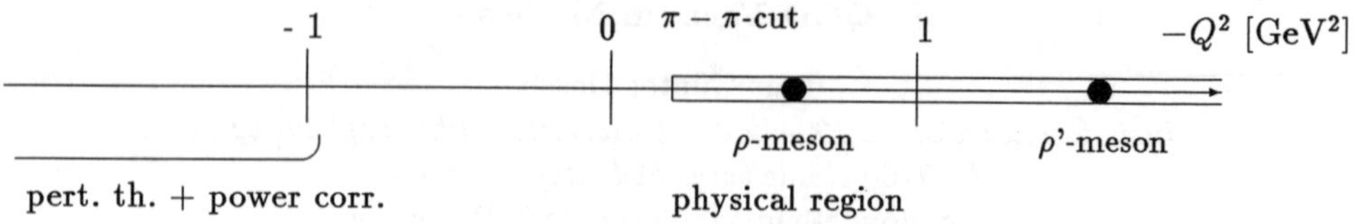

Fig. 1. The analytical structure of the correlator of the vector current with light quarks

QCD sum rules[43–45,28,41] are based on a theoretical evaluation of phenomenologically accessible two or three-point functions in the not so deep Euclidean region, where vacuum condensates make themselves remarked as power corrections to the perturbative contribution. The situation for the correlator of two vector currents with light quarks which is directly accessible in $e^+ - e^-$ annihilation is depicted in Figure 1. The perturbative contribution to the correlator is given to $O(\alpha_s)$ by:

$$\Pi^{pert}(Q^2) = -\frac{1}{8\pi^2}(1 + \frac{\alpha_s}{\pi})\log[\frac{Q^2}{\mu^2}] \tag{1}$$

It is a very good approximation in the deep Euclidean region (i.e. large negative Q^2 values). In the not so deep Euclidean region non-perturbative power corrections as:

$$\Pi^{nonpert}(Q^2) = \frac{\alpha_s}{24\pi Q^2}\langle\Omega|F^C_{\mu\nu}F^C_{\mu\nu}|\Omega\rangle + \ldots \tag{2}$$

become important. The theoretical expression consisting of perturbative and non-perturbative contributions is matched with the phenomenological one which is continued from the physical region to the not so deep Euclidean region via dispersion relations. There are several caveats, some are more puristic, but some also of great phenomenological importance. The most puristic is certainly that the operator product expansion[50,52,53] on which the sum rules are based is only proven in perturbation theory (nevertheless to all orders; for a discussion of objections of that kind see[34]). More serious is the fact that the separation of perturbative and non-perturbative contributions is far from being trivial. If an infra-red cutoff is introduced the perturbative contributions are well defined, but then we have IR-cutoff (and renormalization scale) dependent contributions to the condensates. If on the other hand dimensional regularization is used we have IR renormalons in perturbation theory which also give power corrections and mix with the condensates (see[24,25,8]). Another point to remember is that sum rules are inherently of limited accuracy: If we come too near to the physical region we have to take into account many unknown higher dimensional condensates (I come to that point later) and presumably are also limited by the asymptotic nature of the power series expansion. On the other hand if we stay too far away from the physical region we cannot isolate the contributions of the particular states. In order to isolate the properties of the lowest lying contributions – in Figure 1 the ρ-pole – one has to model the higher resonances. This is done normally by quark hadron duality which then introduces some model dependence.

I now shortly discuss the lowest dimensional condensates: For the quark condensate $< \Omega|\bar{\psi}_q \psi_q|\Omega >$ (mass dimension 3) the second warning about mixing with perturbation theory is not a serious problem since it enjoys chiral protection and hence receives no perturbative contributions in the case of vanishing quark mass, but the other important condensate, the 4-dimensional gluon condensate $< \Omega|F^C_{\mu\nu} F^C_{\mu\nu}|\Omega >$ has no such protection and therefore its definition is much more delicate.

The quark condensate is the most precisely determined one since it plays a dominant role in the two and three-point functions involving heavy quarks where it occurs in the combination $m_Q < \Omega|\bar{\psi}_q \psi_q|\Omega >$. The mixed condensate $< \Omega|\bar{\psi}_q \sigma_{\mu\nu}\, \lambda_C\, F^C_{\mu\nu} \psi_q|\Omega >$is also fairly well determined. The quark condensate is related to to the masses of the up and down quark, the pion mass and the leptonic decay constant of the pion by the Gell-Mann-Oakes-Renner relation[20]

$$(m_u + m_d)(\langle\Omega|\bar{u}u|\Omega\rangle + \langle\Omega|\bar{d}d|\Omega\rangle) + O(m^2_{u,d}) = m^2_\pi f_\pi \tag{3}$$

The determinations of the gluon condensate yield a much wider variety of values than for the quark condensate (see e.g.[28,41] . The gluon condensate is related to the energy density of the vacuum by the trace anomaly (for a discussion of the anomaly and many useful low energy theorems see[31]):

$$\langle\Omega|\Theta^\mu_\mu|\Omega\rangle = \frac{\beta(\alpha_s)}{4\,\alpha_s}\langle\Omega|F^C_{\mu\nu}F^C_{\mu\nu}|\Omega\rangle \quad \frac{\beta(\alpha_s)}{4\,\alpha_s} = -\frac{11}{32\pi^2}g_s^2 + O(g_s^3) \tag{4}$$

We obtain in that way an energy density of the vacuum of circa 1 GeV/fm^3

It should also be noted that there is presumably big difference between the gluon condensate of full QCD and that of the pure gauge theory (i.e. quenched QCD). As estimated by Novikov *et al.*[31] in the latter case the gluon condensate may be a factor two to three larger than in full QCD. This estimate is based on the low energy theorem

$$\frac{d}{dm_q}\langle\Omega|g_s^2 FF|\Omega\rangle = -\frac{96\pi^2}{11 - 2n_f/3}\langle\Omega|\bar{q}q|\Omega\rangle \tag{5}$$

The situation is even worse for the condensates of higher dimensions. There are many different possibilities to form operators of the same dimension and hence also give the same power correction. Usually factorization is applied in this case, but this gives only the order of magnitude. Let us consider the very important four quark condensate. Here the $1/N_C$ expansion is invoked as justification of the factorization hypothesis. But an discrepancy of a factor two to three between the actual and the $N_C \to \infty$ value is certainly compatible with this argument. An example for that is the gluon condensate. In the $N_C \to \infty$ limit the quark loops should have no influence on this quantity but

Condensate	Standard value	realistic limits
$<\Omega\|g_s^2 F^C_{\mu\nu} F^C_{\mu\nu}\|\Omega>$	$(0.82\text{GeV})^4$	stand. value $\times(2\pm 1)$
$<\Omega\|\bar{q}q\|\Omega>$	$(-.230\text{GeV})^3$	
$<\Omega\|\bar{s}s\|\Omega>$	$0.7 <\Omega\|\bar{q}q\|\Omega>$	
$<\Omega\|g_s\,\bar{q}\sigma\mathbf{F}q\|\Omega>$	$0.8\text{GeV}^2 <\Omega\|\bar{q}q\|\Omega>$	
$<\Omega\|g_s\,\bar{s}\sigma\mathbf{F}s\|\Omega>$	$0.8\text{GeV}^2 <\Omega\|\bar{s}s\|\Omega>$	$(1\pm 0.3)\text{GeV}^2\times$ $<\Omega\|\bar{s}s\|\Omega>$
$<\Omega\|\bar{q}q\bar{q}q\|\Omega>$	Factorization normally used	factor. val.$\times(2\pm 1)$

Table 1. The "known" vacuum condensates. If the errors are estimated to be larger than ca 20% realistic limits are given. The renormalization point is 1 GeV

nevertheless it has been argued above that in the case of decoupled quarks (pure gauge) the gluon condensate is a factor two to three larger than in full QCD[28] . This may be due to the high dimensionality. Even if there is a deviation of the $N_c \to \infty$ limit on the 10 to 20 % level for the 1-dimensional quantity, rising to the power four or six leads easily to factors 2 to 3. Another possible source for a violation of factorization of the four quark condensate is the fact that generally the four quark condensate is not chirally protected in contradistinction to the two-quark condensate. So there is no theoretical argument against but some phenomenological evidence for a violation of factorization by a factor two to three . For other high dimensional condensates as for instance the two-gluon two-quark condensates, e.g. $<\Omega|g_s^2\bar{q}F^C_{\mu\nu}F^C_{\mu\nu}q|\Omega>$ the situation is much worse. Here we have not even large N_C arguments for factorization and the result may depend very crucially on the procedure, e.g. if we first take into account the conditions imposed by the equations of motion and then factorize or if we reverse the order. Finally the dependence of the condensates on the renormalization points also tells us, that if at all factorization can be only fulfilled at one scale. In the table 1 I have collected the known condensates , their "standard values" and what I would call reasonable errors.The renormalization point is 1 GeV.

QCD sum rules are not only the most direct source of information on the structure of the physical vacuum but also one of the most important tools to calculate observables in non-perturbative QCD in a fairly model-independent way. It is astonishing how much information can and has been obtained from the few condensates, the theoretical concept of operator expansion[50,51] and another crucial ingredient in most sum rule calculations, namely quark-hadron duality. The treatment of chiral symmetry breaking is very satisfactory since the order parameter of this symmetry, the quark condensate occurs directly in the sum rules.

Sum rules can however not tell us anything on confinement. This is obvious since there is no local order parameter for confinement but the latter manifests itself in the area law of the Wegner-Wilson loop[49,52] which is a non-local quantity. In order to treat confinement we must either have recourse to numerical simulations of the lattice version of QCD[52] or to models.

3. Models

Roughly speaking there are two types of models: One type starts from the QCD-Lagrangian and tries to obtain confinement from it as it stands, the other approach starts from confinement and looks for an *effective theory* which yields confinement. To the first class belong the instanton models (see for instance[38]). Instantons as Euclidean classical solutions of the QCD Lagrangian are certainly there and contribute to the functional integral which describes the matrix elements, but unfortunately they do not seem to yield confinement. My personal view is that instantons contribute essentially to chiral symmetry breaking, but that the confinement properties are a consequence of the very intricate IR behavior of QCD and thus are deeply hidden in the Lagrangian. I shall propose two models which go the second way.

3.1. The model of the stochastic vacuum

The model of the stochastic vacuum, devised and elaborated by Yuri Simonov and myself[14,15] since more than ten years can be formulated in the following way (for reviews see[17,47,27]:

If the functional integral with the QCD-Lagrangian as measure describes a stochastic process with a convergent cluster expansion in the color field strengths then they lead to the area law of the Wegner Wilson loop and hence confinement.

This sounds good and is very general. In a phenomenologically useful form we must be more restrictive and start with the most simple nonlocal quantity in QCD, the gauge invariant non-local gluon correlator $< \Omega | F^C_{\mu\nu}(x, y_R)\, F^D_{\rho\sigma}(0, y_R) | \Omega >$. Here y_R are the coordinates of a reference point to which color is transported via a Schwinger string and which makes this object gauge covariant. If we make now the assumption that the non-perturbative part of QCD can be approximated by a Gaussian stochastic process, i.e. that all n-point correlators of the color field strengths can be factorized into two point correlators the linked cluster expansion converges trivially since all higher clusters vanish, and we have linear confinement. All gauge invariant matrix elements can be evaluated with the help of this correlator.

The essence of the model of the stochastic vacuum can be summarized as follows: *The IR behavior of QCD is approximated by a Gaussian stochastic process.*

A nice feature of this model is that it does not give confinement for an *Abelian* gauge theory like electrodynamics, as it should be the case: The homogeneous Maxwell equations force the string tension to zero. Only non-Abelian gauge theories or Abelian theories with magnetic monopoles yield confinement.

After my warnings with regard to factorization made above this Gaussian assumption seems completely unrealistic. This might be true, but it is not necessarily so. Let me illustrate this with a very simple but nevertheless instructive example. Assume we have a function like $\exp[-x^2 + x^4 - x^6]$. If we want to evaluate this function by a Taylor series around $x = 0$ we have to know the expansion coefficients very precisely in order to reproduce its characteristic feature, namely the rapid fall-off. If we know however that it is a function which falls off to zero quite fast, we may approximate it decently by a simple Gaussian $\exp[-(x/a)^2]$ which has the right value at $x = 0$ and where a is chosen as to give the right second moment.

This is the strategy adopted in the Gaussian approximation: We assume that the linked cluster expansion for QCD is convergent and approximate it by a Gaussian process with the value of the correlator at zero distance, i.e. the gluon condensate, and the appropriate correlation length.

As it is clear from the construction the model only deals with the gauge sector of the QCD and cannot handle, as it stands dynamical quarks. This is certainly a shortcoming of the model but there are several very interesting attempts to relate the model of the stochastic vacuum with chiral symmetry breaking..[48,10,9] But on the other hand a number of other questions of phenomenological relevance can be dealt with it:

- *Heavy Quark interactions.* The central,[14,15] the spin[39,46] and velocity[10] dependent potentials have been evaluated and agree well with those determined from quarkonium spectroscopy.

- *String formation and consistency with low energy theorems.*[12,36,18] One can in the model of the stochastic vacuum evaluate the field strength of a static quark antiquark pair, i.e. the reaction of the vacuum on external sources. Here one observes that the model leads to a string formed by a color-electric flux tube, the full width of which is about 3 correlation length (about one fermi). I only show (Figure 2) a picture of the energy density of the non-perturbative vacuum in

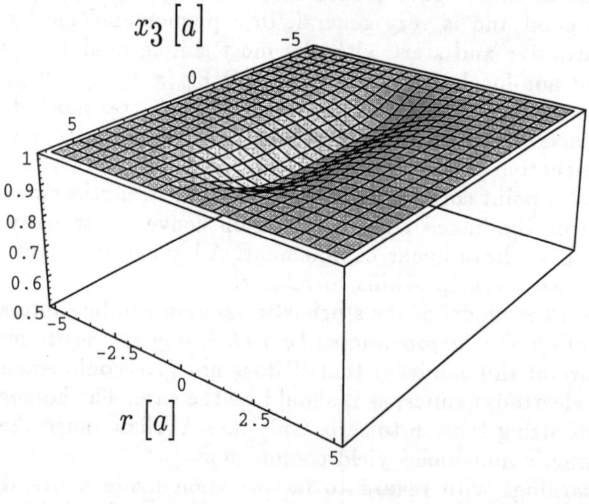

Fig. 2. The squared color-electric field strength in the presence of a static quark-antiquark pair. The presence of the external sources leads to a reduction of the vacuum fluctuations in the Euclidean continuum,which give rise to the flux tube with positive energy density in the Minkowski world.

the presence of a static quark-antiquark pair. Its presence makes itself remarked in Euclidean space as a reduction of the non-perturbative vacuum fluctuations which in term gives rise to a color-electric flux tube in the Minkowski world.

- *High energy diffractive scattering.* This point is closely related to the string formation and I shall come back to it in the last part of my talk.

From the model of the stochastic vacuum we can learn that confinement is quite natural in a non-Abelian gauge theory. It suffices that in the long range quantum fluctuations the higher correlations are sufficiently suppressed. It is however also desirable to pin down a physical mechanism for confinement. This is done in the model of dual QCD with monopole condensation.

3.2. Dual QCD with monopole condensation

The basis of this model is the idea of 't Hooft[19] and Mandelstam[23] that confinement is due to monopole condensation. As in a superconductor the condensation of Cooper pairs confines the magnetic flux lines to magnetic flux tubes so in QCD the condensation of magnetic monopoles leads to currents which squeeze the color-electric field to form a flux tube and hence leads to a linear increase of separation energy with the distance. The concept of non-Abelian monopoles is far from being trivial as we have learned in the contribution of Polykarpov, the most exciting development in this field, namely the appearance of duality in super-symmetric QCD will be treated by Yuval Neeman. So let me just shortly sketch the model and some of the newer phenomenological developments.

The dual theory of pure gauge QCD[3,4] is described by an effective Lagrangian where the field content is described by dual potentials $\mathbf{C}_\mu$ and a triplet of scalar Higgs fields $\mathbf{B}_i$ all in the adjoint representation and coupled minimally to the dual potentials with the color charge $g_m = \frac{2\pi}{g_s}$. Analogously to the Ginzburg-Landau theory the gauge symmetry is broken spontaneously through the Higgs Potential:

$$V_{Higgs}(\mathbf{B}_i) = -\frac{16\lambda}{9}(-F_0^2)(B_1^2+B_2^2+B_3^2)+\frac{14\lambda}{3}(B_1^2+B_2^2+B_3^2)^2+\frac{22\lambda}{3}(B_1^4+B_2^4+B_3^4) \quad (6)$$

This potential has its minimum at a finite value of the vacuum expectation value of the Higgs field:

$$\langle\Omega|B_i|\Omega\rangle = \sqrt{-F_0^2/6}\,/2 \quad (7)$$

The dual gluons thus acquire a mass through the Higgs-Kibble mechanism. Off course this effective Lagrangian for dual theory should not be trusted at distances where the gluon masses become important.

The classical effective Lagrangian so defined has non-trivial solutions which can very well be approximated by those of an Abelian Higgs model familiar from superconductivity, namely the Abrikosov-Nielsen-Olesen strings.[1,30] In this dual QCD the strings are color-electric flux tubes in which the Higgs potential vanishes and which are surrounded by the curls of the dual potential.

The classical solutions can be viewed as flux tubes between color-electric charges at $\pm\infty$. If a static quark-antiquark pair is inserted, the vacuum reacts by forming such a flux tubes which end at the external charges. The model only contains three parameters:

- The dual color charge $g_m = \frac{2\pi}{g_s}$,

- the Higgs coupling λ, and
- the vacuum expectation value of the Higgs fields $\langle\Omega|B_i|\Omega\rangle$.

They can be related to the following phenomenological quantities:

- the string tension σ of the central $Q-\bar{Q}$-potential,
- the Gluon condensate $\langle\Omega|g_s^2\mathbf{FF}|\Omega\rangle$, and
- the Coulombic part of potential: $\frac{4\alpha_s}{3r}$.

The values obtained by a comparison with heavy quarkonium spectra yield[6]:

$$\alpha_s = 0.39,\ \lambda = 1.61,\ F_0^2 = -(0.420\text{GeV})^2 \tag{8}$$

These numbers should be only taken as indications since the comparison with lattice calculations e.g. yields somewhat different values.

The field configuration can be evaluated in detail and related to the heavy quark potentials as has been done in[5] . A detailed comparison with lattice simulations shows a very good agreement.

3.3. Comparison of the models

The potentials of the two models mentioned above have been compared by Nora Brambilla and Antonio Vairo and a remarkable agreement even in details appeared. A comparison on an analytical level[7] showed a particular similarity in the London limit, i.e. the case where the Higgs potential is constant. There the correlator of the gluon field strength which plays the fundamental role in the model of the stochastic vacuum is just the one obtained from a free correlator of the *dual* potential C_μ^D,

$$< \Omega|F^C_{\mu\nu}(x)F^D_{\rho\sigma}(y)|\Omega >= \epsilon_{\mu\nu\mu'\nu'}\epsilon_{\rho\sigma\rho'\sigma'} < \Omega|\partial_{\mu'}C^C_{\nu'}(x)\partial_{\rho'}C^D_{\sigma'}|\Omega > .$$

the correlation length being the inverse mass of the dual potential obtained through the Higgs-Kibble mechanism. An analysis of a string model with "internal strings" leads to the same results in the London limit.[2]

4. Diffractive high energy scattering and confinement

In this section we address the question if the confinement mechanism also leaves its traces in high energy scattering. If one takes the string picture intuitively[35] one may expect that the string between the quarks also participates in the scattering and leads to terms which increase monotonously with the hadron size (string length) and thus hadron-hadron scattering is more than a superposition of quark-quark scattering (if the latter make sense at all). In the model of the stochastic vacuum which leads to a string formation one can also treat high energy scattering and thus answer the question quantitatively.

Using the eikonal approximation one can also reduce the evaluation of scattering amplitudes to the evaluation of Wegner-Wilson loops[26,21,16] and apply factorization. We even have in high energy scattering a big advantage: since the quarks are in the high energy limit frozen on light-like paths they can be treated like external charges and we do not have to bother about chiral symmetry breaking. Again I shall skip all details of the lengthy considerations and calculations and only quote the very intuitive result: The same mechanism which leads to confinement also leads to a string-string interaction in scattering which manifests itself in a cross section which increases (in a quenched theory indefinitely) with the size of the hadrons. One can be more quantitative and evaluate the scattering amplitude of a color-neutral dipole consisting of a quark and an antiquark on a proton without using any information from high energy scattering. This calculation can be compared with an analysis of diffractive electro-production of vector mesons by Nemchik et al.[29] who extracted from experimental data the scattering amplitude of a color dipole on a proton as function of the dipole size. The result is displayed in Fig.3; it is encouraging and to my knowledge the first quantitative relation between non-perturbative QCD in high and low energy physics. The solid curve is our calculation containing no parameters taken from high energy scattering, the dashed one used in the dipole scattering model for diffractive scattering in ref.[29] .

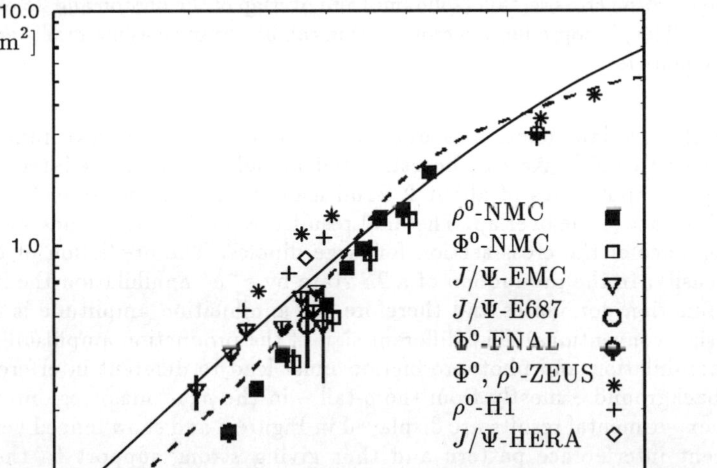

Fig. 3. Comparison of our result for the total cross section for dipole (extension R_D) proton scattering with values extracted from cross sections of lepto-production of vector-mesons by the method of Nemchik et al..[29] The solid line is the result of[37] without any fitting of the parameters to high-energy data. The dashed line is the ansatz of Nemchik et al. for the total cross section; from.[37]

An even more sensitive test of the behavior of the photo-production cross section at large separations of the color charges can be made by calculating the production cross section of excited vector mesons.[22] The production amplitude is obtained by multiplying the dipole proton amplitude with the overlap of the photon and the excited vector meson wave function. Both quantities are displayed in Fig. 4 . The change

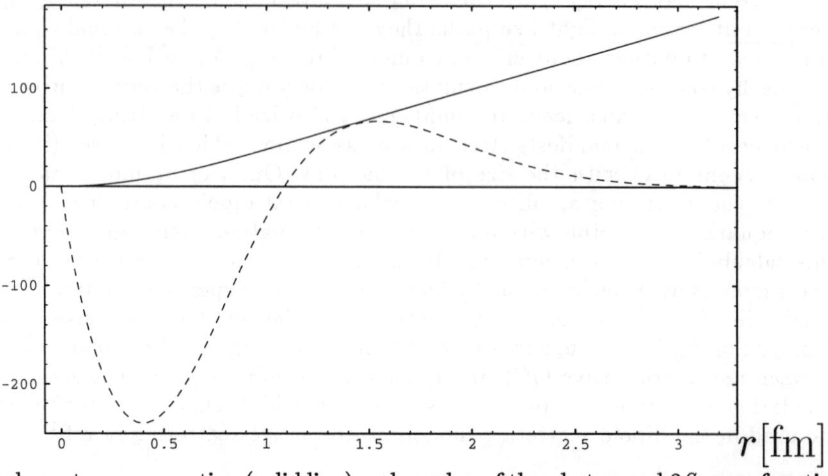

Fig. 4. Dipole proton cross section (solid line) and overlap of the photon and $2S$-wave function (dashed line). The photoproduction cross section can be obtained as integral over the product of the two quantities

of sign in the overlap function is due to the node in the $2S$-wave function, for all details I refer to ref..[22] As can be estimated already from Fig. 4 integration of the product up to dipole sizes of about 2 fermi leads to a compensatiuon of the positive and negative part of the overlap. The final result is a positive amplitude and this result is a consequence of the cross section for large dipoles. The prediction of the sign can be tested easily: In the production of a $2S$ state by $e^+ e^-$ annihilation the inner part of the wave function dominates and therefore the annihilation amplitude is negative (in the usual sign conventions). The different sign of the production amplitude of the ρ', ρ'' in e^+e^- -annihilation and photoproduction must lead to different interference pattern with the background – mostly from the ρ-tail – in the ρ', ρ'' mass region.

The experimental results are displayed in Figure 5 and show indeed very markedly this different interference pattern and thus giving strong support to the amplitude evaluated in the model of the stochastic vacuum.

5. Conclusions

Though we may be still far away from an analytical understanding of confinemet, its existence makes the presence of a relatively simple but non-trivial vaccum structure plausible. From this structure a lot of phenomenological consequences can be drawn.

6. Acknowledgements

I have profitted greatly from discussions with M. Baker, N. Brambilla, D. Gromes, G. Kulzinger, O. Nachtmann, H.J. Pirner, M. Rueter, and A. Vairo. My thanks are also due to the organizers of the workshop for providing a friendly and stimulating atmosphere.

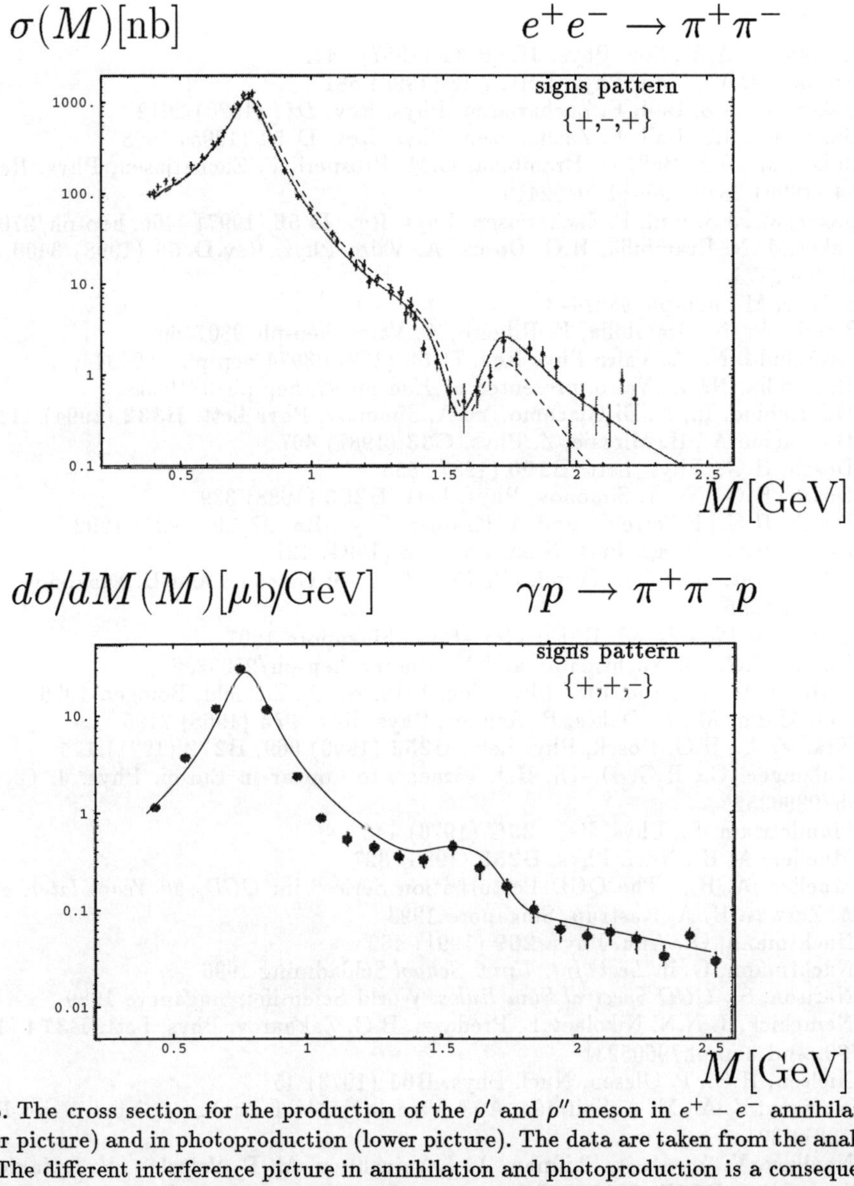

Fig. 5. The cross section for the production of the ρ' and ρ'' meson in $e^+ - e^-$ annihilation (upper picture) and in photoproduction (lower picture). The data are taken from the analysis of.[13] The different interference picture in annihilation and photoproduction is a consequence of the different signs of the production amplitudes, as predicted by the model of the stochastic vacuum. The solid line in the lower picture is the prediction of.[22]

References

1. Abrikosov, A.A., Sov. Phys. JETP **32** (1957) 1442
2. Antonov,D.V., Mod. Phys. Lett. A13 (1998) 581
3. Baker, M., J.S. Ball, F. Zachariasen, Phys. Rev. *D41* (1990) 2612
4. Baker, M., J.S. Ball, F. Zachariasen, Phys. Rev. **D 51** (1995) 1968
5. Baker, M., J.S. Ball, N. Brambilla, G.M. Prosperi, F. Zachariasen, Phys. Rev. **D 54** (1996) 2829. hep-ph/9602419
6. Baker, M., J.S. Ball, F. Zachariasen, Phys. Rev. **D 56** (1997) 4400. hep-ph/9705207
7. Baker,M.,N. Brambilla, H.G. Dosch, A. Vairo Phys. Rev.**D 58** (1998) 3400, hep-ph/9802273
8. Beneke, M., hep-ph/9807443
9. Bicudo, P., N. Brambilla, E. Ribeiro, A. Vairo, hep-ph/9807460
10. Brambilla, N., A. Vairo Phys Rev. **D 55** (1997) 3974 hep-ph/9606344;
11. Brambilla, N., A. Vairo, presented at *Hadron 97*, hep-ph/9710558
12. Del Debbio, L., A. Di Giacomo, Yu.A. Simonov, Phys Lett. **B332** (1994) 111
13. Donnachie A., H. Mirzaie, Z. Phys. **C33** (1987) 407
14. Dosch, H.G., Phys. Lett. **B190** (1987) 555
15. Dosch, H.G., Yu.A. Simonov, Phys. Lett. **B205** (1988) 339
16. Dosch, H.G., E.Ferreira, and A. Krämer, Phys. Rev **D 50** (1994) 1992
17. Dosch, H.G., Progr. Part. Nucl. Phys. 33 (1994) 121;
in Proc. Int. School E. Fermi, CXXX, ed. A. Di Giacomo and D. Diakonov, Amsterdam 1996
in *Hadron Physics*, ed. E. Ferreira *et al.* , Singapore 1997
18. Dosch, H.G., O. Nachtmann ,and M. Rueter, hep-ph/9503386
19. ‘t Hooft, G., in Proc. Eur. Phys. Soc. 1975, ed. A. Zichichi, Bologna 1976
20. Gell-Mann, M., R. Oakes, B. Renner, Phys. Rev. **175** (1968) 2195
21. Krämer,A., H.G. Dosch, Phys.Lett. **B252** (1990) 669, **B272** (1991) 114
22. Kulzinger, G., H.G. Dosch, H.J. Pirner , to appear in Europ. Phys. J. C., hep-ph/9806352
23. Mandelstam. S., Phys. Rep. **23C** (1976) 145
24. Mueller, A. H., Nucl. Phys. **B250** (1985) 327
25. Mueller, A. H., “The QCD Perturbation Series” in: *QCD, 20 Years later*, ed. P. M. Zerwas, H. A. Kastrup, Singapore 1993
26. Nachtmann, O., Ann. Phys. **209** (1991) 436
27. Nachtmann, O. in *Lect. Int. Univ. School* Schladming 1996
28. Narison, S., *QCD Spectral Sum Rules*, World Scientific, Singapore 1989
29. Nemchick, J. N.N. Nikolaev,E. Predazzi, B.G. Zakharov, Phys. Lett. **B374** (1996) 199, and hep-ph/9605231
30. Nielsen, H.B., P. Olesen, Nucl. Phys. **B61** (1973) 45
31. Novikov, V. A., M. A. Shifman, A. J. Vainshtein, V. I. Zakharov, Nucl. Phys. **B191** (1981) 301
32. Novikov, V. A., M. A. Shifman, A. J. Vainshtein, M. B. Voloshin, V. I. Zakharov, Nucl. Phys. **B237** (1984) 525
33. Novikov, V. A., M. A. Shifman, A. J. Vainshtein, V. I. Zakharov, Fortschr. Phys. **32** (1984) 585
34. Novikov, V. A., M. A. Shifman, A. J. Vainshtein, V. I. Zakharov, Nucl. Phys. **B249** (1985) 445

35. Povh, B., J. Hüfner, Phys. Rev. Lett. 58 (1987) 1612
36. Rueter, M., and H.G. Dosch, Z. Phys. **C66** (1995) 245
37. Rueter, M., H.G. Dosch, Phys.Rev. **D 57** (1998) 4097
38. Schäfer, T., E. Shuryak, Rev. Mod. Phys. **70** (1998) 323
39. Schiestl, M., H. G. Dosch, Phys. Lett. **B209** (1988) 85
40. Shifman, M. A., Ann. Rev. Nucl. Part. Sci **33** (1983) 199
41. Shifman, M. A., *Vacuum Structure and QCD Sum Rules*, North Holland, Amsterdam 1992
42. Shifman, M. A., A.I. Vainshtein, V. I. Zakharov, Phys. Lett. **76B** (1978) 471
43. Shifman, M. A., A. I. Vainshtein, V. I. Zakharov, Nucl. Phys. **B147** (1979) 385
44. Shifman, M. A., A. I. Vainshtein, V. I. Zakharov, Nucl. Phys. **B147** (1979) 448
45. Shifman, M. A., A. I. Vainshtein, V. I. Zakharov, Nucl. Phys. **B147** (1979) 519
46. Simonov, Yu. A., Nucl. Phys. **B324** (1989) 67
47. Simonov, Yu. A., Phys. Usp. 39 (1996) 313
48. Simonov, Yu. A., hep-ph/9712248
49. Wegner, F., J. Math. Phys. **12** (1971) 2259
50. Wilson, K., Phys. Rev. **179** (1969) 1499
51. Wilson, K., Phys. Rev. **D3** (1971) 1818
52. Wilson, K., Phys. Rev. **D10** (1974) 2445
53. Zimmermann, W., in *Lect. on Elemeent. Particles and Quantum Field Theory* ed. S. Deser *et al.*, Cambridge 1971

THE FLUX-TUBE MODEL AND ITS SPECTROSCOPY

Jack Paton
Dept of Physics, University of Oxford
Theoretical Physics
1 Keble Rd,
Oxford OX1 3NP
UK

ABSTRACT

I review the flux-tube model[1] and compare its predictions for meson hybrids and glueballs mainly with recent lattice results.

1. Introduction

It would perhaps be more appropriate to alter the title of this talk by inserting quotation marks around "spectroscopy", as it is "spectroscopy" rather than spectroscopy that I shall be talking about most of the time. The difference between "spectroscopy" and the genuine article is that "spectroscopy" is a comparison with computer simulated data rather than with real data. This has two advantages. The real object is difficult to make and is stubborn enough always to remain what it is. The simulated object, on the other hand, may be manufactured more easily and in addition with a variety of different materials, thus allowing different aspects of the construction to be studied independently. This latter facility allows more detailed understanding to be gained of the limitations of the model, and hopefully can lead to ways of modifying and improving the model.

The main part of my talk, Section 3, will be about about mesons and "mesons", the latter being the simulated variety. The first topic in that section is that of the adiabatic limit of states with flavour, the lowest approximation in the model to the states of the quark model and the lowest hybrids. I go on to discuss corrections to the adiabatic limit, how these affect the estimates of hybrid masses and (briefly) the extension of the flux-tube model to predict hadronic decays of hybrids. In the final part of Section 3, I discuss the flux-tube model for "glueballs"and in particular its comparison with lattice "data" on pure gauge theories in two space dimensions. A few conclusions are given in the final Section 4. First, however, in Section 1, I review briefly the foundations of the flux-tube model and a few of its general properties.

2. The flux-tube model

The basic assumption of the flux-tube model[1] is that at low spatial resolution, a, the *quantum field theory* of the gluon fields, $(\mathbf{E}^c, \mathbf{B}^c)$ may be modelled by the *quantum mechanics* of the flux-tube, ie the model involves mechanical degrees of freedom only. The motivation comes from an examination[1] of lattice hamiltonian gauge theory[3] and a suggested reordering of the strong coupling series in this formulation of gauge field theory. In the strong coupling limit the states consist of a number of quarks and anti-quarks at fixed lattice sites attached to "frozen" flux-tube paths along the links of the lattice and connecting the quarks and anti-quarks to each other to form gauge-invariant states. Deviations from the strong coupling limit allow the flux-tube to hop between neighbouring links and the quarks to hop between neighbouring sites. ie they give the flux-tube and quarks kinetic energy. They also allow transitions between patterns of flux-tube paths of different topology and with different numbers of quarks. The flux-tube model is arrived at by imagining the diagonalization of the hamilitonian in two stages: First, one ignores effects of the second type (topological mixing). At this stage the approximate hamiltonian is in block diagonal form with, for example, one block for states with one quark and one anti-quark linked by a continuous flux-tube, and another for states with a single continuous line of flux in a loop. These are the prototypes of the simplest mesons and glueballs respectively of the flux-tube model, as discussed below. Note that the vacuum of the flux-tube model is particularly simple. It is simply the state with no flux on any links and no quarks on any sites- the strong-coupling vacuum.[2]

The model assumes that on scales typical of hadrons the important effects of deviations from the strong coupling limit are taken into account by allowing for the kinetic energies of quarks and flux-tubes (corresponding to transitions within each diagonal block difined above), and that other effects in the strong coupling expansion which allow changes in the topology (transitions between the diagonal blocks) are relatively small and may be treated as small perturbations, at least in some cases.

Of course, for small enough coupling, ie high enough resolution, deviations from this picture based on strong coupling must be large. For example, in strong coupling language the descriptions of the vacuum and of a meson state at high enough resolution are extremely complicated involving an overlapping mess of flux lines and quark-antiquark pairs. It is hoped however that the picture is not too badly wrong at large enough scales $a \approx 0.2$ Fm that are nevertheless small enough to resolve the valence quark structure of hadrons.

We make some remarks on the picture that emerges from these consideration:

• The picture is the same for all numbers of colours, N_C, with two exceptions:

(a) For $U(1)$ gauge theory and for all cases of $SU(N_C)$ gauge theory other than $N_C = 2$ the flux-tube is oriented ie a directional arrow is attached to it. For $N_C = 2$ the flux-tube has no arrow attached to it, leading to some states being disallowed in this case.

(b) The simplest baryons of $SU(N_C)$ gauge theory have N_C flux lines meeting at a junction (three for $SU(3)$, and no junction for $U(1)$ or $SU(2)$).

• The flux-tube has an intrinsic width because of the finite low resolution,

$$a \approx 0.2 \text{ Fm}.$$

There is a further width from quantum fluctuations. This is determined by the string tension, σ and is of order $\sigma^{-\frac{1}{2}} \approx 0.4$ Fm. Since these widths are comparable to the spatial size of some lowest hadron states, we expect that the flux tube model may give a poor representation of such states, but that low-lying excited states with a larger spatial size may be better represented.

• The outcome of these considerations is that the flux-tube model affords a unified picture of the whole of low resolution hadron physics:

1. quark model states
2. non-quark model states, including glueballs and hybrids
3. relatively weak hadron-hadron interactions
4. In addition, high energy low momentum transfer processes are qualitatively identical in their description to that of the dual string (dual resonance model) of the '70s.

I discuss items 1. (for mesons) and 2. in the next section. For baryons the simplest topology (for $SU(3)$) is that of the "Mercedes" shape (sometimes degenerating so that the junction point coincides with the position of a quark). This has been discussed in the literature for quark model states,[4] and, in these Proceedings, Philip Page[5] has described a programme of extending this work to baryon hybrids. Some initial work has been done on 3.[1 6] but for lack of time I shall not describe it here.

The picture afforded by the flux-tube model is likely to be best for excited states whose hadronic decay widths are not too large. (Hadronic decay is an effect of mixing of topologies, which is regarded as a perturbation in the flux-tube model.) Unfortunately, as explained above, it is quite likely to be no more than a caricature for the ground state in any sector, if the physical size of the state is too small.

3. Simplest mesons and "mesons"

3.1. Mesons with flavour: adiabatic limit

The simplest mesons (or "mesons") with flavour consist of a constituent quark and a constituent anti-quark to which a single (fluctuating) flux-tube is attached. To render this model picture tractable requires a simple calculable mechanical representation of the flux-tube. Appropriate to decreasing assumed physical size of the meson the following representations have been used:

- a continuous mechanical string of constant mass per unit length which may be treated either non-relativistically[7] or (presumably better) relativistically[8]
- a "discrete harmonic string" consisting of several "beads" spaced a typical resolution distance a apart and treated non-relativistically[23 24]
- a few "beads" treated as above
- one "bead", undergoing relativistic transverse motion[12]

The picture becomes particularly simple in the *adiabatic limit* where the quarks are a fixed distance R apart, so that the quantum mechanics reduces to that of the flux-tube only. The flux-tube may then exist in various states. The flux-tube energy $E(R)$ for the ground state is identified with the potential $V_0(R)$ of the quark model. A discussion from this point of view of the quark model and its range of validity is given by Godfrey *et al.*[9]
The flux-tube may also be excited. This gives a sequence of excited potentials $V_i(R)$, $i = 1, 2, 3...$ which along with $V_0(R)$ may be compared with those determined on the lattice, most recently by Colin Morningstar and his coworkers.[22] For example, in the continuous string representation, the flux-tube ground state satisfies

$$a_{m\lambda}\Psi_0 = 0$$

for all modes $m = 1, 2, 3...$ and transverse polarizations $\lambda = 1, 2$ where the $a_{m\lambda}$ destroy quanta of transverse vibration. The relativistic string has in this state the energy

$$V_0 = const + \sqrt{\sigma^2 R^2 - \pi\sigma/6}$$

which in the (effectively non-relativistic string) limit $R >> \sigma^{-\frac{1}{2}}$ reduces to

$$V_0 = const + \sigma R - \frac{\pi}{12R}.$$

This form, interpreted as Coulomb plus linear at short distances and linear plus the Luscher correction at large R works "too well" as a phenomenological potential to fit heavy quarkonium. It is really not entirely clear that the exact form of the Luscher correction, which is appropriate to a structureless mechanical string, is really appropriate to the flux-tube which has more and more elaborate structure at scales less than a. However at the scale of present lattice studies its form seems to be reproduced, as explained by Chris Michael in these Proceedings.
The continuous flux-tube model expressions in both the relativistic and non-relativistic forms give that

1. the set of excited potential energy "surfaces" is independent of the number of colours (true for *any* version of the flux-tube model).

2. the potentials $V_i(R)$ depend only on $N = \Sigma n_m m$ where $n_m = 1, 2, 3...$ is the degree of excitation of the mth mode.

3. the first excited potential cooresponds in quantum numbers to a single excitation of the fundamental mode of vibration of the flux-tube.

The first of these results seems to be reasonably well verified on the lattice as far as it has been tested. The second is certainly not valid for $R \leq 1\text{Fm}$. Between 1Fm and 2 Fm it has been claimed to work well[26] but in 2 Fm $\leq R \leq$ 4 Fm, the region of largest R so far explored on the lattice, some discrepancies have been claimed.[22] The result 3. above, that the first excited potential $V_1(R)$ is, in the notation of molecular physics, Π_u, seems well established in the lattice simulations. In fact the form we took for this potential in our original paper,[1] which was based on the non-relativistic string but with a physically motivated cut-off at small R, gives a remarkably good representation of the lattice results. The cut-off is however crucial as otherwise the excited potential is far too large at small R.

These adiabatic potentials may be used, as in molecular physics, to calculate meson energies both for states "living" on the lowest adiabatic surface V_0, which are interpreted as the states of the usual quark model, and, for example, those "living" on the first excited potential surface, V_1. The flux-tube variable representation of the gluonic variables are being treated here as the "fast" variables of the adiabatic approximation, ie the flux-tube variables play an analogous role to that of the electronic variables of molecular physics and are assumed in the lowest approximation to be able to adjust instantaneously to the positions of the quark and anti-quark.

I concentrate henceforth on hybrids. The non-relativistic hamiltonian h_L for the "slow" quark motion in a state of total angular momentum, L, and built on the lowest hybrid surface is

$$h_L = (ke)_{radial} + v_L$$

where

$$v_L = V_0(R) + [V_1(R) - V_0(R)] + \frac{L(L+1)-1}{2m_{red}R^2} + \delta V$$

Here, the first term on the right hand side is fitted to the mass spectrum of established quark model states, the second is given by the model and the third term is the centrifugal barrier which takes the unusual form because the flux-tube, in these states, possesses a non-zero value $\Lambda = \pm 1$ of angular momentum about the quark-antiquark axis. The last term, δV, represents various further corrections to the adiabatic limit for the moment ignored but considered below.

For meson states built on the first excited surface the quantum numbers of the lowest lying states are $L^{PC} = 1^{+-}, 1^{-+}$, the two cases being obtained from symmetric and anti-symmetric combinations of $\Lambda = \pm 1$. This leads, when quark spin is included, to eight degenerate sets of states for each set of quark and anti-quark flavours with J^{PC} quantum numbers

$$J^{PC} = 1^{--};\ 0^{-+},\ \underline{1^{-+}},\ 2^{-+}$$

and

$$J^{PC} = 1^{++};\ \underline{0^{+-}},\ 1^{+-},\ \underline{2^{+-}}$$

where the first state in each case has $S = 0$ and the others $S = 1$. The underlined states all have exotic quantum numbers ie their quantum numbers do not appear in the non-relativistic quark model. Note also[13] that the states with $S = 0$ have "exotic spin" ie states in the quark model with these J^{PC} values have spin one.

Table 1 gives masses as calculated in the adiabatic limit in two different versions of the flux-tube model and compares them with values obtained from lattice Monte Carlo potentials. There is good agreement for $c\bar{c}$ and $b\bar{b}$ the only cases quoted by the lattice workers.

3.2. Mesons with flavour: corrections, δV, to the adiabatic limit.

Four corrections to the above adiabatic masses may be identified all of which involve transitions of the system between adiabatic potential surfaces. ie They are

	lattice	non-rel harmonic with small R cut-off[1]	1 bead, standard parameters[12]
$b\bar{b}$	10.56(30),[19] 10.81(25)[20] 10.8[21]	10.8	
$c\bar{c}$	4.04,[19] 4.19(15)[20]	4.2	4.18-4.19
$s\bar{s}$		2.0	1.91-1.94
$s\bar{l}$,$\bar{s}l$		1.8	
$l\bar{l}$		1.7	1.80-1.83

Table 1. *Masses of the lowest hybrid mesons in the adiabatic limit [GeV].*

analogous to the Born-Oppenheimer corrections of molecular physics. Their effects are summarised in Table 2. They are:
(a) shifts of the masses of states built on the lowest potential surface (the quark model states)
(b) shifts of the lowest hybrids relative to the quark model states
(c) splitting between states of $L^{PC} = 1^{+-}, 1^{-+}$
(d) spin splittings
The last two of these corrections lead to a lifting of the eight-fold degeneracy. The first effect may be of interest as it would show up as an apparent quark mass dependence of the effective quark model potential. However to get the best flux-tube model estimate of hybrid masses it may simply be incorporated into the parameters of the effective quark model potential V_0 which is adjusted to fit the spectrum of established quark model states, and contributes then to the column labelled (a)(b) in Table 2. The next column is my (unpublished) estimate of the effect (c). A proper calculation of this effect in the flux-tube model requires an improvement in the representation of the flux-tube which allows the system with fixed quark and anti-quark sources to be spherically symmetrical in the limit of $R \to 0$. This was not the case in either the non-relativistic harmonic oscillator representation[7] or the relativistic one bead representation.[12] The present estimate is based on using the lattice Σ_u^- potential at small R which becomes degenerate with Π_u as required by spherical symmetry, and incorporating this into a more elaborate centrifugal barrier potential which interpolates to that of V_L above at large R. The transition region was taken to be 0.5-1.0 Fm.
Arguments[16 27] suggest that the only spin interactions of the flux-tube model should be Thomas-like spin-orbit interactions. The last column of the table gives an estimate[16] of these for hybrids where because of the inertia of the flux-tube the calculation of the Thomas effect is more elaborate than the standard one. It gives rise to equal spacing δ in the states of $L^{PC} = 0^{+-}, 1^{+-}, 1^{++}, 2^{+-}$ in order of decreasing mass and a smaller equal spacing δ' of the same sign in the four states of opposite P and C.
The best flux-tube model estimates of the lowest hybrid masses, combining Tables 1 and 2, are given in Table 3. Where they exist, corresponding recent lattice results, with dynamical quarks, for the lowest states with quark model exotic quantum numbers

	(a)(b) non-rel. harmonic[7] (rel. 1 "bead"[12])	(c) $L^{PC} = 1^{+-}$[17] $L^{PC} = 1^{-+}$	(d) "Thomas"[16]
$b\bar{b}$	+0.02	+0.03 -0.03	$\delta = .001$ $\delta' = 0.000$
$c\bar{c}$	+0.06 (+0.01-0.02)	+0.05 -0.05	$\delta = .006$ $\delta' = 0.004$
$s\bar{s}$	+0.14 (+0.06-0.09)	+0.08 -0.14	$\delta = .065$ $\delta' = 0.015$
$l\bar{l}$	+0.19 (+0.10-0.13)	+0.09 -0.13	$\delta = .140$ $\delta' = 0.020$

Table 2. *Corrections to the adiabatic limit for hybrid masses [GeV]. The columns labelled (a)(b), (c) and (d) refer to the corresponding effects discussed in the text.*

are given in the rows below the corresponding model results. The brackets () indicate lattice error estimates. Of course, error estimates in a model are always hard to make. A crude guess might be a factor of two uncertainty in the *corrections* to the adiabatic mass values.
Note the substantial agreement of the model with lattice "data" but also note the substantial disagreement with the hybrid interpretation of the phenomenon (no quotation marks) seen in the quark model exotic channel 1^{-+} at about 1.4 GeV with a width of about 0.4 GeV.[31]
Hybrids have also been invoked[13 14] in the interpretation of a number of light quark states of non-exotic quantum numbers.
The exact mass positions of the $c\bar{c}$ hybrids could be crucial to their observation since the threshold for their preferred decay channel (see below) to $D\bar{D}^*$ in which one of the charmed mesons is orbitally or radially excited is in the region of 4.3 GeV.

3.3. Mesons with flavour: Hadronic decays of hybrids.

Hadronic decay of hybrids requires a slight extension of the flux tube model, since the creation of a quark-antiquark pair necessary for decay is a topological mixing effect. This was originally treated by Isgur, Kokoski and Paton.[24] There has recently been proposed an alternative extension[28] to compare with the the "canonical" extension given originally by us and developed by Close and Page.[13] I prefer the canonical extension for three reasons. First, it seems to have a very plausible basis in the same analysis of lattice hamiltonian theory that gives rise to the flux-tube model itself. Second, the hybrid parameters are fixed by the decays of quark model states[1 25] where it substantially reproduces the results of the $^3P_0$ model, known to be accurate overall to some ±50% in widths. Finally, as dealt with in detail by Geiger and Isgur[15] it gives a very plausible picture of quark loop effects in spectroscopy. Nevertheless, it is good to have alternative decay models to compare with and to see whether results are reasonably model-independent. In fact, though the alternative decay model differs in some details, such as the ratio of decays in a given channel to different partial waves,

J^{PC}	ll	$s\bar{s}$	$c\bar{c}$	bb
0^{+-}	$1.9+.37$	$2.1+.21$	$4.3+.06$ $4.61(.11)_{st}(.2)_{qu}$[29]	$10.8+.03$ $11.08(.11)$[30]
1^{++}	$1.9+.23$	$2.1+.14$	$4.3+.06$	$10.8+.03$
1^{+-}	$1.9+.09$	$2.1+.08$	$4.3+.05$	$10.8+.03$
2^{+-}	$1.9-.05$	$2.1+.15$	$4.3+.04$	$10.8+.03$
0^{-+}	$1.9-.09$	$2.1+.06$	$4.3-.04$	$10.8-.03$
1^{-+}	$1.9-.11$ 1.90[30] $1.97(.09)_{st}(.30)_{qu}$[29]	$2.1-.12$ $2.00(.2)$[30] $2.17(.08)_{st}(.08)_{qu}(.2)_{latt}$[29]	$4.3-.05$ $4.39(.08)_{st}(.20)_{qu}$[29]	$10.8-.03$
1^{--}	$1.9-.13$	$2.1-.14$	$4.3-.05$	$10.8-.03$
2^{-+}	$1.9-.15$	$2.1-.15$	$4.3-.05$	$10.8-.03$

Table 3. *Best flux-tube model estimates of lowest hybrid masses ($l = u, d$). The added or subtracted pieces are estimates of the combined effect of the different corrections of Table 2 [Recent lattice estimates with dynamical quarks in subsequent rows, with quoted error estimates in brackets.]*

it agrees with the two main qualitative results of the canonical decay mechanism: the suppression of decay to two quark model ground states[11] and the forbidding of the decay $S = 0 \rightarrow S = 0 + S = 0$. Of course it has, as any decay model must have, very strong phase space dependence. Because of the latter it is very much more useful to give model predictions as function of assumed mass as in eg the work of Page[10] rather than for a single fixed assumed mass, as is usually done. By the way, the normalization of the canonical decay model would imply that a hybrid state at as low a mass as 1.4 GeV would be very narrow. So in addition to the mass, also the width of the state at 1.4 GeV would be hard to understand in the flux-tube model.

3.4. Flux-tube model for "glueballs" in 2+1 dimensions

The simplest and lowest lying glueballs in the flux-tube model consist of a continuous loop of flux ("glueloops"). The original work on the spectrum of "glueloops"[1] was done for the realistic case of three spatial dimensions, in which case an *ad hoc* extra adiabatic assumption was made *viz* that the average radial coordinate of the loop is a slow variable compared to the variables describing the deviation of the loop from the circular shape. The assumptions are not so severe in two space dimensions, and in addition lattice studies for theories in two space dimensions[33 32 35] give more detailed information than for their $3+1$ dimensional counterparts. Thus it makes sense to compare first with the two-dimensional "glueball" spectrum. The work of Agostini *et al*[32] is on Z_2 gauge theory. Like $SU(2)$ the flux-tube has no arrow associated with it. In addition to results for $SU(2)$, Teper[35] also gives results for $N_C = 3, 4, 5$ and considers the extrapolation to the limit of an infinite number of colours. The glue-loops of the

M	n_2	n_3	n_4	J^P
0	0	0	0	0^+
2	1	0	0	$2^\pm$
3	0	1	0	$3^\pm$
4	2	0	0	$0^+, 4^\pm$
4	0	0	1	$4^\pm$
5	1	1	0	$1^\pm, 5^\pm$

Table 4. $C = +1$ *states of lowest value of* M *(see text) in the 2 + 1 dimensional flux-tube model*

flux-tube model should be the same for all cases except that the $C = -1$ states should be absent for Z_2 and $SU(2)$. Further, the spectrum for all N_C should be identical in the model, with degeneracy between states of positive and negative C where the latter exist.
Taking the continuous string realization of the flux-tube, the loop is specified by the polar coordinate parametrization:

$$r(\theta) = r_0 + \Sigma_{m=2}^{\infty}(q_m \exp im\theta + q_{-m} \exp -im\theta)$$

and the states of the quantum system may be classified by

1. the number of quanta n_m corresponding to each oscillation mode m, with $m = 2, 3, 4, ...$
2. the degree of radial excitation n_r

The radial ground states are then ordered by

$$M = \Sigma n_m m$$

Table 4 gives the quantum numbers of the lowest flux-tube model states of $C = +1$ where the mass of each state should be a monotonic increasing function of M. I show in Figure 1 the lattice results[35] for $N_C = 2$. The orderings for Z_2 and for the $C = +1$ states for $N_C = 3, 4, 5$ are all identical to this and the masses are also not too different at least for the SU(N_C) sequence. The quantum numbers of the first two states are correctly given by the flux-tube model corresponding to $M = 0$ and $M = 2$, but the next two, given as $J^P = 0^-, 1^\pm$ would correspond in the flux-tube model to $M = 8$ and $M = 5$ respectively and would be very much higher in mass than the lattice results appear to make them. However, it is amusing to note that on a square lattice there is an ambiguity of the assignment of J *modulo* 4. It is conceivable therefore that the lattice assignment of the lowest 0^- state is wrong and that it is really the lowest $4-$ state and similarly that the $1^\pm$ states are really $3^\pm$. In this case it would not be difficult to envisage that one of the two degenerate $M = 4$ states with $J^P = 4^\pm$ is pushed down mass (J^P)

– 10.8 ($1^{\pm}$)
– 9.9 (0^-)

– 7.8 ($2^{\pm}$)

– 4.7 (0^+)

– vacuum

Fig. 1. *Level diagram of lowest mass states of given quantum numbers, as determined for $SU(2)$ lattice gauge theory (masses in string tension units)*[35]

just below the $M = 3$ state, and that the flux-tube model is essentially correct for the first four sets of quantum numbers.
The situation[35] for excited states on the lattice of these $C = +1$ states more-or-less bears out flux-tube model expectations: see for example Table 5 compiled from the lattice data of Agostini[32] and Teper[35] for $J^{PC} = 0^{++}$ states. A simple version of the relativistic string gives spacing in squared mass, ΔM^2, of $8\pi\sigma = 25.13\sigma$ for radial recurrences.[1] The corresponding lattice values are given in the table for what would be the first and second radial excitation in the flux-tube model. It may be seen that the values of $\Delta M^2/\sigma$ cluster around a value not far from $8\pi = 25.13$ but, using the lattice statistical errors, most are outside 1σ of this value. I have not attempted a least squares linear fit to the lattice data for M^2/σ.

gauge group	Z_2	$SU(2)$	$SU(3)$	$SU(4)$	$SU(5)$
mass	3.08	4.718(43)	4.329(4)	4.236(50)	4.184(55)
mass	5.80(4)	6.83(10)	6.52(9)	6.38(13)	6.20(13)
$\Delta M^2/\sigma$	24.15	24.40	23.77	22.76	20.93
mass	7.97(11)	8.15(15)	8.23(17)	8.05(22)	7.85(22)
$\Delta M^2/\sigma$	29.88	19.77	25.22	24.10	23.18

Table 5. *Lattice masses and spacing in squared mass, ΔM^2, both in string tension units, of 0^{++} states in various gauge theories. In its relativistic string realization the flux-tube model gives $\Delta M^2 = 8\pi\sigma = 25.13\sigma$.*

gauge group	$SU(3)$	$SU(4)$	$SU(5)$
mass	6.48(9)	6.27(9)	6.03(18)
mass	8.15(16)	7.86(20)	7.87(27)
$\Delta M^2/\sigma$	24.43	22.47	25.57
mass	9.81(26)	9.21(30)	9.54(41)
$\Delta M^2/\sigma$	29.81	23.04	29.07

Table 6. *Lattice masses and spacing in squared mass, ΔM^2, both in string tension units, of 0^{+-} states in various gauge theories.*

Turning now to the $C = -1$ states for gauge groups $SU(N_C)$ with $N_C = 3, 4, 5$[35] Teper's results for eg the $J^{PC} = 0^{+-}$ states are reproduced in Table 6 along with the values of $\Delta M^2/\sigma$ calculated from them. Note that the lowest $C = -1$ states are very much more massive than the lowest $C = +1$ states of Table 5, being degenerate within errors with the *first excited* $C = +1$ states. The same pattern is seen also for $J = 2$ states of $C = -1$ in comparison to those of $C = +1$.[35] It is clear that *either* the $C = \mp 1$ degeneracy of the flux-tube model is badly broken *or else* the $C = -1$ states corresponding to the 0^{++} and 2^{++} ground states are for some reason missing in all cases. I favour the latter interpretation. The $C = \pm 1$ degeneracy of "glue-loop" states follows from the assumption that the kinetic energy operator derived from the "hopping" of flux links on the lattice cannot change the orientation of the loop. This follows if, for all important configurations of the glue-loop, the deviation of $r(\theta)$ from its mean value r_0 is small enough. This assumption is clearly most dangerous for the smallest "glue-loops". In the opposite extreme, fluctuations in the shape of the flux-tube loop produced by the kinetic energy operator would completely mix configurations of opposite orientation. This would have the effect of removing the lowest $C = -1$ state entirely from the spectrum leaving the $C = \pm 1$ degeneracy of excited states undisturbed. This appears to be the case.

It is possible therefore that this comparison of the flux-tube model with "spectroscopy" is teaching us that a modification to the model is required in exactly the place (states of smallest physical size) where it might be expected, and that the modification required is simple and limited. It would be worthwhile to determine whether the lowest $C = -1$ states of $J = 0, 2$ are missing in a more sophisticated version of the oriented closed relativistic string.

4. Conclusions

As advertised in the Introduction, my conclusions are primarily about "spectroscopy" rather than about spectroscopy. In the case of "hybrids" it appears that the flux-tube model and the lattice agree on the quantum numbers of the lowest hybrids, and that corrections to the adiabatic limit mass values in the flux-tube model agree at the moment with preliminary lattice dynamical calculations where these are available. The one conclusion on genuine spectroscopy with no quotation marks appears to be negative. It would be difficult to accomodate a hybrid $J^{PC} = 1^{-+}$ state of mass as low as 1.4 GeV and width as wide as 400 MeV.

On the "spectroscopy" of glueballs, the flux-tube model exhibits several intriguing qualitative agreements with the lattice in $2+1$ dimensional gauge theories. There is also an intriguing disagreement which suggests that the original flux-tube model assumption that the orientation of the flux-tube loop (its arrow) is well defined, and implying the $C = \pm 1$ degeneracy of states, breaks down for the ground states of a given set of quantum numbers. It is also worth noting[33] that the flux-tube model also puts into question in some cases the common lattice assumption that the lowest state of a given lattice symmetry should be assigned to the lowest compatible J value. It will be worth bearing both these points in mind in going to the more realistic case of three space dimensions.

References

1. N. Isgur and J. Paton, Phys. Rev. **D 31**, 2910 (1985)
2. The vacuum is "empty" in the sense that it is in an eigenstate of $\mathbf{E^2}$ on every link with eigenvalue zero. Since the components of **E** and **B** are conjugate variables, it is not an eigenstate of $\mathbf{B^2}$ and indeed it has a non-zero expectation value of $\mathbf{B^2}$. Topological mixing would allow this vacuum to be modified to include a non-zero density of virtual $q\bar{q}$ mesons and glueballs.
3. J. Kogut and L. Susskind, Phys. Rev. **D 11**, 395 (1975)
4. J. Carlson, J. Kogut and V. R. Phandaripande, Phys. Rev. **D27**, 233 (1983), S. Capstick and N. Isgur, Phys. Rev. **D34** 2809 (1986)
5. P. R. Page, these Proceedings, S. Capstick and P. R. Page, work in progress
6. A. M. Green and J. E. Paton, Nucl. Phys. **A492**, 595 (1989)
7. J. H. Merlin and J. Paton, J. Phys. **G 11**, 439 (1985)
8. See eg M. B. Green, J. H. Schwarz and E. Witten *Superstring Theory, vol 1 Introduction*, Cambridge, 1987
9. S. Capstick, S. Godfrey, N. Isgur and J. Paton, Phys. Let **175B**, 457 (1986)
10. See talk by A. Szczepaniak in these Proceedings; also, E. S. Swanson and A.Szczepaniak hep-ph/9704434 and P. R. Page, hep-ph/9711241
11. This selection rule follows also in some other models eg A. Le Yaouanc, L. Oliver, O. Pene and J.-C. Raynal, Phys. Rev. **D8**, 2233 (1973)
12. T. Barnes, F. E. Close and E. S. Swanson, Phys. Rev. **D 52**, 5242 (1995)
13. F. E. Close adn P. R. Page, Nucl. Phys **B443**, 233 (1995)
14. A. Donnachie, Y.-S. Kalashnikova and A. B. Clegg, Proceedings of the Jefferson Lab/NCSU Workshop on Hybrids and Photoproduction, November,1997
15. P. Geiger and N. Isgur, Phys. Rev. **D55**, 299 (1997)
16. J. Merlin and J. Paton, Phys. Rev **D35**, 1688 (1987)
17. J. Paton (unpublished)
18. M. Luscher, Nucl. Phys. **B180 [FS2]**, 317 (1981)
19. S. Perantonis and C. Michael, Nucl. Phys. **B 347**, 854 (1990)
20. Previous reference, but allowing for an estimate of the effect of quenching.

21. K. Juge, J. Kuti and C. Morningstar, Nucl. Phys B (Proc. Suppl. 63, (1998) *in press* (LAT97)
22. C. Morningstar, these Proceedings, and Juge *et al*[21]
23. N. Dowrick, J. Paton and S. Perantonis, J. Phys. **G13**, (1987)
24. N. Isgur, R. Kokoski and J. Paton, Phys. Rev. Lett. **54**, 869 (1985)
25. R. Kokoski and N. Isgur, Phys. Rev. **D35** 907, (1987)
26. T. J. Allen and M. G. Olsson, hep-ph/9804452 and T. J. Allen, these Proceedings
27. E. S. Swanson, Phys. Rev. **D55** 3987 (1997)
28. E. S. Swanson and A. P. Szczepaniuk, Phys.Rev. **D56** 5692, (1997) See also P.R. Page in HADRON97, Brookhaven, Aug. 1997
29. MILC Collaboration, hep-lat/9707008 Nucl. Phys B (Proc. Suppl. 63, 1998) (LAT97)
30. UKQCD Collaboration, P. Lacock, C. Michael, P. Boyle and P. Rowland,Nucl. Phys B (Proc. Suppl. 63, 1998) (LAT97)
31. D. R. Thompson et al (E852 Collaboration) Phys. Rev. Lett. **79** 1630 (1997). The present experimental status of the quark model exotic channel 1-+ is summarised by D. P. Weygand (these Proceedings) who notes some corroboration of this state from VES, D. V. Amelin (VES Collaboration), HADRON97, Brookhaven,Aug.1997 and the Crystal Barrel, W. Dunnweber (Crystal Barrel Collaboration) HADRON97, Brookhaven,Aug.1997. There is, in addition to the possibility of a wide state at around 1.4 GeV, also some evidence, also from BNL-E852, D. P. Weygand and A. I. Ostrovidov (E852 Collaboration) HADRON97, Brookhaven,Aug.1997 for a state at around 1.6 GeV.
32. V. Agostini, G. Carlino, M. Caselle and M. Hasenbusch. Nucl. Phys. **B484**, 33 (1997)
33. T. Moretto and M. Teper, hep-lat/9312035
34. R. Johnson and M. Teper, hep-lat/9709083
35. M. Teper, hep-lat/9804008

QCD CONFINEMENT PARAMETERS FROM DUAL SUPERGRAVITY

Yuval Ne'eman
Raymond and Beverly Sackler Faculty of Exact Sciences Tel-Aviv University, Israel
and
Center for Particle Physics, University of Texas, Austin, USA

1. A string-based gravitational algorithm

Though the notion of *color confinement* has been with us almost from the moment of the discovery of asymptotic freedom and the birth of QCD, it should still be considered as a very plausible but as yet mathematically unproven hypothesis. In my report to Confinement I [1], I reviewed the two approaches which had been used by theorists in devising models for the confinement mechanism:

(a) *Dual Superconductivity*, i.e. the emergence of a finite *penetration length* (a Meissner-like expulsion from the hadron vacuum outside, into the inside of the hadron), as a result of the formation of a *chromomagnetic monopole condensate* [2-3]. Monopoles, however, appear only as solutions of *spontaneously broken* Yang-Mills equations and it has accordingly been suggested that such a breakdown of local $SU(3)_{color}$ symmetry might be due to either *effective* Higgs mechanisms, such as gluonic condensates, behaving like $\underline{8}$ and leaving a $Z(3)$-gauge unbroken, the opposite (i.e. $Z(3)$-breakdown) at the quark level, or also 't Hooft's 'abelian projection', yielding $U(1) \times U(1)$ - or to physical Higgs fields, whose presence might, for instance, be imposed by *supersymmetry*.

(b) *Chromogravity*, mechanisms emulating gravity, based on (b1) the emergence of an $sl(3, R)$ symmetry or spectrum-generating algebra, either through (b1,1) the identification of a diffeomorphisms-like subgroup within the $SU(3)_{color}$ gauge group [4], or (b1,2) by the application of a Hamiltonian quantization procedure to QCD [5] or again (b2) through the identification of Schwarzschild-like solutions to the Yang-Mills equation for QCD [6].

In my report to Confinement II [7], I described the successful mathematical proof of the emergence of the monopole condensate and consequent confinement, following mechanism (a), due to N. Seiberg [8] and E. Witten [9]. Supersymmetry was essential in those proofs in two ways: *physically*, by the provision of a supermultiplet including a scalar field - which then plays the role of a Higgs field, breaking the color group and thus providing for monopoles - and *mathematically*, through BPS saturation and the algebraic fixation of the complete energy spectrum, including the monopole condensate.

At this Confinement III conference, I again have an encouraging result to report

[10], and once again it is tied to supersymmetry, though in a rather different way. The calculation I am reporting consists in the obtention of the parameters of QCD relating to confinement - in a rather convoluted procedure - from *supergravity* in a superspace whose even submanifold is 5-dimensional anti-de Sitter spacetime AdS_5, i.e. a (-1,-1,1,1,1) 5-manifold. The linkage between QCD and sGR(AdS_5) is provided by a version of (10-dimensional) *Superstring Theory* [11] known as "Type IIB" and the (11-dimensional) *supermembrane* extension of superstring theory [12], known as "M-theory", the present incumbent candidate "Final Theory". Supersymmetry has been a fundamental ingredient in strings, ever since, in its role as tachyon-remover, it guaranteed the unitarity of the theory. This is then how it enters here, but the conclusions are the same as in the 1994/95 results: we are dealing with a version of QCD in which the gluons are accompanied in the supermultiplet by scalar fields in the adjoint representation of the color group; moreover, the construction utilizes the notorious [13] *renormalization-invariant* $\mathcal{N} = 4$ supersymmetric $D = 4$ Yang-Mills ($\mathcal{N}$ is the number of supersymmetries) so that supersymmetry provides also for self-duality in the *weak* $\leftrightarrow$ *strong* coupling context. Before going into a detailed survey and analysis of Maldacena's derivation [10], I shall sketch the logical sequence and steps it involves:

A) Back in 1974 't Hooft introduced [13] the $N_c \to \infty$ limit (N_c is the number of "colors") and proved that in this Yang-Mills model, the *planar* diagrams dominate, in a quantum field theory's perturbative expansion and that their sum *is equivalent to a string.* Furthermore, in the 1998 version, one first embeds this "QCD($N_c \to \infty$)" within its $D = 4, \mathcal{N} = 4$ renormalization-free supersymmetric extension [14]. This yields a model which can be reobtained - in a dimensional reduction process - from a $\mathcal{N} = 1$ super-YM in 10 dimensions [15]. This 10-dimensional quantum field theory itself can be reproduced as the sub-Planck-mass truncation of Type IIB superstring theory. The latter can further be extended, by adding one spatial dimension to both the string (turning it into a membrane, or turning the superstring into a supermembrane) and to the embedding manifold (now 11-dimensional) - into "M-theory". Our "link A" thus prescribes how to get to M-theory and its Type IIB superstring truncation, from $\mathcal{N} = 4$ superYM($N_c \to \infty$), as an *exact* statement. From this point back to the physical $N_c = 3$ QCD, we are dealing with an approximation, namely - plagiarizing Gamow's title [16], "One, Two, Three, Infinity". Recapitulating, we have the inclusion sequence (sMT is the supermembrane in "M-theory") ,

$$\begin{aligned}[QCD] \equiv [YM(D=4, N_c=3)] \to [YM(D=4, N_c \to \infty)] \subset \\ [\mathcal{N}=4sYM(D=4, N_c \to \infty)] \in [\mathcal{N}=1sYM(D=10, N_c \to \infty)] \\ \subset [sSIIB(D=10)] \subset [s'M'T(D=11)] \qquad (1)\end{aligned}$$

where we have used the inclusion $\in$ to represent a dimensional reduction. The M-theory last step is in fact somewhat redundant here, and we situate the system with respect to M-theory mostly for the sake of completeness, in this era in which that theory "dominates" the scene.

B) The next link in the logical chain is provided by *compactification.* M-theory is best known in a mode in which the supermembrane in it, when sub-Planck mass truncation is applied, begets the Cremmer-Julia $D = 11, \mathcal{N} = 1$ supergravity [17], which has a natural *spontaneous compactification* [18] into $AdS(D=4)\otimes S^7$ (anti de Sitter is "$-+++$") yielding $sGR(D=4,\mathcal{N}=8)$, the maximal supergravity theory in $D=4$. The art of Kaluza-Klein compactification has, however, undergone great advances since [19], and in the present case, another (though similar) reduction is selected, a compactification of type IIB superstring, to the direct product of the above-mentioned supergravity in 5-dimensional anti-de Sitter spacetime and a 5-sphere S^5.

$$[sSIIB(D=10)] \doteq [sGR(D=5,AdS)]\otimes S^5 \tag{2}$$

C) The third and last step has been realized in two different but mathematically equivalent ways. Juan Maldacena embeds a series of equally-spaced *D3-branes* (extendons - 3-dimensional 'lumps' - with Dirichlet boundary conditions) - or better, their 4-dimensional world-hypervolumes - in AdS_5. In the limit of zero-spacing, this becomes a 4-dimensional Minkowski spacetime, except that a *duality* transformation has thereby meanwhile occured. Thus, we close the triangle by reidentifying $\mathcal{N}=4sYM(D=4,N_c\to\infty)$, but only after inverting $g\leftrightarrow 1/g$ and making the corresponding changes in spacetime and energy-momentum variables.

Witten, on the other hand, has shown that, using polar coordinates for the three space-like dimensions in $AdS_5(-1,-1,1,1,1)$, this is equivalent to taking $r\to\infty$ - an adaptation of the well-known Wigner-Inönü group contraction. With an infinite radius-vector, a rotation becomes a translation and the $r\to\infty$ boundary of AdS_5 is Minkowski spacetime, in which we again identify our $\mathcal{N}=4sYM(D=4,N_c\to\infty)$.

$$\begin{aligned}&[sGR(D=5,AdS)]\supset[\mathcal{M}(p=3,D=4)]^{\otimes\infty}\to\\&\qquad[sYM(\mathcal{N}=4,D=4,N_c\to\infty)],\\&[sGR(D=5,AdS](r\to\infty)\to[sYM(\mathcal{N}=4,D=4,N_c\to\infty)],\end{aligned}\tag{3}$$

This then is the "geometrodynamical" outline, with the final result consisting in a new gravity-inspired algorithm: a *duality* transformation, relating $sGR(D=5,AdS)$ - a purely gravitational arena whose characteristic features are *horizons* - with our QCD color confinement issues (though in the " 't Hooft-Gamow" approximation). We shall now follow this path in some detail.

2. The Yang-Mills end: $\mathcal{N}=4$, $N_c\to\infty$, the spacetime loop representation

First, a brief explanation of the $\mathcal{N}=4$ choice. For those who are less familiar with the highlights of supersymmetry in its 27 years' history as theorists' pet (and as yet no experimental confirmation) let me mention that it was originally believed that systems with $\mathcal{N}=4j^z_{max}$ should display improved renormalization features. In such systems, supersymmetry spans the entire helicity spread, from $+j^z_{max}$ to $-j^z_{max}$, i.e. $[2j^z_{max}/(1/2)]$ steps (with antisymmetrization at each), in one representation - whereas for the Poincaré group, the positive and negative helicities decouple, with only CPT relating them. This is a powerful algebraic constraint and the case of $j^z_{max}=+1$, $\mathcal{N}=4$ indeed fulfilled these hopes beautifully, allowing for zero radiative corrections to the Yang-Mills coupling [14]. The $j^z_{max}=1/2,\mathcal{N}=2$ case, known as the Wess-Zumino model, displays an economy of one renormalization constant. When we discovered the constraint, Gell-Mann and I in 1976, we immediately thought of quantum gravity and called for the construction of the $j^z_{max} = 2, \mathcal{N} = 8$ case. Cremmer et al [17] indeed duly achieved it, in a *tour de force* performance (via $\mathcal{N} = 1, D = 11$ supergravity). By 1984, however, hopes were very low and the model was practically abandoned, in the great string goldrush. Considering that this model is contained as the 'beneath Planck mass truncation' of the presently enshrined 'M'-theory, there is perhaps some justification for new hopes..

Returning to our $\mathcal{N} = 4$ sYM, we note the helicity composition of the super-multiplet: $1 \times +1, 4 \times +1/2, 6 \times 0, 4 \times -1/2, 1 \times -1$. We note the presence of 6 scalar fields, capable of playing Higgs field roles. Everything is in the adjoint representation of the color-gauge group.

The dimensionality of the $\mathcal{N} = 1$ embedding from which the same system can be obtained by dimensional reduction is determined by the constraint requiring an equal number of boson and fermion components [15]. For a vector ($J = 1$) field, the number of physical (transverse) bosonic components (per color) is $D-2$; the number of physical spinorial components is $2^{[D/2]-\epsilon(D)}$, where $[D/2]$ denotes the integer part of $D/2$ and $\epsilon(D) = 0, 1, 2$ according to the type of spinor available in that dimensionality [20] (and selected, if there is more than one possibility), namely Dirac ($\epsilon = 0$), Weyl or Majorana ($\epsilon = 1$), or Weyl-Majorana ($\epsilon = 2$). $D = 10$ obviously solves the sYM case.

Now to the $N_c \to \infty$ limit. This is taken [13] together with $\alpha_s = (g^2/4\pi) \to 0$, holding the product constant

$$g^2 N_c = (g_0)^2 = const \tag{4}$$

The last point relates to the *encoding* of the dynamics in spacetime. This is essential in an approach in which the "analog computer" is gravitational. It is achieved by the (gauge invariant) Wilson loop integral,

$$W[c(s)] = (1/N_c) < TrPe^{\oint_{c(s)} A^i_\mu \{\lambda_i\} dx^\mu} > \tag{5}$$

where W is the path-integral, P indicates *normal ordering*, A_μ is the $SU(N_c)$ connection (i.e. the gluon), the trace - in the present case - is taken over the fundamental representation matrices λ_i (i.e. quarks, because we are going to evaluate the potential energy between quark and antiquark) and $c_\mu(s)$ is a parametrized ($0 \leq s \leq 1$) contour. The Wilson loop is the prime calculational tool in *Lattice Gauge Theory*; for an intuitive picture, consider the quark-antiquark pair as very heavy, allowing for a non-relativistic treatment. The S-matrix will be given by $S = lim_{t' \to -\infty, t" \to +\infty} e^{i(t'-t")H}$, H being the Hamiltonian. This contains kinetic and potential energy and in the interaction picture reduces to the latter, $H_I = A$ and taking a long but finite time T, with the "i" incorporated by working in a Euclidean 4-space, taking the matrix element for the interaction between the pair and completing the integration path to a closed rectangle by adding the spatial distance Δx between them ($\Delta x << cT$) twice, we conclude that this is $W([c(s)]$. Notice that confinement results here from the breaking of the $Z(3)$ subgroup in this *nonzero triality* representation.

In his construction, Maldacena gives us an approximative method of describing a heavy quark: start with $U(N+1)$, with spontaneous symmetry breakdown (Higgs mechanism) to $U(N)$. The gauge fields of the quotient (broken symmetry) $U(N+1)/U(N)\otimes U(1)$ are in the N-dimensional fundamental representation of $U(N)$ and have acquired masses (proportional to the Higgs mass). Working at a much lower energy, the additional $U(1)$ decouples and the massive mesons can play the role of "quarks" in the calculation. We shall describe and analyze the results in the next section.

Note that the dynamics have thus been related to a loop, marrying spacetime with the YM gauge-potential. In *loop space* the second variational derivative provides a Schwinger-Dyson equation,

$$\begin{aligned} \hat{L}W(c) &= g_0^2 \int dsdu[\ c(s)\ c(u)\delta(c(s)-c(u))W('c)W('c) \\ \hat{L} &\equiv lim_{\epsilon\to 0} \oint ds \int_{-\epsilon}^{+\epsilon} dt \frac{\delta^2}{\delta c(s+t)\delta c(s-t)} \end{aligned} \qquad (6)$$

Here $W(c)$ is a sum over surfaces with disk topology. It is also the amplitude for an open string with ends moving on the contour c. Polyakov and collaborators have been developing this aspect since the early eighties [21]. I shall not dwell here upon various interesting features in the detailed analysis, such as Polyakov's *"zigzag"* symmetry, in *winding* around the contour, somewhat in the manner of basket weavers..

3. Kaluza-Klein compactification of supergravity models

The 11-dimensional *maximal* supergravity of Cremmer et al [17] supplied a geometrical model from which one obtains the (maximal) $\mathcal{N}=8$ supergravity in $D=4$ by Kaluza-Klein reduction. Once again, the dimensionality is determined by the constraint of having an equal number of boson and fermion components, from 4-dimesional supersymmetry. Starting with the positive helicity $+2$ component of the graviton as *highest weight* and acting with the 8 Majorana spinor "spin lowering" supersymmetry generators, we get 8 gravitinos with helicity $+3/2$, 28 vector states with $+1$, 56 spinors with $+1/2$, $70\times j_z=0$, $56\times j_z=-1/2$, $28\times j_z=-1$, $8\times j_z=-3/2$, $1\times j_z=-2$ state; altogether a total of 128 boson and 128 fermion components. Our listing also makes it obvious that $\mathcal{N}=8$ is maximal, since 9, for instance, would have created a $J=5/2$ spinor, etc..

In constructing the higher-dimensionality D $\mathcal{N}=1$ "originator", we note that the gravitational field will have $[(D-2)(D-2+1)/2]-1=D(D-3)/2$ transverse (and thus physical) components (a symmetric tensor in $D-2$ dimensions, minus the trace). The gravitino satisfies a constraint $\gamma^M\psi_M=0$ and will thus have $(D-3)\times 2^{[D/2]-\epsilon_D}$ components (see previous section for the notation). In D=11, this gives indeed 128 fermion states, but the bosons supplied by the gravitational field only number 44. The difference, as discovered by Cremmer et al., is provided by a new *3-form potential* A_{PQR} with $(D-2)!/3!(D-5)!=84$ components, making up the necessary 128. Note that a point has a world line - making a 1-form in an integration, a string has a world-surface - making a 2-form, a membrane has a world-volume - making a 3-form; alternatively, remember that a (point) "local" gauge field is a 1-form, a string gauge field is a 2-form (the Kalb-Ramond A_{PQ}), and a membrane is gauged by a 3-form. This is why it was possible for Bergshoeff et al. [22] to show that $D = 11$ supergravity, with the A_{PQR} potential is the truncation of a supermembrane.

In the Lagrangian, the field strength $F_{PQRS} = 24\partial_{[P}A_{QRS]}$ appears both in a Maxwell-like $F \wedge F$ and in a Chern-Simons-like $F \wedge F \wedge A$. The spontaneous compactification discovered by Freund and Rubin [18] exploits this fact by solving, $< F_{\mu\nu\rho\sigma} >= 3m\epsilon_{\mu\nu\rho\sigma}$ and $< F_{pqrs} >= 0$, i.e. it is the existence of a membrane as a starting point which makes spacetime have $D = 4$. The other 7 dimensions then compactify into a sphere S^7. With "M-theory" thus defined as the quantum supermembrane embedded in 11 dimensions, one thus has a natural compactification and a "stable vacuum" for the "Theory of Everything" - something one was desperately searching for in 1984-1994 with superstring theory, by looking for a *string field theory* "meta-equation". Here "it comes for free"..

This was $AdS_4 \otimes S^7$ compactification, perhaps (?) the true vacuum for the TOE. However, once this is understood, it becomes easy to have other compactifications, just by putting in a $D - 1$-form as a potential. Thus $AdS_4 \otimes S^7$ can be replaced by the $AdS_5 \otimes S^6$ we need here as a computational device - or $AdS_5 \otimes S^5$ if we start already from the Type II superstring, instead of from M-theory.

4. $AdS_5 \otimes S^5$ and Maldacena's result

G. 't Hooft's YM $\to$ string transition, when applied to $[\mathcal{N} = 4, sYM(D = 4] \in [\mathcal{N} = 1, sYM(D = 10)]$ yields a TypeII superstring with action,

$$I = (1/2\pi\alpha') \int d^2\zeta \sqrt{detG_{MN} g^{\beta\gamma} \partial_\beta X^M \partial_\gamma X^N} \tag{7}$$

and the metric G_{MN} describes the embedding 10-dimensional spacetime which will correspond to $AdS_5 \otimes S^5$ after truncation (for some specific steps one takes a Euclideanized G_{MN}). To get some intuition about de Sitter and Anti de Sitter spaces, note that Einstein's equation reduces for constant curvature (and with an effective cosmological constant, as a consequence) to

$$R_{\mu\nu} - (1/2)Rg_{\mu\nu} = (1/4)Rg_{\mu\nu} \tag{8}$$

$R = 0$ is Minkowski spacetime, $R > 0$ is de Sitter spacetime and $R < 0$ is 'anti de Sitter', with $\Lambda = R/4$ changing from repulsion to attraction.

Type II superstring ($D = 10$), taking the (quantum field theory) limit of zero slope $\alpha' \to 0$, is compactified as $sGR(AdS, D = 5) \otimes S^5$ then has the G_{MN} metric

$$\begin{aligned} ds^2 &= f^{-1/2} dx_\mu^2 + f^{1/2}(dr^2 + r^2 d\Omega_5^2) \\ f &= 1 + (4\pi g N \alpha'^2)/r^4, \end{aligned} \tag{9}$$

where x_μ represents the 4 coordinates of the D3-branes world volume and $d\Omega_5$ is the metric of a unit 5-sphere. One defines a variable $u \equiv (r/\alpha')$, taking the double limit

$\alpha' \to 0$ (local field theory) and $r \to 0$ (merger of the D3-branes), with u fixed.

$$ds^2 = \alpha' \left[\frac{u^2}{R^2}(dt^2 + \sum_1^3 dx_i^2) + R^2 \frac{du^2}{u^2} + R^2 d\Omega_5^2 \right] \tag{10}$$

where $R = (4\pi g N_c)^{1/4}$ is now the "radius" of S^5 and also of AdS_5. Inserting this metric in the string action enabled Maldacena to calculate the string energy, obtaining

$$E = -\frac{4\pi^2 (2g_{YM}^2 N_c)^{1/2}}{\Gamma(1/4)^4 L} \tag{11}$$

where L is the distance between the two "quarks". Here we get no confinement - this corresponds in $\mathcal{N} = 4sYM(D = 4)$ to a zero-temperature superconformal theory. Going to nonzero temperature and breaking the superconformal symmetry, Maldacena found the energy to be proportional to the area enclosed by the contour c_μ on the 4-dimensional boundary of AdS_5. Note that the "Poincaré" group in AdS_5 can be derived as a Wigner-Inönü contraction of $SO(4,2)$ - and this is the conformal group in sYM(D=4).

For the nonzero temperature case, the AdS_5 metric chosen corresponds to a Schwarzschild solution, a black hole: the minimal radius (the horizon) will translate into a minimal energy density, a temperature. This is the metric (this time we leave out the 5-sphere S^5),

$$ds^2 = \left(\frac{r^2}{b^2} + 1 - \frac{w_4 M}{r^2} \right) dt^2 + \frac{dr^2}{\frac{r^2}{b^2}} + 1 - \frac{w_4 M}{r^2} + r^2 d\Omega^2. \tag{12}$$

b^{-2} is the scalar curvature, so that b is a 'radius of curvature'. w_D (here w^4) is the coefficient ($2G/c^2$ in $D = 4$) multiplying the mass of the black hole, yielding the Schwarzschild radius, $w_D = \frac{16\pi G_N}{(D-1)vol(S^{D-1})}$ (G_N is Newton's constant in $D+1$ and vol(S^{D-1}) is the volume of a $D-1$ sphere). The horizon is given, for AdS_{D+1} by the larger solution of

$$\frac{r^2}{b^2} + 1 - \frac{(w_D M)}{r^{(D-2)}} = 0 \tag{13}$$

After some further adaptations and simplifications one has an AdS_5 metric

$$ds^2 = \left(\frac{r^2}{b^2} - \frac{b^2}{r^2} \right) dt^2 + \frac{dr^2}{\left(\frac{r^2}{b^2} - \frac{b^2}{r^2} \right)} + r^2 \sum_{i=1}^{i=3} dx_i^2 \tag{14}$$

and we can see that after dividing by r^2, we recover Minkowski spacetime for the boundary $r \to \infty$, normalizing by setting $b = 1$. Note that the supergravity picture is most appropriate when gN_c is very large, making the AdS_5 almost flat. This is the strong coupling regime for QCD. When AdS_5 becomes nonperturbative (small gN_c) it is QCD which becomes perturbative.

The transformation $[AdS_5 \otimes S^5] \to sYM_4(N_c \to \infty)]_{r \to \infty}$ has now been exploited and validated in verifying a large number of examples. It thus provides a curious link between gravity and the Strong Interaction (QCD), a new and more sophisticated version of "Chromogravity" (rather than of "Strong Gravity", since one does not add a new force to QED, only a new "analog computer").

References

1. Y. Ne'eman, in *Quark Confinement and the Hadron Spectrum*, edited by G.M. Prodperi and N. Branbilla (World Scientific, Singapore, 1995).
2. S. Mandelstam, Phys. Rep. **23C**, 245 (1976).
3. G. 't Hooft, in *High Energy Physics* (EPS Palermo 1975 Conf.), edited by A. Zichichi.
4. Y. Ne'eman and Dj. Šijački, Int. J. Mod. Phys. **A10**, 4399 (1995); Modern Phys. Lett. **A11**, 217 (1996).
5. D.Z. Freedman and R.R. Khuri, Phys. Lett. **B329**, 263 (1994); P.E. Haagensen and K. Johnson, Nucl. Phys. **B439**, 597 (1995).
6. D. Singleton, Phys. Rev. **D54**, 5911 (1995); F.A. Lunev, Modern Phys. Lett. **A9**, 2281 (1994).
7. Y. Ne'eman, in *Quark Confinement and the Hadron Spectrum II*, edited by G.M. Prosperi and N. Branbilla (World Scientific, Singapore, 1997).
8. N. Seiberg, Phys. Rev. **D49**, 6857 (1994).
9. N. Seiberg and E. Witten, Nucl. Phys. **B426**, 19 (1994); **B431**, 484 (1994).
10. J.M. Maldacena, "The large N limit of superconformal field theories and supergravity", hep-th/9711200; "Wilson loops in large N field theories", hep-th/9803002; N. Itzhaki, J.M. Maldacena, J. Sonnenschein and S. Yankielowicz, "Supergravity and the large N limit of theories with sixteen supercharges", hep-th/9802042; E.Witten "Anti-de Sitter space, thermal phase transition, and confinement in gauge theories", hep-th/9803131; "Anti-de Sitter space aand holography", hep-th/9802150.
11. M.B. Green, J.H. Schwarz and E. Witten, *Superstring Theory*, 2 vols. (World Scientific, Singapore, 1987). J.H. Schwarz, *Superstrings (The first 15 years)*, 2 vols., (World Scientific, Singapore, 1985).
12. E. Bergshoeff, E. Sezgin and P.K. Townsend, Phys. Lett. **B189**, 75 (1987). See Y. Ne'eman and E Eizenberg, *Membranes and other Extendons (p-branes)*, (World Scientific, Singapore, 1995), 178 pp., Chapter 5 and section 4.6.
13. G. 't Hooft, Nucl. Phys. **B72**, 461 (1974).
14. S. Mandelstam, Nucl. Phys. **B213**, 149 (1983). See ref. 15, p.84 for further references.
15. P.G.O. Freund, *Supersymmetry*, (Cambridge Univ. Press, 1986).
16. G. Gamow, *One Two Three ...Infinity*, (Mentor Books, New York, 1946).
17. E. Cremmer and B. Julia, Phys. Lett. **B80**, 48 (1978); Nucl. Phys. **B159**, 141 (1979).
18. P.G.O. Freund and M.A. Rubin, Phys. Lett. **B97**, 233 (1980).
19. M.J. Duff, H. Lu and C.N. Pope, Physics Rep. **130**, 1 (1986).
20. P. Budinich and A. Trautman, *The Spinorial Chessboard*, (Springer Verlag, Berlin - New York, 1985).
21. A. Polyakov, Nucl. Phys. **B164**, 171 (1980); Yu. Makeenko and A. Migdal, Nucl. Phys. **B188**, 269 (1981).

ADIABATIC SURFACES FROM THE LATTICE: EXCITED GLUONIC POTENTIALS

Chris Michael

Dept. of Math. Sci., University of Liverpool, Liverpool L69 3BX, UK

ABSTRACT

I describe the results from lattice studies of probing the vacuum with static colour sources. Topics include

- Twang the flux tube: string models and hybrid mesons.
- Charge up the flux tube: adjoint and sextet sources.
- Several sources: how does the colour flux get arranged?

1. Introduction

The non-perturbative sector of QCD is a very interesting area of study. Experimental investigation is difficult since it is not possible to make many controlled changes to explore the response. Lattice QCD allows a much more comprehensive set of questions to be asked. In particular, we can vary the quark màsses. Moreover the simplifying approximation of removing quark contributions to the vacuum (the so-called quenched approximation) allows phenomenological models to be tested in this context also.

The origin of string models in particle physics is from the hadronic string that grew out of dual models. It is worthwhile to explore the region of applicability of hadronic string-like models. The lattice study of the response of the vacuum to static colour sources is an ideal testing ground. For a pair of sources in the fundamental representation, there has been much progress in extracting the continuum spectrum from the lattice. As well as the ground state which has been well known for a long time, there are now comprehensive results for excited gluonic levels. I compare these lattice results with string models and also discuss the small separation limit.

Hybrid mesons are states beyond the quark model in which the gluonic degrees of freedom contribute in a non-trivial way. One of the clearest ways to study this area theoretically is from the excited gluonic potentials. This gives predictions for hybrid meson spectroscopy with heavy quarks. A brief aside will cover the situation with hybrid mesons containing light quarks: this is of topical relevance since there are new experimental data on these hybrid mesons.

I cover a range of other situations which have been studied using static sources. These include the bound state spectrum of a single adjoint source (the gluelump), the potential between adjoint sources and between sextet sources and string-breaking.

Some preliminary results are described of relevance to baryons: with 3 static quarks. In this case one can compare models with minimal flux length to a simple sum of two-body interactions. Leading to a study of hadron-hadron interactions, I present a summary of results for the energy of four static sources ($Q\bar{Q}Q\bar{Q}$) and some new results for the interaction energy between two B mesons.

It is worth recalling the weakness as well as the strength of the lattice approach to QCD. Any quantity which can be expressed as a vacuum expectation value of fields can be extracted straightforwardly in lattice studies. Thus masses and matrix elements of operators can be determined. What is not so easy is to explore hadronic decays. This is difficult because the lattice, using Euclidean time, has no concept of in and out states. About the only feasible strategies are to evaluate the mixing between states of the same energy - so giving some information on on-shell hadronic decay matrices, or to make a model dependent analysis[1] of lattice results.

2. Gluonic Excitations between Static Quarks in the Fundamental Representation

In a lattice regularisation, static sources in the fundamental colour representation are readily implemented. In the spirit of the heavy quark effective theory, they correspond to heavy quarks or anti-quarks. Thus in the experimental realm they give insight into b quark physics. Because of the lattice regularisation, the self energy of these static sources is unphysical. Energy differences, however, are physical and it is possible to obtain the continuum limit (ie extrapolate the lattice spacing to zero).

Before focussing on excitations of the colour flux between static quarks, I recall briefly the great insights that have come from the ground state potential itself. Some of these topics are also covered by other presentations.[2] The ground-state potential energy $V(R)$ between a static quark and antiquark at separation R has been measured accurately on the lattice: the continuum limit can be obtained with precision in the quenched approximation when the quark contributions to the vacuum are neglected. These results enable discussion of confinement at large R, of heavy quark bound states and provide one way to determine the running coupling non-perturbatively. The distribution of colour electric and magnetic fields in the flux tube have also been determined from a lattice.[3] This again has impact on models of confinement. Another insight obtained from the ground state potential is its spin dependence - in particular for the long range confining component which proves to be 'scalar' exchange.[4,5]

We now consider gluonic excitations of the potential between static quarks: they will play a rôle in hybrid mesons, for example. An analysis of the colour representation of the quark and antiquark is not useful when they are at different space positions since their combined colour is not gauge invariant. A better criterion is to focus on the spatial symmetry of the gluonic flux. The ground state distribution is symmetric - it is rotationally symmetric about the separation axis, and is symmetric about end to end inversion. As well as this symmetric ground state of the colour flux between two static quarks, there will be excited states with different symmetries.

These can be classified, in the continuum, by the group $D_{\infty h}$. This classification is well known in molecular applications. The representations are labelled for $J_z = 1$, 2, 3 as Σ, Π, Δ respectively where z is the axis of separation of the fundamental sources which are R apart. The $J_z \neq 0$ representations are two-dimensional. The other labels of the representations are g, u for $CP = \pm 1$ and, for the one-dimensional Σ states only, an additional $\pm$ label indicating whether the sate is even/odd under reflection in the plane containing the z-axis.

Note that this discussion of excited gluonic configurations between heavy quarks is very similar to that of electronic wavefunctions in a di-atomic molecule. Indeed the oxygen molecule has contributions from non-trivial representations of $D_{\infty h}$.

On a lattice with the separation axis along a lattice axis, the relevant symmetry group is D_{4h} and its representations are labelled such that A, E, and B are related to the $J_z = 0$, 1 and 2 excitations., etc. The lattice data for the lowest gluonic excited states have been determined by the Liverpool group using the Wilson gauge action for SU(2) colour[6–8] and for SU(3) colour.[9] The lowest non-trivial excitation was found to be in the E_u (Π_u) representation (corresponding to flux states from an operator which is the difference of U-shaped paths from quark to antiquark of the form ⊓ − ⊔). Although on a lattice, the self energy of the static sources is unphysical, the energy differences between excited gluonic potentials and the ground state are physical and have a continuum limit. An illustration of this first gluonic excitation is shown in fig.1.

Recently, a comprehensive study of this area has been undertaken using an asymmetric lattice.[10] The advantage of using a relatively large spatial lattice spacing is that large R can be reached, while a smaller temporal lattice spacing allows a more precise determination of the energy values. The disadvantage of a coarse spatial lattice spacing is that lattice corrections are increased but this can be offset by using improved lattice discretisations with smaller corrections than the Wilson lattice action which has order a^2. Recent results extend out to R-values of 4fm which is most impressive. By using several different lattice spacings and also by considering source separations in other directions than the lattice axes, it is possible to extract continuum results for the excited gluonic potentials. Because of the lattice dependence of the self-energy, only energy differences can be obtained.

In order to set the scale in lattice calculations, it is now conventional to use r_0. This is defined[11] in terms of the force between static sources implicitly as $r_0^2 f(r_0) = 1.65$ where $f(r) = dv(r)/dr$. From relating the interquark potential to that needed phenomenologically for heavy quark bound states, we know that $r_0 \approx 0.5$fm. Since this potential can be very accurately measured on a lattice, this is an appropriate observable to use to set the energy scale for comparisons of lattice results. Thus the lattice excited potential $\tilde{V}(R)$ expressed as $R_0(\tilde{V}(R) - V(R_0))$ where $R_0 = r_0/a$ is dimensionless and, in the limit of small lattice spacing, will be equal to the continuum expression $r_0(\tilde{v}(r) - v(r_0))$ up to corrections of order a^2 for the Wilson lattice discretisation. This is thus a suitable way to present and compare results from different lattice studies. We present some of the recent comprehensive results[12] in fig. 2.

It is worth exploring what is to be expected for these gluonic excitations between static sources at separation R.

Stability: The most obvious criterion is that if the excited gluonic energy lies below the energy of all states with the same symmetry made from a glueball together with a lower-lying potential, then the excited gluonic potential will be stable, at least

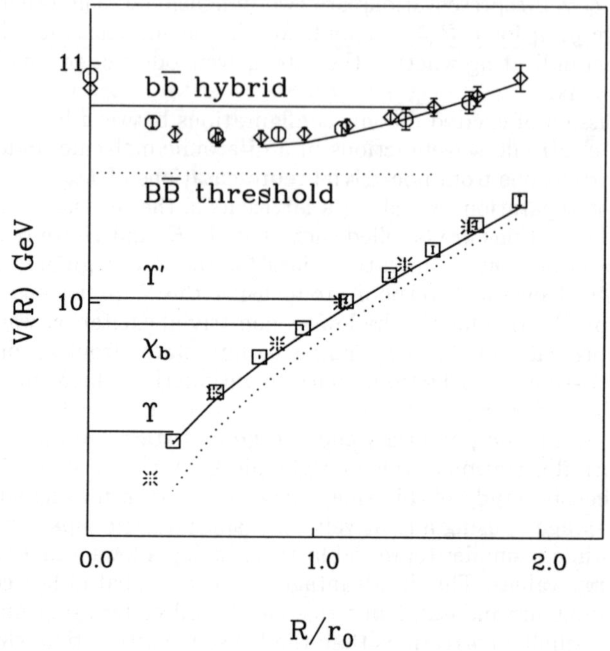

Fig. 1. Potentials $V(R)$ between static quarks at separation R (in units of $r_0 \approx 0.5$fm) for the ground state (□ and ∗) and for the Π_u (E_u) symmetry which corresponds to the first excited state of the gluonic flux (octagons and diamonds). Results from quenched calculations[9] with SU(3) colour are shown by symbols corresponding to different lattice spacings. The $R=0$ data points[19] for the excited state are discussed in the text. For the ground state potential the continuous curve is an interpolation of the lattice data while the dotted curve with enhanced Coulomb term fits the spectrum and yields the masses shown. For the excited gluonic potential, the continuous curve shows the string excitation expression of eq. 1. The hadronic bound states of b quarks in these potentials are shown, including the lightest hybrid level in the excited gluonic potential.

in the quenched approximation. Since the lightest glueballs are found, in quenched QCD, to have energies[13,14] $m(0^{++})r_0 = 4.33(5)$ and $m(2^{++})r_0 = 6.0(6)$, this criterion is satisfied in most cases, especially at larger R.

String models: We consider now simple phenomenological models for the excited gluonic potentials: one such model is the hadronic string. A bosonic string fixed at two points R apart and with no intrinsic width will give a parameter-free prediction although the tachyonic problem implies that this prediction will be unphysical for the lowest energy excitation at small string length R. The lattice excited potentials can be compared with this bosonic string model provided an appropriate expression[7] is used

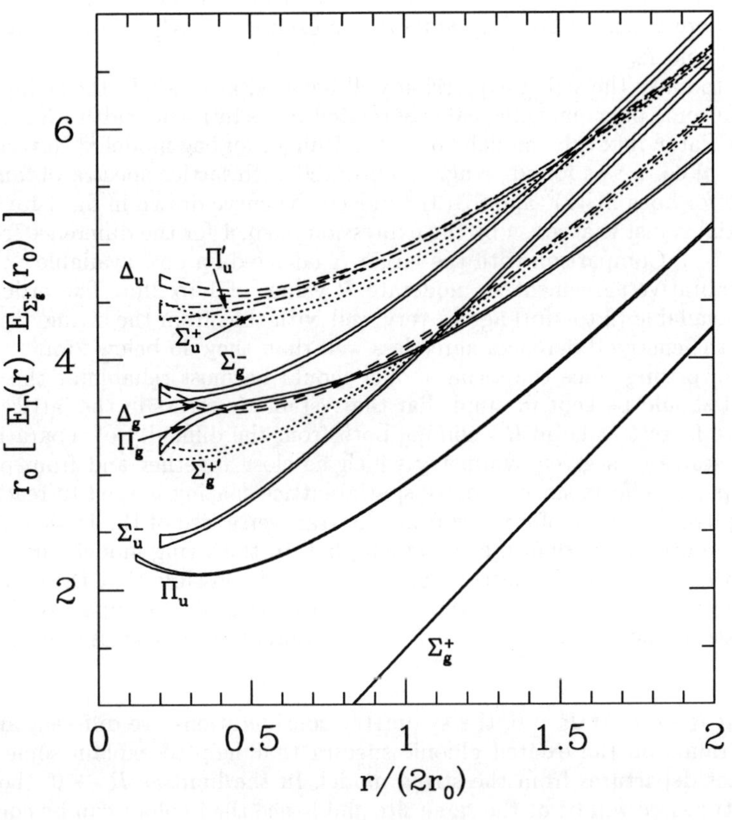

Fig. 2. Potentials $E(r)$ between static quarks at separation r (in units of $2r_0 \approx 1.0$fm) for different symmetry representations from Juge et al.[12] The two lines shown in each case represent an estimate of the error on the determination of the continuum value.

for excited level n:

$$V_n(R) = (\sigma^2 R^2 - \frac{\pi\sigma}{6} + 2\pi n\sigma)^{\frac{1}{2}}. \tag{1}$$

If one uses this expression for the energy differences of excited levels from the ground state, there are no free parameters since the string tension σ is determined by the ground state potential with $n = 0$. At large R, this expression has the simple consequence that gluonic excitations are at multiples of π/R in energy higher than the ground state. The relationship[7,9,12] between the excitation symmetry and n is that the $n = 0$, 1 levels are given by Σ_g^+, Π_u, while $n = 2$ excites $\Sigma_g^{+'}$, Π_g and Δ_g, and $n = 3$ excites $\Sigma_u^\pm$, Π_u' and Δ_u.

In order to make the string expression well behaved at small R, the string has to be regulated in some phenomenological way related to its intrinsic width - for instance as in the Isgur-Paton flux tube model[15] or in the Hungarian bag model.[16] Nevertheless, the expression of eq. 1 was found to agree quite well with lattice spectra obtained for R above 0.5fm for both SU(2)[7] and SU(3).[9] Indeed the curve drawn in fig. 1 for the Π_u excited potential is just that given by the expression of eq. 1 for the difference from the ground state (Σ_g^+). Comparison with the larger R lattice data now available[12,17] again gives good qualitative agreement at moderate R values of 1 to 2fm. The ordering at the largest R available (2 to 4fm) agrees very well with that from the string excitation approach but the energy differences agree less well than they do below 2fm.

This is surprising since the string model should be most reliable at the largest R. However, it should be kept in mind that the systematic errors in the lattice determinations[12] are largest at large R - coming both from the difficulty of separating the many excited states of a given symmetry which lie close together and from possible finite lattice spacing effects since a coarse spatial lattice spacing is used to reach large R. A further possible source of error is from the transverse size of the lattice which is assumed to be much larger than the string length R in the string model whereas it is comparable to R in some of the lattice analyses. So, it is possible that the systematic errors in the lattice results are somewhat underestimated. This comparison between the hadronic string model and the large R excited potentials is discussed in detail by Morningstar.[12]

The limit $R \to 0$: At $R = 0$, the symmetry considerations are different and this imposes constraints on the excited gluonic spectra that help to explain some of the more significant departures from the string model. In the limit as $R \to 0$, the static source and anti-source will be at the same site and hence their colour can be combined in a gauge invariant way - creating an adjoint colour source and a singlet (glueball correlator). Pictorially this is shown in figure 3. This former situation has been studied previously[18,19]: the gluelump is an adjoint source in the presence of a gluonic field - it is rotationally invariant and is described by J^{PC}. Thus the Wilson loop correlation satisfies

$$\lim_{R\to 0} W(R,t) = ce^{-M_{\text{gluelump}}t} + \frac{1}{3}c'e^{-M_{\text{glueball}}t} \tag{2}$$

In the large t limit, the lighter of the two states will dominate the correlation function. In most cases of present interest, the gluelump state is lighter than the glueball. Thus, we can obtain a relationship between the gluonically-excited states of the generalised Wilson loop, in the limit $R \to 0$, and those measured in the gluelump spectrum. This relationship in the continuum is obtained by subducing the rotation group representations appropriate to the gluelump to the $D_{\infty h}$ representations appropriate to the generalised Wilson loop when $R \neq 0$. By subducing the irreducible representation of the gluelump with J^{PC} we will find $D_{\infty h}$ representations with $J_z = -J, \dots, J$; labels

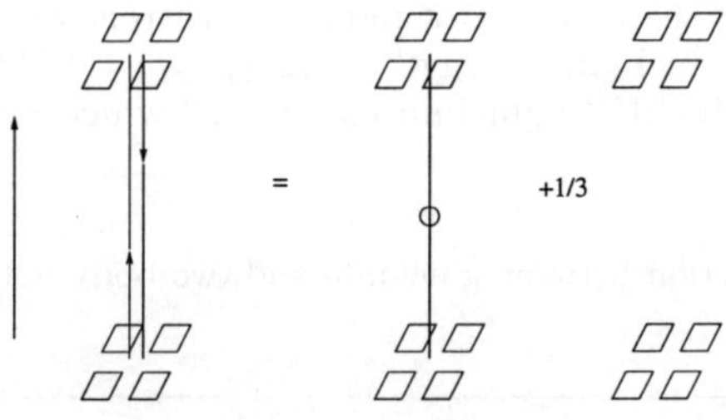

Fig. 3. Relation of $R \to 0$ Wilson loop with adjoint source (gluelump) and glueball correlations

g, u given by CP and, for any $J_z = 0$ states an additional label given by $P(-1)^J$. These relationships are given in Table 1.

Similar identities as $R \to 0$ also apply to the lattice discretisation. Then the O_h representations appropriate for the gluelump can be subduced into the D_{4h} representations appropriate for the generalised Wilson loop with $R \neq 0$. Thus, as has been emphasised previously,[18] the ground state gluelump with 1^{+-} (T_1^{+-}) implies that as $R \to 0$ there must be a degeneracy of the two-dimensional Π_u state (E_u) and a Σ_u^- state (A_{1u}). This explains why the Σ_u^- state which is high lying at large R (since it is a $n = 3$ string excitation) is seen in fig. 2 to decrease in energy rapidly at small R since it has to be degenerate at $R = 0$ with the Π_u state which is a $n = 1$ string excitation and so is lower lying in energy. The lowest lying gluelump states[19] have been determined to be 1^{+-}, 1^{--} and 2^{--} with energy differences in units of r_0 of 0.93(2) and 1.44(3) from the 1^{+-} respectively. This implies that the $R = 0$ limit of the excited gluonic potentials should have degenerate states as given in Table 1 with energy differences as quoted above. These considerations explain why the Σ_g^-, which is a $n = 4$ string level, also departs significantly (see fig. 2) from the string-like expectation since it has to be degenerate at $R = 0$ with the $n = 2$ string level Δ_g.

One gluon exchange: Although the above group-theoretical identities are a good guide to the behaviour of the excited gluonic potentials at small R, the limit as $R \to 0$ of the excited gluonic potential is not trivial to extract from lattice data with $R = a$, $R = 2a, \ldots$. A guide is to consider the gluon exchange contributions perturbatively. One way to investigate this is to consider the self energies of the contributions: $2E_F$ at $R \neq 0$ and E_A at $R = 0$, where F and A label fundamental and adjoint colours. Since, to lowest order, $E_A = 9E_F/4$ for SU(3) of colour, there will be a mismatch and one might expect the energy to increase as $R \to 0$ since the adjoint self-energy is larger. Another way to investigate this, is to imagine that as $R \approx 0$, there is a gluonic field in the adjoint representation, so that the heavy quark and anti-quark are also in an adjoint and hence will have a Coulombic interaction energy given by $-1/8$ of the Coulombic energy between a quark and antiquark in the fundamental representation (which is approximately given by $-0.25/R$ in lattice quenched studies). This again suggests that the excited gluonic potentials should rise as $R \to 0$, here as $0.03/R$.

Lattice data for the E_u representation for small R from SU(2) colour studies[8] at $\beta = 2.4$ with values of $aV_{Eu}(R) = 1.31$, 1.32 and 1.38 for $R = 3a$, $2a$ and a respectively do qualitatively support these estimates and are consistent with a limit as $R \to 0$ which agrees with the lattice gluelump energy[20] of $aE_{\rm gluelump} = 1.50$. Also in fig. 1, we show the $R = 0$ point from the SU(3) gluelump analysis[19] which fits in well with the above considerations.

Table 1. Connection between gluelump and two-body potential as $R \to 0$.

Gluelump J^{PC}	Two-body potential states
1^{+-}	Σ_u^-,Π_u
1^{--}	Σ_g^+,Π_g
2^{--}	Σ_g^-, Π_g, Δ_g

It is possible to measure the distribution of the colour flux around an excited gluonic state. Results have been presented[21] for the Π_u and for the first excited Σ_g^+ states. The excited states show a wider transverse distribution than for the ground state, as would be expected. There is some evidence for a node in the transverse distribution of the first excited Σ_g^+ state.

2.1. *Hybrid Mesons with Heavy Quarks*

The potential $V(R)$ between static sources obtained from the lattice (it is approximately of the form $V(R) = -e/R + \sigma R$) can be used to determine the spectrum of $b\bar{b}$ mesons by solving Schrödinger's equation since the motion is reasonably approximated as non-relativistic. The result from quenched lattices is similar to the experimental Υ spectrum: see fig.1. The main difference is that the Coulombic part (e) is effectively too small (0.28 rather than 0.4). This produces[9] a ratio of mass differences $(1P - 1S)/(2S - 1S)$ of 0.71 to be compared with the experimental ratio of 0.78. This difference is understandable as a consequence of the Coulombic force at short distances which would be increased by $33/(33 - 2N_f)$ in perturbation theory in full QCD compared to quenched QCD. Indeed lattice studies[22] including sea-quark effects in the vacuum do see evidence for an increase of the Coulombic component of the potential, although they have not yet reached sea-quark masses as small as those in nature.

Consider now the hadrons corresponding to bound states of the excited gluonic potentials. The lightest such gluonic excitation (Π_u) corresponds to a component of angular momentum of one unit along the quark-antiquark axis. Since the energy scale associated with the gluonic excitation is much larger than the energy scale associated with orbital or radial excitations, it is a reasonable approximation to use the adiabatic approximation. Thus one solves for the spectrum of hybrid mesons in the excited potential using the Schrödinger equation. For those excitations with a non-zero angular momentum J_z about the separation axis, an extra centrifugal term has to be added. The spatial wave function necessarily has non-zero angular momentum and corresponds[6] to $L^{PC} = 1^{+-}$ and 1^{-+}. Combining with the quark and antiquark spins then yields[6] a set of 8 degenerate hybrid meson states with $J^{PC} = 1^{--}$, 0^{-+}, 1^{-+}, 2^{-+} and 1^{++}, 0^{+-}, 1^{+-}, 2^{+-} respectively. These contain the spin-exotic states with $J^{PC} = 1^{-+}$, 0^{+-} and 2^{+-} which will be of special interest since they do not occur as $\bar{q}q$ states and so any experimental evidence for a resonance with these quantum numbers is a strong suggestion for the existence of a hybrid meson.

Since the lattice calculation of the ground state and hybrid masses allows a direct prediction for their difference, the result for this 8-fold degenerate hybrid level is illustrated in fig. 1 and corresponds[9] to masses of 10.81(25) GeV for $b\bar{b}$ and 4.19(15) GeV for $c\bar{c}$. Here the errors take into account the uncertainty in setting the ground state mass (i.e, Υ or ψ) using the quenched potential as described above. Using the recent comprehensive results[12] confirms the results above and the preliminary values quoted for the lightest hybrid meson is 10.8 GeV for $b\bar{b}$ with no error estimate given.

The quenched lattice results show that the lightest hybrid mesons lie above the open $B\bar{B}$ threshold and hence are likely to be relatively wide resonances. This could also be checked by comparing with quenched masses for the B meson itself,[23] but at present there are quite large uncertainties on that mass determination. The very flat potential implies a very extended wavefunction: this has the implication that the wavefunction at the origin will be small, so hybrid vector states will be weakly produced from e^+e^-. An explicit evaluation[12] of the hybrid wavefunction shows that it is very extended: being significant out to a radius of 1fm.

It would be useful to explore the splitting among the 8 degenerate J^{PC} values obtained. This could come from different excitation energies in the $L^{PC} = 1^{+-}$ (magnetic) and 1^{-+} (pseudo-electric) gluonic excitations, from spin-orbit terms, as well as from mixing between hybrid states and $Q\bar{Q}$ mesons with non-exotic spin. One way to study this on a lattice is to use the NRQCD formulation which describes non-static heavy quarks which propagate non-relativistically. Preliminary results for hybrid excitations from several groups[24,25,12] give at present similar results to those with the static approximation as described above, but future results may be more precise and able to measure splittings among different states.

2.2. *Hybrid Mesons with Light Quarks*

Unlike very heavy quarks, light quark propagation in the gluonic vacuum sample is very computationally intensive — involving inversion of huge ($10^7 \times 10^7$) sparse matrices. Current computer power is sufficient to study light quark physics thoroughly in the quenched approximation. The state of the art[26] is the Japanese CPPACS Collaboration who are able to study a range of large lattices (up to about 64^4) with a range of light quark masses. Qualitatively the meson and baryon spectrum of states made of light and strange quarks is reproduced with discrepancies of order 10% in the quenched approximation.

Here I will focus on hybrid mesons made from light quarks. In the quenched approximation, there will be no mixing involving spin-exotic hybrid mesons and so these are of special interest. The first study of this area was by the UKQCD Collaboration[27] who used operators motivated by the static ($Q\bar{Q}$) studies referred to above. Using non-local operators, they studied all 8 J^{PC} values coming from $L^{PC} = 1^{+-}$ and 1^{-+}

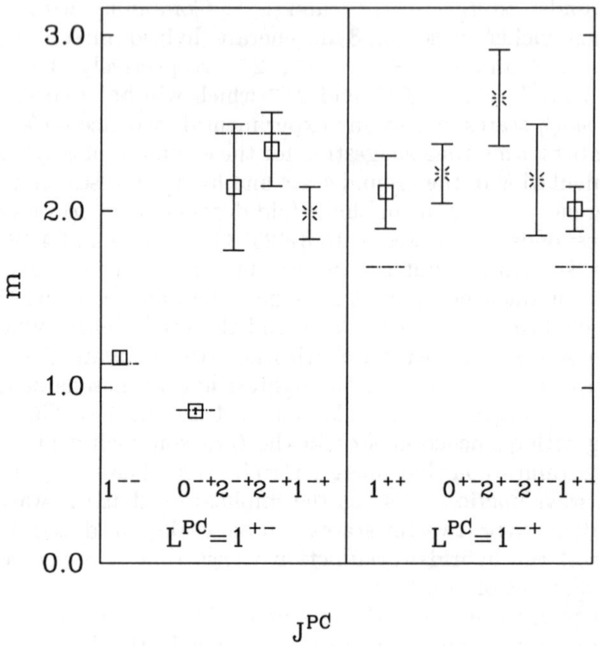

Fig. 4. The masses in GeV of states of J^{PC} built[27] from hybrid operators with strange quarks, spin-exotic (∗) and non-exotic (□). The dot-dashed lines are the mass values found for $s\bar{s}$ operators.

excitations. The resulting mass spectrum is shown in fig. 4 where the $J^{PC} = 1^{-+}$ state is seen to be the lightest spin-exotic state with a statistical significance of 1 standard deviation. The statistical error on the mass of this lightest spin-exotic meson is 7% but to take account of systematic errors from the lattice determination, a mass of 2000(200) MeV is quoted for the $s\bar{s}$ meson. Although not directly measured, the corresponding light quark meson would be expected to be around 120 MeV lighter. In view of the small statistical error, it seems unlikely that the 1^{-+} meson in the quenched approximation could lie as light as 1.4 GeV where there are experimental indications for such a state.[28] There are also recent experimental claims[29] for a second 1^{-+} spin exotic meson at 1.6 GeV. Beyond the quenched approximation, there will be mixing between such a hybrid meson and $q\bar{q}q\bar{q}$ states such as $\eta\pi$ and this may be significant in explaining the apparent discrepancy between experiment and lattice.

One feature clearly seen in fig. 4 is that non spin-exotic mesons created by hybrid meson operators have masses which are very similar to those found when the states are created by $q\bar{q}$ operators. This suggests that there is a significant coupling between hybrid and $q\bar{q}$ mesons even in the quenched approximation. This lattice result is difficult to quantify in terms of decay widths but does imply that the $\pi(1800)$ is unlikely to be a pure hybrid, for example.

A second lattice group has also evaluated hybrid meson spectra from light quarks. They obtain[30] masses with statistical and various systematic errors for the 1^{-+} state of 1970(90)(300) MeV, 2170(80)(100)(100) MeV and 4390(80)(200) MeV for $n\bar{n}$, $s\bar{s}$ and $c\bar{c}$ quarks respectively. For the 0^{+-} spin-exotic they have a noisier signal but evidence that it is heavier. They also explore mixing matrix elements between spin-exotic hybrid states and 4 quark operators.

3. The String Self-Energy

The expression for the energy of the ground state string mode ($n = 0$) will have a contribution from string fluctuation as well as the linear string tension component, namely[31] $V(R) \approx \sigma R - \pi/(12R)$ as $R \to \infty$. In practice this $\pi/(12R)$ string fluctuation term for a bosonic string is very hard to disentangle from the Coulomb term e/R in the potential. One way to get round this in lattice studies is to consider a hadronic string that encircles the periodic spatial boundaries of length L. Then there are no sources and hence no Coulomb component. The appropriate string fluctuation term in the energy of this state, called the torelon, is then given by $E(L) = \sigma L - \pi/(3L)$ as $L \to \infty$. By plotting $E(L)/L$ against L, see fig. 5, lattice studies[32] have confirmed the presence of this string fluctuation term with its coefficient determined to agree with the expected value and to be determined quite accurately (as $(\pi/3) \pm 0.03$). This is impressive evidence that the hadronic string is a good model of the energy of the colour flux tube at large distances.

4. Excited Colour Sources

4.1. Adjoint Sources

Table 2. Energy differences from the ground (1^{+-}) state of gluelump states with J^{PC}.

Excited State	$\Delta(M_0 r_0)$	Energy (MeV)
1^{--}	0.933(18)	368(7)
2^{--}	1.438(25)	584(10)
0^{++}	2.771(72)	1092(28)

One adjoint source: The 'hydrogen atom' of pure gauge QCD has a static adjoint source with a gluon field making it an overall colour singlet. This is the gluelump whose spectrum has already been mentioned above. It has been explored in SU(2)[18]

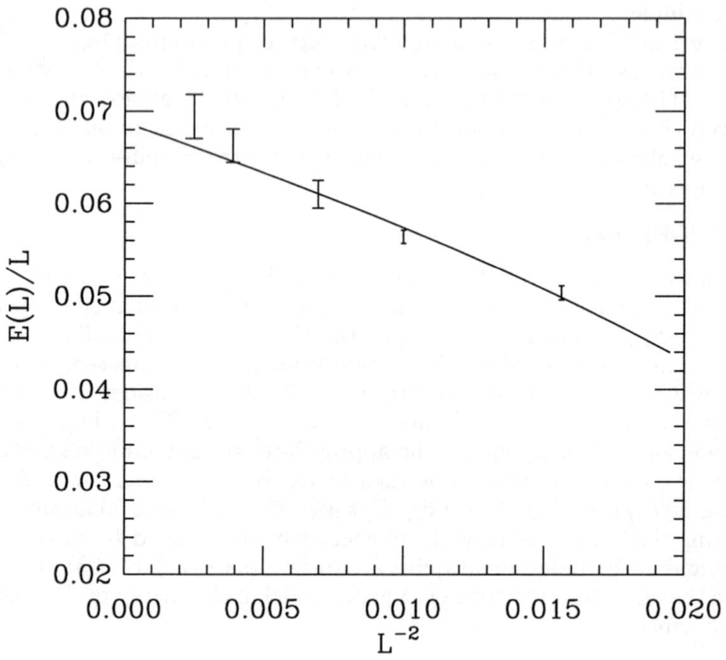

Fig. 5. Relationship between energy E and length L around the periodic spatial boundary for a torelon in lattice units[32]. The continuous line shows the bosonic string result.

and SU(3)[19] quenched lattice studies. The 'magnetic gluon' with $J^{PC} = 1^{+-}$ is found to be the ground state with the 1^{--} state as the first excited state. See table 2.

This spectrum would be accessible experimentally should a gluino exist which is sufficiently heavy for the static approximation to be valid and which lives long enough to allow spectroscopic energy levels to be determined.

As well as exploring the spectrum, the distribution of colour fields has been measured.[18] They are found to extend out to a radius of about 0.5fm. This is consistent with the result (see below) that the adjoint string breaks into two gluelumps at a separation of around 1fm - that is when the two gluelumps are just touching. More detailed results give some evidence of the nature of the colour fields in the low-lying states: disk-like for the 1^{+-} and toroidal for the 1^{--}.

Two adjoint sources: Another route to explore confinement is to measure on a lattice the potential energy between two static sources in the adjoint representation of the colour group. Here three regions are expected as R varies. At small R one gluon exchange should give a Coulombic region with strength proportional to C_A/R where C_A is the Casimir appropriate to the colour representation; at intermediate R an effective string tension may be discernible, while at large R the adjoint potential $V_A(R)$ must become independent of R because each adjoint colour source can be screened by a

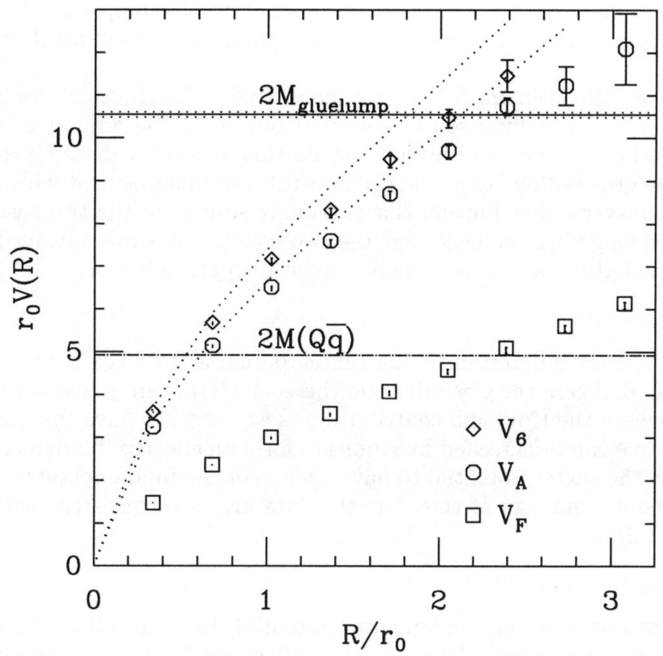

Fig. 6. The potentials in units of r_0 between static sources for quenched SU(3) with fundamental, adjoint and sextet sources. From quenched $\beta = 5.7$ $12^3 24$ lattices, setting the scale using $r_0 \approx 0.5$fm. Also shown[19,43] are $2M_{\text{gluelump}}$ and $2M(Q\bar{q})$ and the Casimir ratios $\frac{9}{4}V_F(R)$ and $\frac{5}{2}V_F(R)$.

gluonic field. Indeed at large R, $V_A \to 2m_{\text{gluelump}}$ where the gluelump is the ground state hadron with a gluon field around a static adjoint source discussed above. Of interest to model builders is the adjoint potential at moderate R values: the effective string tension region. Some theories of confinement[33] have suggested that the effective adjoint string tension is given by the Casimir ratio (i.e. $\sigma_A/\sigma_F = C_A/C_F$): for instance from assuming that confinement in four-dimensional QCD acts like two-dimensional QCD - which has Casimir scaling exactly.

Precise data exist[20] for $SU(2)$ of colour and they do show a region of linear rise, although with a slope less steep than that given by the Casimir ratio (namely $\frac{8}{3}V(R)$ where $V(R)$ is the fundamental colour source potential discussed previously). For SU(3) of colour, I have determined the adjoint potential to large R and the result is shown in fig.6. Again the Casimir ratio, here $\frac{9}{4}V(R)$, is an underestimate except at very small R where the Coulomb contribution is expected to have this ratio. A rule that $V_A(R) \approx 2V(R)$ seems to be a better guide to this intermediate R region.

The R value where the two gluelump state becomes degenerate in energy with the measured adjoint potential is found for both 2 and 3 colours to be near 1.2fm. For

SU(2) colour, a variational basis[20] with both colour-flux and two-gluelump basis states was used which allowed the mixing in this string breaking region to be studied more readily.

Lattice studies have also been made[34] of the colour field distributions for the adjoint potential. This allows a comparison of the distribution of the adjoint colour field with the fundamental case. The longitudinal distribution at intermediate r-values shows a ratio of field strengths falling below the Casimir ratio so in agreement with the discussion above. The transverse distribution is surprisingly similar in the two cases - which may be interpreted as giving evidence that the two colour flux tubes involved in the adjoint case are attracted to each other - as in a type I superconductor.

4.2. Sextet Sources

For SU(3) of colour, I have determined the sextet potential to large R and the result is also shown in fig. 6. Again the Casimir ratio (here $\frac{5}{2}V(R)$) is an underestimate except at very small R where the Coulomb contribution is expected to have this ratio. At large R, the sextet source can be screened by gluons to form an effective fundamental source. Thus one expects the sextet potential to have a slope of the fundamental string tension σ at large R. There is no sign of this, but the data are also consistent with a change of slope beyond 1.2fm.

4.3. Fundamental string breaking

In passing, we point out that the fundamental potential itself, in full QCD, will show string breaking at R-values where $V(R) > 2M(Q\bar{q})$ where by $Q\bar{q}$ we mean the ground state meson with a light quark q bound to a static heavy quark Q. This comparison of the energies of the static potential and the two meson system can be explored in the quenched approximation - although there will be no actual mixing between these systems in that case. The situation is illustrated in fig. 6 and again breaking would be expected to occur at an R-value near 1.2fm where there is a degeneracy. Attempts to look for this mixing explicitly using $V(R)$ determined from full QCD simulations[22,35] have not yet achieved sufficient precision to address this topic.

5. Multiple Sources

5.1. Baryonic Applications

Early attempts to study the potential with three static quarks were rather exploratory.[36] Indeed at moderate R values, the baryonic potential was found to be given quite well by an average of the three two-body potentials. In contrast, in a flux tube approach, one would expect an important contribution from a three-body term with a flux tube of minimal length (ie with a star-like configuration): see fig. 7. This would give a different behaviour of the energy levels as a function of the position of the three sources than the case of a two-body sum. The above studies had rather small interquark separations of 0.5fm or less which may be an explanation for the apparent dominance of the two-body terms. Some recent preliminary results[37] give some support to the three-body flux-tube picture for the special case of 3 sources at the corners of an equilateral triangle.

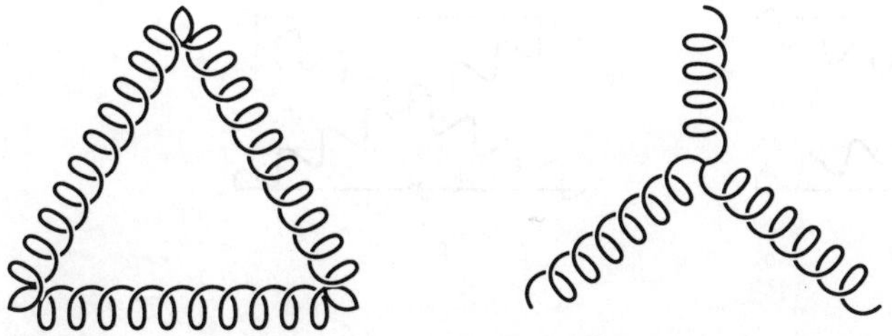

Fig. 7. Diagrams showing two possible assignments of colour flux between three static quarks in a baryonic configuration: as a sum of two-body terms or as a minimal length string.

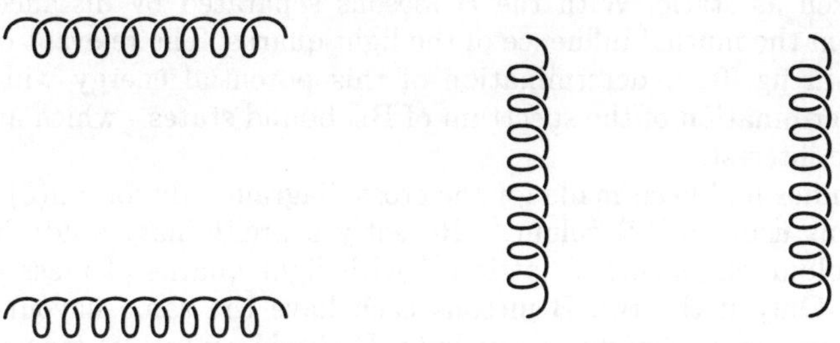

Fig. 8. Diagrams showing two possible assignments of colour flux between static $Q\bar{Q}Q\bar{Q}$.

5.2. $Q\bar{Q}Q\bar{Q}$

To study the residual strong force between hadrons on a lattice is difficult since it is so much weaker than the colour force between quarks. For this reason, preliminary studies have either used static quarks and/or SU(2) of colour to simplify the computations. There has been extensive work[38] using static quarks in SU(2) colour lattice studies - this has a bearing on the meson-meson force. Of special interest is whether the measured 4-body potential can be expressed as a sum of 2-body components. It turns out that a description in terms of the mixing of different 2-body assignments as shown in fig. 8 is possible but that the mixing coefficient is geometry-dependent. In other words there is a four-body force.

As well as the energy levels for a wide range of geometries, the spatial distribution of the colour flux has been determined for the case of 4 sources at the corners of a square.[39]

5.3. The BB interaction

The $Q\bar{q}Q\bar{q}$ system with the heavy quarks Q treated as static at separation R is a step towards studying the strong force between realistic hadrons. This case can be

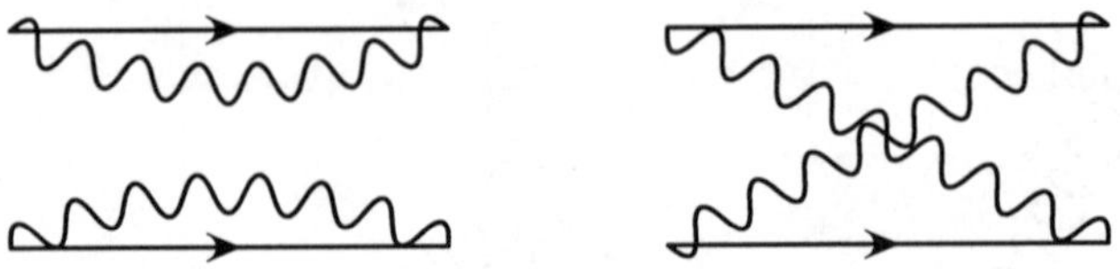

Fig. 9. Diagrams showing the interaction between two B mesons: the light quarks are shown as wiggly lines.

considered as the interaction between two B mesons in the heavy quark limit - where the b quarks are taken as static. With the B mesons separated by distance R, the interaction comes from the mutual influence of the light quarks. The relevant diagrams to study are shown in fig. 9. A determination of this potential energy will allow a non-perturbative determination of the spectrum of BB bound states - which are exotic states of considerable interest.

Exploratory studies had been made for the cross diagram only for SU(3) colour[40] and for both diagrams using SU(2) colour.[41] Recently a preliminary study has been made using SU(3) colour for quenched lattices[42] with light quarks of mass near the strange quark mass. Only if the two B mesons each have the same flavour of light quark will the additional cross diagram contribute. Preliminary results for same light flavour and different light flavour are shown in fig. 10. Binding is seen in each case, but it is stronger in the unlike flavour case. This suggests that a $B - B_s$ exotic dimeson might exist. It will also be possible to study the spin-dependence of these potential energy differences on a lattice.

Note that again, at $R = 0$, the system is related to one previously studied. Thus the colour of the two static quarks can be combined to a triplet or sextet and the lightest state will have them in a colour triplet - creating a system like the Λ_b baryon but with a static b quark. These baryonic states with one static quark have been studied on the lattice previously and their mass gives the BB energy on the lattice at $R = 0$. Lattice estimates[43,42] give around 0.3 GeV for this binding at $R = 0$. For light quarks of the same flavour, there is no Λ_b state - the lightest baryon will be the Σ_b which is somewhat heavier so the binding in this case will be around 0.2 GeV.

In the continuum, in the heavy quark limit, one would expect that the binding of the BB system at $R = 0$ for unlike light quark flavours is given by $2(M_B - m_b) - (M_{\Lambda_b} - m_b)$ where m_b is the b-quark mass.

The interaction energy between a B and $\bar{B}$ can be studied similarly. However, this system at separation R has a coupling to a state consisting of the flux tube joining the static sources (ie the state which occurs in the static potential $V(R)$) plus a light quark-antiquark meson (or vacuum when the light quark and antiquark have identical flavour). This reflects the physical process that $B\bar{B} \to \Upsilon + q\bar{q}$ and since the Υ is known

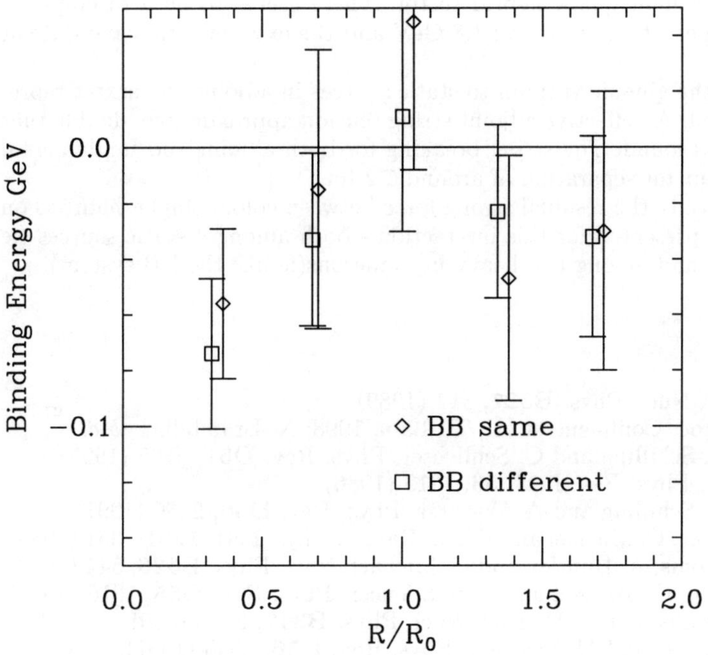

Fig. 10. Preliminary results[42] for the binding energy between two B mesons with either same flavour of light quarks, or different flavours at separation R in units of $R_0 \approx 0.5$fm. The light quark mass used corresponds to strange quarks. At $R = 0$, the binding energy is given from lattice results for baryons with static quarks[43] as around -0.3 GeV for different flavours and -0.2 GeV for same flavour- see the text.

to be tightly bound, this latter channel may have a lower energy at small to moderate R values. Conversely, at large R, this transition is essentially that of fundamental string breaking: where it is energetically favourable for the string of length R to break leaving a B and $\bar{B}$. A thorough study of this area thus needs to use a basis with both string (flux) states and meson-antimeson states.

6. Conclusions

The lattice has a lot to offer to phenomenological model builders and those wishing to unravel the mysteries of QCD.

String-like features are seen in the excited gluonic potentials for $R > 1$fm. The departure from this behaviour at moderate and small R, especially for the Σ^- states, can be understood as a consequence of the symmetry requirements at $R = 0$. There is even lattice evidence from closed string studies for the self energy appropriate to a bosonic string.

As well as predictions for hybrid mesons with heavy quarks, recent lattice results give predictions for the light quark hybrid sector. There seems to be a discrepancy between the lattice expectation of around 1.8 GeV and the experimental candidate at 1.4 GeV.

The response of the gluonic vacuum to static sources in adjoint and sextet representations was discussed. An effective adjoint string tension approximately double that of the fundamental was found. The string breaking for both adjoint and fundamental sources appears to occur for separation of around 1.2 fm.

It is possible to study the residual strong force between colour singlet hadrons on a lattice. Results were presented for this interaction - both among 4 static sources (ie like the Υ - Υ system) and among two heavy-light mesons(ie like the BB system).

References

1. C. Michael, Nucl. Phys. **B327**, 517 (1989).
2. G. Bali, Proc. Confinement III, TJ Lab., 1998; N. Brambilla, *ibid.*
3. G. Bali, K. Schilling and C. Schlichter, Phys. Rev. **D51**, 5165 (1995).
4. C. Michael, Phys. Rev. Lett. **56**, 1219 (1986).
5. G. Bali, K. Schilling and A. Wachter, Phys. Rev. **D56**, 2566 (1997).
6. L.A. Griffiths, C. Michael and P.E.L. Rakow, Phys. Lett. **129B**, 351 (1983).
7. S.J. Perantonis, A. Huntley and C. Michael, Nucl. Phys. **B326**, 544 (1989).
8. A. M. Green, C. Michael and P. S. Spencer, Phys. Rev. **D55**, 1216 (1997).
9. S.J. Perantonis and C. Michael, Nucl. Phys. **B347**, 854 (1990).
10. C. Morningstar and M. Peardon, Phys. Rev. **D56**, 4043 (1997).
11. R. Sommer, Nucl. Phys. **B411**, 839 (1994).
12. K. Juge , J. Kuti and C. Morningstar, Nucl. Phys. B (Proc. Suppl.) **63**, 326 (1998); Hadron Spectroscopy, Seventh International Conference, AIP New York 1998, ed. Chung and Willutzki, 137, hep-ph/9711451; contribution to Confinement III.
13. UKQCD collaboration, G. Bali et al., Phys. Lett. **B309**, 378 (1993).
14. H. Chen et al., Nucl. Phys. B (Proc. Suppl.) **34**, 357 (1994).
15. N. Isgur and J. Paton, Phys. Lett. **124B**, 247 (1983); Phys. Rev. **D31**, 2910 (1985); Phys. Rev. Lett. **54**, 869, (1985).
16. P. Hasenfratz, R. Horgan, J. Kuti and J. Richard, Phys. Lett. **95B**, 299 (1980); K. Juge, J. Kuti and C. Morningstar, Nucl. Phys. B (Proc. Suppl.) **63**, 543 (1998).
17. T. Allen, M. Olsson and S. Veseli, hep-ph/9804452.
18. I.H. Jorysz and C. Michael, Nucl. Phys. **B302**, 448 (1988).
19. M. Foster and C. Michael, Nucl. Phys. B (Proc. Suppl.) **63**, 724 (1998); and in preparation.
20. C. Michael, Nucl. Phys. B (Proc. Suppl.) **6**, 417 (1992).
21. P. Pennanen, A. Green and C. Michael, Phys. Rev. **D56**, 3903 (1997).
22. S. Güsken, Nucl. Phys. B (Proc. Suppl.) **63**, 16 (1998).
23. R. Sommer, Phys. Rep. **275**, 1 (1996).
24. UKQCD Collaboration, T. Manke et al., Nucl. Phys. B (Proc. Suppl.) **63**, 332 (1998).

25. UKQCD Collaboration, S. Collins et al., Nucl. Phys. B (Proc. Suppl.) **63**, 335 (1998).
26. T. Yoshie, Nucl. Phys. B (Proc. Suppl.) **63**, 3 (1998).
27. UKQCD Collaboration, P. Lacock, C. Michael, P. Boyle, and P. Rowland, Phys. Rev. **D54**, 6997 (1996); Phys. Lett. **B401**, 308 (1997); Nucl. Phys. B (Proc. Suppl.) **63**, 203 (1998).
28. D. Thompson, et al., Phys. Rev. Lett. **79**, 1630 (1997); A. Ostrovidov, Hadron Spectroscopy, Seventh International Conference, AIP New York 1998, ed. Chung and Willutzki, 263.
29. D. Weygand, Confinement III, TJ Lab, 1998.
30. C. Bernard et al., Nucl. Phys. B (Proc. Suppl.) **53**, 228 (1996); Phys. Rev. **D56**, 7039 (1997).
31. M. Lüscher, Nucl. Phys. **B180[FS2]**, 317 (1981).
32. C. Michael and P. Stephenson, Phys. Rev. **D50**, 4634 (1994).
33. J. Ambjorn, P. Olesen and C. Peterson, Nucl Phys. **B240**, 189 (1984); For a recent discussion of Casimir scaling: see M. Faber, J. Greensite and S. Olejnik, Phys. Rev. **D57**, 2603 (1998) and references therein.
34. H. Trottier, Nucl. Phys. B. (Proc. Suppl.) **47**, 286 (1996).
35. UKQCD Collaboration, C. Allton et al., hep-lat/9808016.
36. R. Sommer and J. Wosiek, Nucl. Phys. **B267**, 531 (1986); H. Thacker, E. Eichten and J. Sexton, Nucl . Phys. B (Proc. Suppl.) **4**, 234 (1988).
37. G. Bali, private communication.
38. A. M. Green et al., Phys. Lett. **B280**, 11 (1992); Nucl. Phys. **A554**, 701 (1993); Int. J. Mod. Phys. **E2**, 476 (1993); Z. Phys. **C67**, 291 (19950; Phys. Rev. **D53**, 261 (1996),
39. P. Pennanen, A. Green and C. Michael, Phys. Rev. **D** (in press), hep-lat /9804004.
40. D. Richards, D. Sinclair and D. Sivers, Phys. Rev. **D42**, 3191 (1990).
41. C. Stewart and R. Koniuk, hep-lat/9803003.
42. P. Pennanen and C. Michael, in preparation.
43. C. Michael and J. Peisa, Phys. Rev. **D** (in press), hep-lat/9802015.

25. UKQCD Collaboration, S. Collins et al., Nucl. Phys. B (Proc. Suppl.) 63, [illegible] (1998).
26. T. Yoshié, Nucl. Phys. B (Proc. Suppl.) 63, 3 (1998).
27. UKQCD Collaboration, P. Lacock, C. Michael, P. Boyle and P. Rowland, Phys. Rev. D54, 6997 (1996); Phys. Lett. B401, 308 (1997); Nucl. Phys. B (Proc. Suppl.) 63, 203 (1998).
28. [illegible] et al., Phys. Rev. Lett. 79, 1630 (1997); A. [illegible], Hadron Spectroscopy, Seventh International Conference, AIP, New York, 1998, ed. Chung and Willutzki, 708.
29. P. Rowland, [illegible] thesis (1998).
30. C. Bernard et al., Nucl. Phys. B (Proc. Suppl.) 53, 228 (1997); Phys. Rev. D56, 7039 (1997).
31. [illegible], Nucl. Phys. B[illegible] (198[illegible]).
32. [illegible] and [illegible], Phys. Rev. [illegible] (198[illegible]).
33. J. Ambjørn, P. Olesen and C. Peterson, Nucl. Phys. B240, 189 (1984). For a recent discussion of Casimir scaling see M. Faber, J. Greensite and Š. Olejník, Phys. Rev. D57, 2603 (1998) and references therein.
34. [illegible], Nucl. Phys. B (Proc. Suppl.) 47, 286 (1996).
35. UKQCD Collaboration, C. Michael et al., hep-lat/9808016.
36. R. Sommer and J. Wosiek, Nucl. Phys. B267, 531 (1986); H. [illegible] and J. [illegible], Nucl. Phys. B (Proc. Suppl.) 4, 284 (1988).
37. G. Bali, private communication.
38. A. M. Green et al., Phys. Lett. B280, 11 (1992); Nucl. Phys. A554, 701 (1993); Int. J. Mod. Phys. E2, 479 (1993); Z. Phys. C67, 291 (1995); Phys. Rev. D58, [illegible] (1998).
39. [illegible] and C. Michael, Phys. Rev. D (in press), hep-lat/98[illegible].
40. D. Richards, D. Sinclair and D. Sivers, Phys. Rev. D42, 3191 (1990).
41. [illegible] Stewart and R. Koniuk, hep-lat/98[illegible].
42. P. Pennanen and C. Michael, in preparation.
43. [illegible] and [illegible], Phys. Rev. D (in press), hep-lat/98[illegible].

FIELD THEORY BASED CONFINING MODELS

John A. Tjon
Institute for Theoretical Physics, University of Utrecht,
3508 TA Utrecht, The Netherlands

ABSTRACT

We review some of the work done on relativistic confining models for the quark-antiquark system. The full Dirac structure is accounted for. As a result issues about the scalar or vector nature of confinement have to be considered. Confining models based on the Dyson-Schwinger formalism are also discussed.

1. Introduction

Nonrelativistic constituent quark models have been highly successful in describing the level structure of mesons and baryons and their electromagnetic decay properties.[1] The one-gluon-exchange (OGE) and QCD inspired (linear) confining force are main ingredients in these models. Although such a description is á priori expected to be especially suitable for the heavy quark situation, it has proven to be also applicable for the light quark systems. In extending the modeling to the latter case effects of special relativity have to be included. A simple way to include some relativity is to relativize the kinetic operator in the Hamiltonian.[2] In so doing spectra are remarkably well reproduced.

In recent years attempts have been made using the more basic approach of relativistic field theory to search for appropriate relativistic equations to describe the nonperturbative dynamics of hadronic systems. The full Dirac structure of the quark degrees of freedom are maintained in these approaches, which give rise to certain complications. Self consistent treatment of the propagation of the quarks is important to allow for dynamical chiral symmetry breaking. Attempts have been made to simplify the Dyson-Schwinger equations to study the problem of confinement and dynamical chiral symmetry breaking. We review in this talk some of the developments.

2. Relativistic Quasipotential Equations

In contrast to the nonrelativistic case, the basis for a dynamical scheme is not unique. Relativistic extensions have been investigated along the lines of relativistic quantummechanical particle description or using field-theoretical based approaches. For definiteness, confining ourself to the mesonic sector, in the field theory based approach the starting point is the four-point function of the $q\bar{q}$ system. Assuming an underlying field theory like QCD, one may formally integrate out the gluonic degrees of freedom to yield an effective force between the valence quark and antiquark and where the dynamics is described through the Bethe-Salpeter equation. For the existence of a bound state with mass $M = \sqrt{s}$ of the $q\bar{q}$ system the four-point function T has to satisfy the homogeneous Bethe-Salpeter equation

$$T(s) = V(s)G(s)T(s), \tag{1}$$

V being the set of all 2-particle irreducible graphs and G the free *dressed* two-particle Green function, i.e.

$$G(s) = S(p_1)S(p_2), \tag{2}$$

where S is the dressed quark propagator

$$S(p) = [\not{p} - m_q + \Sigma(\not{p})]^{-1} \tag{3}$$

with Σ being the self-energy contribution.

In many studies one for simplicity approximates the quark propagators by free particle ones and V by some effective phenomenological interaction, containing certain essential features like confinement. Moreover, usually multiple $q\bar{q}$ channels are assumed to play a minor role and their contributions to V are neglected. In this socalled quenched approximation one arrives at a natural extension of the nonrelativistic equations, in which relativistic covariance is satisfied. In such a framework there is the essential difference with the nonrelativistic description, that the particles in the intermediate state are not only off-the energy shell, but also off the mass-shell. As a consequence one has to deal with the complexity of four-dimensional equations.

In most of the studies carried out in the Bethe-Salpeter equation framework one makes further use of 3-dimensional reductions. The resulting quasipotential equations are constructed by choosing the free particle Green function with certain constraints on the relative energy of the intermediate state so that relativistic covariance is satisfied. In so doing we effectively go from a two time formalism of the Bethe-Salpeter equation to the more familiar single time description of the two-particle system. The resulting equations have the same mathematical structure as the well known nonrelativistic Lippmann-Schwinger equation. In momentum space we have

$$T(\vec{p},\vec{p}') = \frac{1}{(2\pi)^3} \int d\vec{q}\, V(\vec{p},\vec{q})\; g(\vec{q})\; T(\vec{q},\vec{p}') \tag{4}$$

Amongst some considered choices[3–11] for the quasipotential Green function $g(\vec{q})$, the instantaneous approximation by Salpeter[3] and the spectator approximation by Gross[6] have widely been used in the literature. In the special case of one light and one heavy quark, one can convince oneself that the heavy quark can be considered to act like a potential source, provided that we asssume the validity of the quenched approximation.[6,12,13] As a consequence a Dirac equation description should hold for the light quark. The ladder theory of the Bethe-Salpeter equation, i.e. when only the one-meson-exchange diagram is kept in V, doesnot satisfy this socalled one-body limit. To get the proper limit the sum of all crossed ladders have to be included. In some of the quasipotential equations the choice of the Green function is such that the one-body limit is recovered. This is true for the spectator model, while the Salpeter Green function can readily be extended to yield the proper one-body limit.[8] Clearly, it is of interest to see for other mass ratios how well the various quasipotential approximations do.

Nonpertubative solutions have recently numerically been constructed for the generalized ladder theory in the case of the scalar QED and the $\phi^2\chi$ interaction. This is done in Euclidean space. Considering the above φ^3 theory as an example the action is

$$S = \int \mathrm{d}^4x \left[(\partial_\lambda \phi)^2 + m^2\phi^2 + \frac{1}{2}(\partial_\lambda \chi)^2 + \frac{1}{2}\mu^2\chi^2 + g\phi^2\chi \right]. \tag{5}$$

According to Ref. 12, the Feynman-Schwinger representation can be written down for the 'quenched' (i.e. the possible occurence of vacuum fluctuation ϕ-loops is neglected) two-body Green function G in terms of a path integral

$$G = \int_0^\infty \mathrm{d}s \int_0^\infty \mathrm{d}\bar{s} \int (\mathcal{D}z)_{xy} \int (\mathcal{D}\bar{z})_{\bar{x}\bar{y}} \exp\left[-K[z,s] - K[\bar{z},\bar{s}] + V[z,\bar{z},s,\bar{s}]\right], \tag{6}$$

where the functionals K and V are defined as

$$K[z,s] = m^2 s + \frac{1}{4s}\int_0^1 \mathrm{d}\tau\, \dot{z}^2(\tau), \tag{7}$$

$$V[z,\bar{z},s,\bar{s}] = g^2 s\bar{s} \int_0^1 \mathrm{d}\tau \int_0^1 \mathrm{d}\bar{\tau}\, \Delta\left(z(\tau) - \bar{z}(\bar{\tau})\right). \tag{8}$$

In this representation the χ fields have been replaced by quantummechanical particle degrees of freedom, characterized by the trajectories z and $\bar{z}$. The functional integrations are over all possible paths, subject to the boundary conditions: $z(0) = x$, $z(1) = y$ and similarly for $\bar{z}$. In (3+1) dimensions the free two-point function Δ is given by

$$\Delta(x) = \frac{\mu}{4\pi^2|x|} K_1(\mu|x|), \tag{9}$$

where K_1 is the modified Besselfunction of order 1.

The description in terms of particle trajectories amounts to a considerable reduction in number of degrees of freedom as compared to for example putting the field action (5) on a discrete 4-dimensional lattice. As a result accurate calculations can be carried out with this approach even for considerably larger times T than realized in lattice calculations. The mass spectrum can be extracted from the study of the logarithmic derivative of the four-point function, i.e.

$$L(T) = \frac{d}{dT} log[G(T)], \tag{10}$$

where $T = (y_4 - x_4 + \bar{y}_4 - \bar{x}_4)/2$. Physically, Eq. (6) resums all ladder and crossed-ladder graphs that contribute to G. In Ref. 12 it was shown that it satisfies the correct one-body limit. In addition, the formalism has the advantage that an explicitly gauge invariant formulation can be found and also that confinement can be accounted for in a natural way via the area-law.[14,15] The numerical calculations were done using standard Monte Carlo methods to evaluate the path integrals.

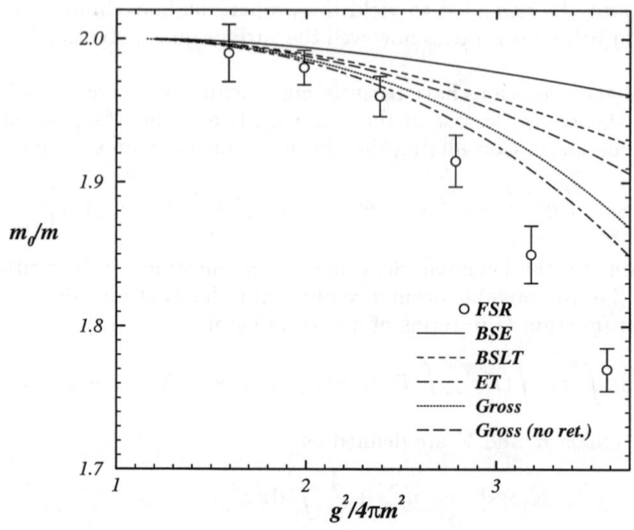

Fig. 1. Ground state mass m_0 of the ϕ^3-theory as a function of the dimensionless coupling constant $g^2/4\pi m^2$ for $\mu/m = 0.15$. The ladder Bethe-Salpeter (BSE) and various quasipotential results are also shown. BSLT, Gross and ET are from Refs. 4-5, 6 and 8 respectively.

In Fig. 1 are shown the results for the ground state of a $\phi^2\chi$ theory in (3+1) dimension, together with various quasipotential predictions. The ladder Bethe-Salpeter prediction gives a considerably less binding as compared to the exact answer, whereas the quasipotential results are in closer agreement. In particular, the equal time (ET) prescription, corresponding to neglecting the retardation in the one-meson-exchange contribution and in addition modifying the Green function to yield the proper one-body limit,[8] leads to predictions comparable to the exact results. Dropping the modification of the Green function yields results which are more repulsive. This approximation is in the case of scalar particles equivalent to the minimal relativity prescription by Logunov and Tavkhelidze[4] and Blankenbecler and Sugar.[5]

All the ET-like approximations should basically be reasonable provided that the time scale is such that the interaction can be taken to be instantaneous. In Fig. 2 contour plots are shown of calculations of the interaction density as a function of the two proper times, characterizing the two particle trajectories in the case of (2+1) scalar QED. From this we see, that there is indeed a relative sharp fall-off in the direction of the relative time $t = \tau - \bar{\tau}$ as compared on the time scale of $T = (\tau + \bar{\tau})/2$.

The discussed quasipotential equations have been used to study the spectroscopy of both the light and heavy quark systems. As the driving force one usually assumes a

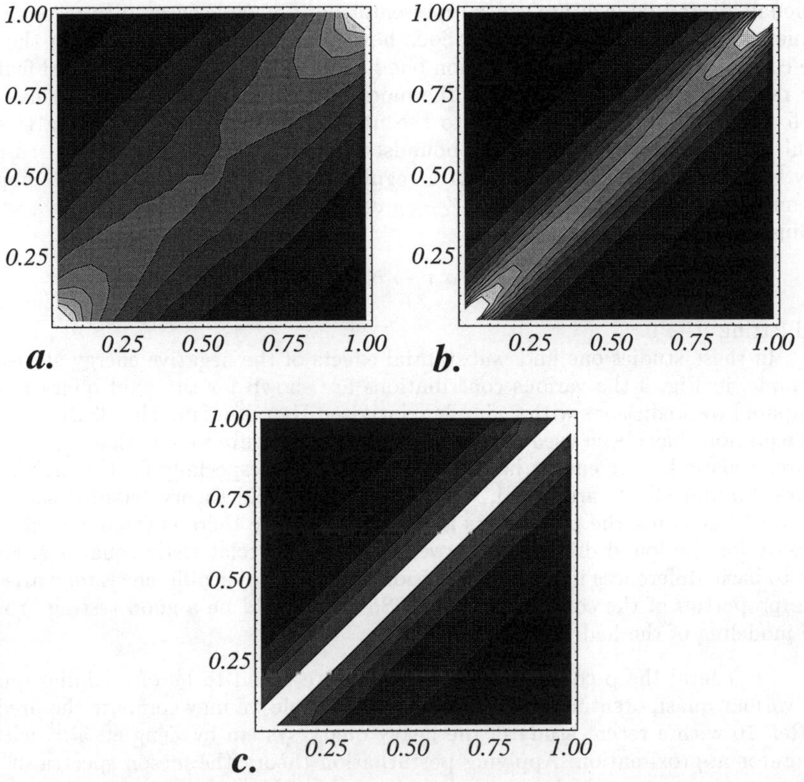

Fig. 2. The contourplot of the interaction density $E(\tau, \bar{\tau})$ of scalar QED_{2+1} as a function of the two proper times characterizing the particles in the four-point function. The subplots (a-c) are for $m = \bar{m} = 1, 10, 100$ respectively.

linear confining interaction V_{CONF} and the one-gluon-exchange interaction

$$V = V_{CONF} + V_{OGE} \tag{11}$$

In most of the studies a running OGE coupling is used, while the string tension is assumed to be independent of the constituent masses of the system. Both the configurational[10,11] and momentum space representation[3–9] have been used to reconstruct the numerical solutions. Numerical methods have been devised to do this. In the case of the configuration space representation one has to worry about the nonlocal features of the relativistic kinetic operator. In the momentum space formulation one has to deal with numerical difficulties related to the unboundedness of the confining force. The resulting noncompact kernel of the boundstate equation has to be regularized in some way.[7,9,16,17] As a typical example of a regularization procedure, the $1/k^4$ singularity for small momentum transfer k in V_{CONF} can be removed by introducing a screening radius $1/\mu$

$$V_{CONF} \approx \sigma\, r \rightarrow \sigma\, \frac{d^2}{d\mu^2}[\frac{exp(-\mu r)}{r}] \tag{12}$$

and letting $\mu \rightarrow 0$.

In these studies one finds substantial effects of the negative energy states. As an example, in Fig. 3 the various contributions are shown for the light quark system as compared to predictions with a simple relativized Hamiltonian. The BSLT quasipotential equations have been used in this study. From the figure we see, that relativizing the nonrelativistic kinetic energy has an important effect, especially for the higher excited states. Similar effects are found, when we use of a field theory based description. In general by refitting the parameters in the driving force there is room enough to compensate for the found differences between the various relativistic equations. However, due to basic differences in the wavefunction content, major differences may arise in the e.m. properties of the composite system. So this should be a good testing ground for the modeling of the hadron system.

In general the predictions of the spectra are found to be of a similar quality in the various quasipotential approaches. As an example we may compare the predictions of Ref. 10 with a recent study of the heavy quark system by Zeng et. al.[18] within the spectator approximation. Applying perturbation theory the meson spectra of various quark systems were calculated up to order $1/M_Q^2$, M_Q being the heavy quark mass. The predictions of the lowest levels of the $c\bar{c}$ are shown in Table 1, together with the full calculations of Ref. 10 using the BSLT and ET approximation. From this we see, that quantitatively similar results are found for the various quasipotential descriptions, except possibly in the case of the p wave levels.

In the various investigations substantial sensitivities on both the interaction modeling and quasipotential descriptions are found. Multi-gluon contributions are expected to play also a role in the effective interaction between the quarks. Some studies have recently been carried out by using instead of the OGE an instanton induced NJL-like interaction.[19,20] These are done within the framework of the Salpeter equation. The

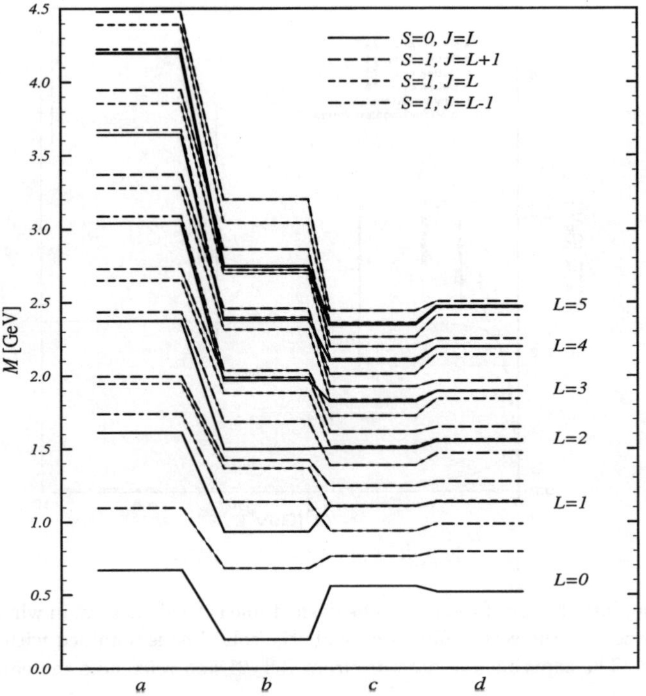

Fig. 3. Spectrum of the light meson system. The levels (a) are nonrelativistic predictions. Relativizing the kinetic energy we find (b), while (c) and (d) are the BSLT quasipotential predictions with only positive energy and all ρ spin states included.

Meson	J^P	$^{2S+1}L_J$	Exp	SM	BSLT	ET
D	0^-	$^1S_0$	1.87	1.85	1.91	1.86
D^*	1^-	$^3S_1$	2.01	2.02	2.03	2.06
D_1	1^+	$^1P_1$	2.42	2.40	2.40	2.42
D_0^*	0^+	$^3P_0$		2.47	2.33	2.63
D_1'	1^+	$^3P_1$		2.41	2.42	2.49
D_2^*	2^+	$^3P_2$	2.46	2.46	2.46	2.47

Table 1. Predictions of the lowest levels in the $c\bar{c}$ system. The meson masses are in GeV. The various quasipotential results are using the BSLT,[4,5] spectator model[6] (SM) and ET[10,11] prescriptions. The spectator model predictions have been obtained using perturbation theory.[18]

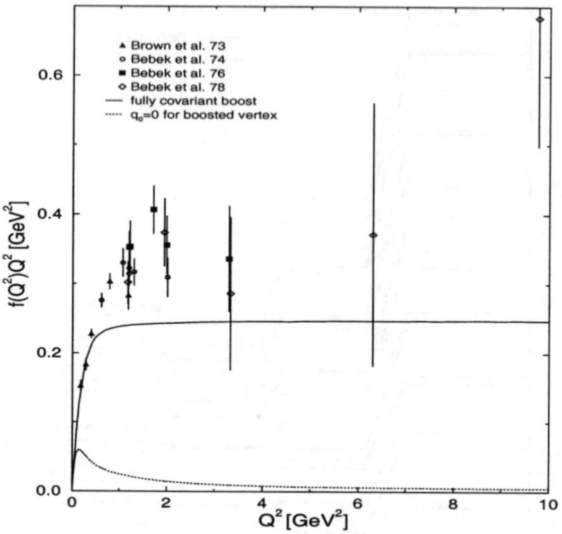

Fig. 4. The e.m. form factor of the pion. The dashed line is a calculation, in which the relative energy is kept zero in the vertex function, while the solid line is obtained with the complete boost included. The experimental data are from Ref. 21 and references therein.

NJL interaction has the virtue of having the ingredients of chiral symmetry and as a result the empirical small mass of the pion can be reproduced in a natural way in such a force model. Moreover, the calculated e.m. properties like the pion form factor are found to be in good agreement with the data. In Fig. 4 the predictions are shown for the elastic pion form factor. To calculate the current matrix elements the Lorentz transformation properties of the $\pi q\bar{q}$ vertex function are needed. From Fig. 4 we see, that it is crucial to include the proper boost in the calculation.

3. Some Issues

Within the Bethe-Salpeter approach we in general need the full operator structure between the positive and negative energy spinor states, i.e. the full ρ-spin structure of the interaction has to be given. Relativistic covariance restricts its form to a limited set of Dirac operators. For the confining interaction we may write

$$V_c = v_c\, O^{(1)} O^{(2)} \tag{13}$$

with

$$v_c = C + \sigma r \tag{14}$$

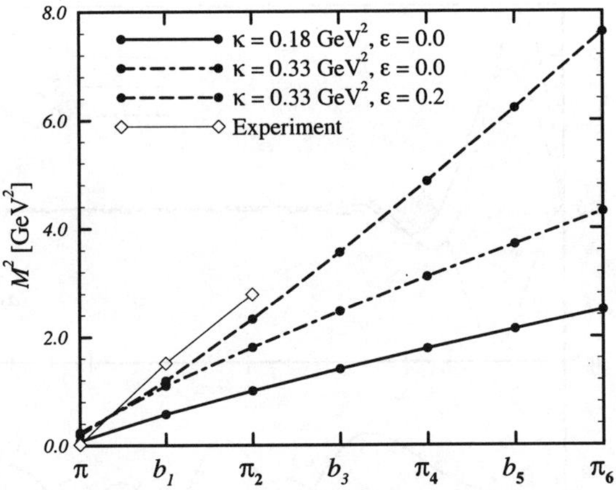

Fig. 5. The Regge trajectory for various string tension and admixture of vector confinement. Also the relativized Hamiltonian prediction is shown. The constituent quark mass is taken to be 0.25 GeV

and where $O^{(n)}$ are Dirac operators in the n-th quark space. There exist two forms of covariants which have been studied in some detail. These are the vector and scalar confinement or a combination of both. We have

$$O^{(1)}O^{(2)} = \epsilon\gamma_\mu^{(1)}\gamma_\mu^{(2)} + (1-\epsilon)I^{(1)}I^{(2)}. \tag{15}$$

For a given string tension σ the quasipotential solutions with a pure scalar confining interaction ($\epsilon = 0$) are found to yield a considerably smaller slope β_1 of the Regge trajectory of the bound state mass M

$$M^2 = \beta_0 + \beta_1 J^2 \tag{16}$$

as compared to the nonrelativistic prediction. In Fig. 5 are shown the predictions for the BSLT equations.[10] Similar results were found in the Salpeter approximation by Durand et al.,[22] keeping only the $(+,+)$ ρ-spin states in the calculations. As a result the value of the string tension needed to accomodate for the experimental data for the light quark system has to be chosen much larger than the nonrelativistic $\sigma \approx 0.18$ GeV^2, leading to some difficulties in the fitting of the heavy quark spectrum assuming a pure scalar confinement. From Fig. 5 we see that a small admixture parameter ϵ the slope increases significantly. This clearly facilitates the fitting of the various quark systems with a single choice of the confining force. A calculation of the Regge trajectory using the relativized Hamiltonian shows that the results are similar to the nonrelativistic prediction, which

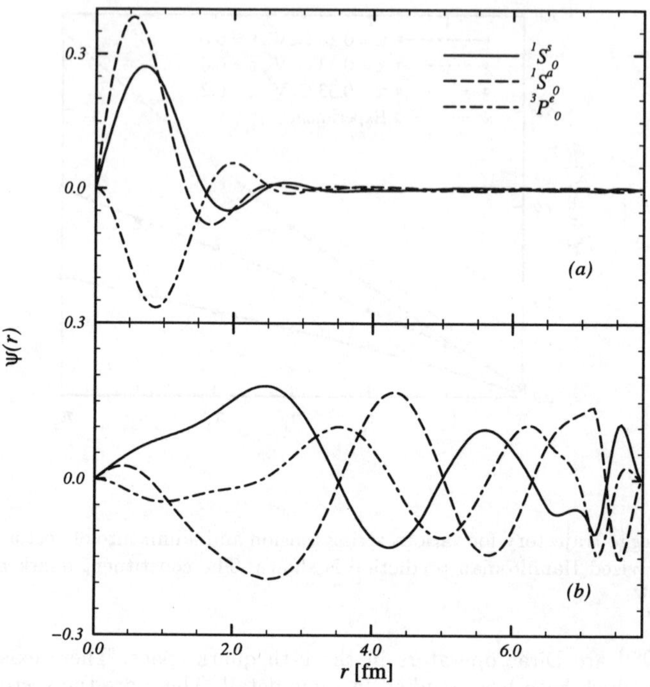

Fig. 6. A bound state solution (upper panel) with mass M = 1.23 *GeV* and an oscillatory continuum solution (lower panel) at M = 1.30 *GeV* for the BSLT equation.

yields a substantial larger Regge slope as compared to the relativistic quasipotential calculations. Clearly, the major difference with the quasipotential results are due to the presence of the lower Dirac components in the confining interaction and which are neglected in the relativized treatment.

In the ET approximation the admixture parameter ϵ cannot unfortunately be increased above a value of $\epsilon \approx 0.22$.[10,11] Beyond this critical value the ET equations are found to become unstable, i.e. solutions of a not square integrable type are found. In particular, we find above this value in addition to bound state solutions also oscillatory solutions which are extended to infinity. A similar problem arises for the case of the BSLT equation.[4,5] In Fig. 6 an example of the two possible types of solutions are shown.

This instability is due to the coupling of the positive energy states to the negative ones and is related to the Klein paradox. The found instability can readily be seen to also occur in the Dirac equation. For definiteness let us consider a Dirac particle with mass m in the presence of a confining potential

$$V = \sigma r[1 - \epsilon + \epsilon\gamma_0]. \tag{17}$$

The Dirac equation

$$[E\gamma_0 - \vec{\gamma}.\vec{p} - m + V]\Psi = 0 \tag{18}$$

can be written, after partial wave decomposition for total angular momentum J as

$$\begin{aligned} (m-E)\Psi_1 + D_{J+3/2}\Psi_2 &= -\sigma r\Psi_1 \\ D_{-J+1/2}\Psi_1 + (m+E)\Psi_2 &= -\sigma r(1-2\epsilon)\Psi_2 \end{aligned} \tag{19}$$

where $D_l = \frac{d}{dr} + \frac{l}{r}$ and Ψ_1 and Ψ_2 are the upper and lower 2-components of Ψ respectively. Introducing $\Psi_1 = F/r$ and $\Psi_2 = G/r$ and eliminating the lower 2-component G one readily finds an effective Schrödinger equation. For large r we get

$$\left[-\frac{d^2}{dr^2} + \frac{1}{r}\frac{d}{dr} + \frac{1}{r^2}(J+1/2)(J-3/2) + (1-2\epsilon)\sigma^2 r^2\right] F = 0. \tag{20}$$

For pure scalar confinement ($\epsilon = 0$) Eq. (20) supports only bound state solutions. Moreover, from Eq. (20) we see that the effective Schrödinger potential changes sign, when $\epsilon > 1/2$. As a result we get an inverted harmonic well. Hence deconfinement takes place in that case. Whether such an instability occurs depends strongly on the type of equation.

Various studies have been made of the stability problem of the Salpeter equation with some conflicting results. They in general are predominantly based on numerical analysis. Kopashvilli et al.[23] have noticed that for a timelike vector confining interaction of the form $\gamma_0^{(1)}\gamma_0^{(2)}$ the Salpeter equation is stable, whereas for small quark masses a pure scalar confinement leads to unstable solutions. Clearly, this finding is different from the Dirac equation result. Moreover, it is found that for increasing quark masses the solutions become stable. This is in disagreement with the momentum space stability analysis of Parramore and Piekarewicz.[24] The latter in particular argue, that an effective potential V_{eff} exists in the pure scalar confining case which has the infrared singular behaviour of

$$V_{eff}(p,p') = \sigma(p)/(p-p')^4, \tag{21}$$

but with a momentum dependent string tension σ which changes sign for large momentum p. As a result they deduce that the Salpeter equation should be unstable independent of the quark mass. Using the same method they do find that the timelike vector confinement is stable, in agreement with Ref. 23.

More recently the stability of the ET[25] and spectator solutions[26] have been investigated using the momentum space respresentation. Evidence is presented of also the occurrence of similar difficulties in these cases. Again results differ from the previous analysis. Uzzo and Gross[26] have studied the Salpeter equation and find that for sufficiently heavy quark mass pure scalar confinement doesnot support unstable solutions as suggested by Parramore et al.[24] Moreover, Wallace and collaborators[25] find for pure

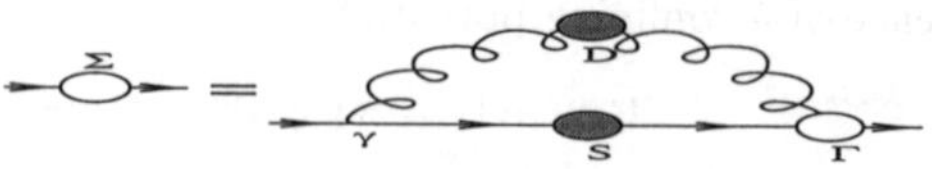

Fig. 7. The Dyson-Schwinger equations for the quark self-energy. Both the gluon and quark propagators are dressed

scalar confinement also instabilities in the ET case, to be contrasted with the findings of Tiemeijer.[10,11] Since many of the studies are of numerical nature, a more mathematical rigorous analysis would be welcome to settle some of these issues. It is in particular not clear whether the regularization scheme of the singular interaction in momentum space could also be a source of the introduction of additional (unphysical) solutions.

4. Chiral Symmetry Breaking and Confinement

The nonperturbative solutions as discussed in Section 2 do not contain the dynamical chiral symmetry breaking mechanism. To realize the breaking mechanism the dressed quark propagator should be solved in some consistent way. A simple example has been studied by Gross and Milana[7] in an extended spectator model with a confining interaction, satisfying a $SU(2) \times SU(2)$ symmetry. The dressing and generation of the constituent mass takes place through the confining interaction.

More generally, the Dyson-Schwinger formalism is especially suitable to implement the chiral symmetry breaking. In the case of QCD the self-energy part Σ of the quark propagator S can be expressed in terms of the dressed gluon and quark progators as depicted in Fig. 7. From this we see, that the set of equations for S are in general nonlinear. Moreover, since the quark-gluon vertexfunction Γ enters in Σ one has an infinite tower of coupled equations.

Some truncation scheme is needed to arrive at a finite number of coupled equations. Roberts and collaborators[27] have studied a lowest order type of approximation. The infinite tower can essentially be terminated by making some appropriate ansatz for the quark-gluon vertex function and the gluon propagator. To do this it is assumed that the vertex function can be approximated simply by the bare vertex, the socalled rainbow approximation. In some studies more sophisticated choices have been made for Γ by satisfying the Ward-Takahashi (WT) identity

$$(p-k)^{\mu}\Gamma_{\mu} = S^{-1}(p) - S^{-1}(k), \tag{22}$$

p and k being the initial and final quark momenta. The WT identity clearly doesnot determine Γ uniquely. An often used vertex function, satisfying Eq. (22) is due to Ball and Chui.[28] It is given by

$$\Gamma_\mu(p,k) = \frac{A(p^2)+A(k^2)}{2}\gamma_\mu + \frac{(p+k)_\mu}{p^2-k^2}\left[[A(p^2)-A(k^2)]\frac{\not p + \not k}{2} + [B(p^2)-B(k^2)]\right] \tag{23}$$

where A and B are the invariants of the quark propagator

$$S(p) = [A(p^2)\,\not p + B(p^2)]^{-1} \tag{24}$$

Inspired by the studies of Refs. 29 and 30, showing that for certain gauge choices the dressed gluon propagator $G_{\mu\nu}$ the existence of an infrared singularity

$$G_{\mu\nu}(k) = P_{\mu\nu}/k^4, \tag{25}$$

forms of the gluon propagator are taken, which reflects the singular infrared behaviour. Using the Landau gauge one has $P_{\mu\nu} = \delta_{\mu\nu} - \frac{k_\mu k_\nu}{k^2}$. Various forms for $G_{\mu\nu}$ have been studied. In particular, Roberts et al. chose, following Munzcek and Nemirovsky[31]

$$G_{\mu\nu} = P_{\mu\nu}\delta_4(k). \tag{26}$$

For both choices (25) and (26) of $G_{\mu\nu}$ one finds that the Dyson-Schwinger solutions for S exhibits the 'confining' property, that it doesnot have a free mass pole.[32,33] Using the above quark propagator and $G_{\mu\nu}$ as driving force the Bethe-Salpeter equations are solved. This is done in Euclidean space, where the kernel is well behaved.

It should be noted, that for the considered approximation of the Dyson-Schwinger equations, confinement is basically due to the confinement of each quark individually. Since the quark propagator doesnot have a free mass pole, the individual quarks cannot exist as asymptotic free objects. In view of the δ-function in Eq. (26) the effective potential arising from the gluon propagator becomes a constant and hence doesnot grow unbounded for large quark separations. As a result this naive argument shows that no infinite rising Regge trajectories should be present in this model. Moreover, the Dirac structure is clearly vector-like due to the OGE.

Using the above truncation scheme the Argonne group[27,34] has studied in a series of papers the structure of the $q\bar{q}$ system and its e.m. properties. Since the quark propagators are dressed in the Dyson-Schwinger formalism, bare e.m. current operators like γ_μ do not suffice, but should be chosen similar as Eq. (23) in order to satisfy gauge invariance and as a consequence current conservation. Properties like the momentum transfer dependence of the e.m. form factor of the pion are well reproduced.

There are certain indications that the confinement mechanism is of a nonperturbative nature arising from multi-gluon effects. Recently encouraging results have been obtained to describe this using the vacuum correlator method. Considering the case of a light-heavy quark system Simonov[35] and Brambilla[36] have derived a Dyson-Schwinger like equation, where the effective $q\bar{Q}$ interaction is essentially nonlocal and of a NJL like type. Consider the gauge-invariant four-point function of the light-heavy quark system

$$G_{q\bar{Q}} = N\int \mathcal{D}\Psi\,\mathcal{D}\Psi^\dagger\,\mathcal{D}A\;exp(-S)\;\Psi^\dagger(x)S_{\bar{Q}}(x,y)\Psi(y), \tag{27}$$

where S_Q is the heavy quark propagator, Ψ the light quark field and S the QCD action

$$S = \int d_4x \left[\frac{1}{4}F_{\mu\nu}^2 + \Psi^\dagger[-i\,\not{\partial} - im_q - \not{A}]\Psi\right]. \tag{28}$$

Treating the heavy quark as static and located at the origin, the heavy quark propagator can be written as

$$\begin{aligned} S_{\bar{Q}} &= \frac{1+\gamma_4}{2} Pexp[ig\int_0^T A_4(\vec{r}=0,\tau)d\tau] \\ &= \frac{1+\gamma_4}{2}. \end{aligned} \tag{29}$$

Using the lowest order contribution in the cumulant expansion of the Wilson loop[14,15] the four-point function can be expressed in terms of the vacuum field strength correlator, which leads to the minimal area law and the Nambu-Goto string for large separations of the quarks. In the considered static approximation the light particle moves as a Dirac particle in the presence of a potential field due to the presence of the heavy quark. The interaction obtained in this way is nonlocal. Furthermore the dressing of the light quark propagator is such, that the 'self-energy' is dependent on the presence of the heavy quark, yielding only confinement when the quarks are asymptotically far separated from each other. This should be contrasted to the previously discussed Dyson-Schwinger truncation of Roberts et al., where the quark propagator is 'confining' independent of the presence of the other quark. The effective Lagrangian for the light quark is found to be

$$\mathcal{L}_{eff} = \int d_4x\; d_4y\; \Psi^\dagger(x)\gamma_\mu\Psi(x)\; J_{\mu\nu}(x,y)\; \Psi^\dagger(y)\gamma_\nu\Psi(y), \tag{30}$$

where $J_{\mu\nu}$ is the vacuum field correlator

$$J_{\mu\nu}(x,y) = < A_\mu(x)A_\nu(y) > . \tag{31}$$

The derived Dyson-Schwinger equations satisfy chiral symmetry in the limit of vanishing bare quark mass. Moreover, the kernel has the property that for large separation between the quarks it becomes linear confining.

5. Concluding Remarks

To study systems with light quarks it is in general important to have a relativistic treatment available. The general philosophy is that one may hope that there exist only a few explicit degrees of freedom to describe the system, while the remaining ones can be integrated out to yield an effective interaction. Moreover, the theory of the effective interaction should be transparant and simple enough. Attempts have been made to construct field theory based equations, where quark degrees of freedom are singled out and the effects of gluons are reflected in the effective interaction. Such a description has met with some success. Some basic problems arise which partially are of a mathematical nature but it also may signal that something is missing in the description.

Relativistic treatments lead for example to the need of a large string tension. This may indicate that the effects of gluon degrees of freedom may not simply be described by only the confining force and the OGE, but some additional contributions like for example effects due to the existence of a rotating string[37,38] have to be included. This need not invalidate the framework as long as the missing pieces can be incorporated in the effective interaction. Obviously, it would be interesting to see how this can be done in the relativistic quasipotential approach. It may however be that the string degrees of freedom have also to be included explicitly in the dynamics. More work is also needed to understand the covariant structure of the confining force.

Investigations using the Dyson-Schwinger formalism have recently been carried out to include chiral symmetry breaking and at the same time the confining properties. In the various calculations distinct different mechanisms of confinement are put in. Depending on whether it is due to a dressed OGE force with infrared behaviour or the gluon field vacuum correlator the confining interaction will be vector or mostly scalar. A better understanding of the nature of confinement is clearly needed.

References

1. W. Lucha, F.F. Schöberl and D. Gromes, Phys. Rept. **200**, 127 (1991).
2. S. Godfrey and N. Isgur, Phys. Rev. D **32**, 189 (1985).
3. E. Salpeter, Phys. Rev. **87**, 328 (1952).
4. A. A. Logunov and A. N. Tavkhelidze, Nuove Cimento **29**, 380 (1963).
5. R. Blankenbecler and R. Sugar, Phys. Rev. **142**, 1051 (1966).
6. F. Gross, Phys. Rev. C **26** (1982) 2203.
7. F. Gross and J. Milana, Phys. Rev. D **43**, 2401 (1991).
8. S. J. Wallace and V. B. Mandelzweig, Nucl. Phys. **A503**, 673 (1989).
9. J.R. Spence and J. P. Vary, Phys. Rev. C **47**, 1282 (1993).
10. P. Tiemeijer and J. A. Tjon, Phys. Rev. C **49**, 44 (1994).
11. P. Tiemeijer, PhD thesis, University of Utrecht (1993).
12. Yu. A. Simonov and J. A. Tjon, Ann. Phys. **228**, 1 (1993).
13. C. Itzykson and J. B. Zuber, 'Quantum Field Theory' (McGraw-Hill, 1985).
14. H. G. Dosch, Phys. Lett. **B190**, 177 (1987).
15. Yu. A. Simonov, Nucl. Phys. **B307**, 512 (1988); Yad. Fiz. **54**, 192 (1991).
16. K. M. Maung, D.E. Kahana and J. W. Norbury, Phys. Rev. D **47**, 1182 (1993).
17. H. Hersbach, Phys. Rev. D **47**, 3027 (1993).
18. J. Zeng, J.W. VanOrden and W. Roberts, Phys. Rev. D **52**, 5229 (1995).
19. C. R. Münz et. al., Nucl. Phys. **A578**, 418 (1994).
20. C. R. Münz et. al., Phys. Rev. C **52**, 2110 (1995).
21. C. J. Bebek et al., Phys. Rev. D **17**, 1693 (1978).
22. B. Durand and L. Durand, Phys. Rev. D **28**, 396 (1983); A. Gara et al., Phys. Rev. D **40**, 843 (1989); Phys. Rev. D **42**, 1651 (1990).
23. M. Chachkunashvili and T. Kopaleishvili, Few-Body Systems **6**, 1 (1989); A. Archvadze, M. Chachkunashvili and T. Kopaleishvili, Few-Body Systems **14**, 53 (1993).
24. J. Parramore and J. Piekarewicz, Nucl. Phys. **A585**, 705 (1995).

25. M. Ortalano et al., preprint (1998)
26. M. Uzzo and F. Gross, preprint (1998).
27. For a review and references, see C. D. Roberts and A. G. Williams, Prog. Part. Nucl. Phys. **33**, 477 (1994).
28. J. S. Ball and T. W. Chui, Phys. Rev. D **22**, 2542 (1980).
29. M. Baker et al., Nucl. Phys. **B186**, 531 (1981); Phys. Rept. **209**, 73 (1991).
30. N. Brown and M. R. Pennington, Phys. Rev. D **39**, 2723 (1989).
31. H. J. Munczek and A. M. Nemirovsky, Phys. Rev. D **28**, 181 (1983).
32. J. S. Ball and F. Zachariasen, Phys. Lett. **B106**, 133 (1981).
33. C. J. Burden, C. D. Roberts and A. G. Williams, Phys. Lett. **B285**, 347 (1992).
34. See C.D. Roberts, 11th Physics Summer School on Frontiers in Nuclear Physics, Australian National University, Canberra, Australia (1998), nucl-th/9807026, and references therein.
35. Yu. A. Simonov, Phys. Atom. Nucl. **60**, 2069 (1997); Few-Body Systems **24** (1998), in press.
36. N. Brambilla, this conference.
37. A. Yu. Dubin, A. B. Kaidalov and Yu. A. Simonov, Phys. Atom. Nucl. **56**, 1745 (1993).
38. T.J. Allen, M.G. Olsson and S. Veseli, this conference.

PARALLEL SESSION PRESENTATIONS

SECTION "A"

MECHANISM FOR CONFINEMENT: FLUX TUBE CONFIGURATIONS AND STRINGS

A PROOF OF QUARK CONFINEMENT IN QCD

Kei-Ichi Kondo

Department of Physics, Faculty of Science, Chiba University, Chiba 263, JAPAN

ABSTRACT

I propose to reformulate the gauge field theory as the perturbative deformation of a novel topological quantum field theory. It is shown that this reformulation leads to quark confinement in QCD_4. Similarly, the fractional charge confinement is also derived in the strong coupling phase of QED_4. As a confinement criterion, we use the area decay of the expectation value of the Wilson loop.

1. Introduction

An idea of magnetic monopole and its condensation was used in the analytical proofs of "quark" confinement so far:

1975: Polyakov[1] for 3D compact U(1) gauge theory

1977: Polyakov[2] for 3D Georgi-Glashow model

1994: Seiberg-Witten[3] for 4D N=2 SUSY YM/QCD

How to prove quark confinement in QCD_4 analytically? Recent numerical simulations have established that also in QCD the magnetic monopole plays the dominant role in quark confinement (monopole dominance) under the Abelian projection.[4] To explain this fact analytically and then to prove quark confinement, we propose to use a novel type of topological quantum field theory (TQFT)[5] which describes 4D magnetic monopole, and to reformulate QCD as a perturbative deformation of the TQFT.

We can show:[6–9]

1. The TQFT is derived from QCD (without any approximation) in the maximal Abelian gauge (MAG).
2. The calculation of the Wilson loop in 4D YM theory is reduced to that in 2D nonlinear sigma model (NLSM) due to the dimensional reduction. This is an exact result.
3. Area decay of the Wilson loop is derived from 2D Instanton calculus in NLSM.

2. Basic Idea

We take into account the topological nontrivial background field Ω_μ which corresponds to 4D magnetic monopole current k_μ. The current k_μ obeys the topological conservation law $\partial_\mu k_\mu = 0$ and hence k_μ denotes a loop in 4D spacetime. Usually, Ω_μ denotes an arbitrary, but fixed background (as a solution of classical field equation). In our case, we sum up all topological nontrivial background Ω_μ after integrating out quantum fluctuations Q_μ,

$$Z_{QCD} = \sum_{\Omega} \int [dQ_\mu] \exp\{-S_{QCD}[\Omega_\mu + Q_\mu]\}, \tag{1}$$

where

$$S_{QCD} = \int d^D x \left[-\frac{1}{2g^2} \mathrm{tr}_G(\mathcal{F}_{\mu\nu}\mathcal{F}_{\mu\nu}) + \bar{\psi}(i\gamma^\mu \mathcal{D}_\mu[\mathcal{A}] - m)\psi \right]. \tag{2}$$

We must regard this background to be different from the 4D instanton discussed in the topological YM theory by Witten[5] and 3D magnetic monopole by Polyakov.[2]

3. Quantization of YM theory based on BRST formalism

Using the nilpotent BRST transformation δ_B, it is well known that the gauge fixing and the Faddeev-Popov (FP) ghost parts is written in the form,

$$\mathcal{L}_{gf} = -i\delta_B G_{gf}, \quad G_{gf} := \mathrm{tr}_G \left\{ \bar{C} \left(F[\mathcal{A}] + \frac{\alpha}{2}\phi \right) \right\}, \tag{3}$$

where α is the gauge-fixing parameter and ϕ is the Lagrange-multiplier field. For the manifestly covariant Lorentz gauge, we choose $F[\mathcal{A}] = \partial^\mu \mathcal{A}_\mu$.

In the maximal Abelian gauge (MAG),[6,7] a suitable choice of gauge fixing parameter $\alpha = -2$ leads to

$$G_{gf} := \bar{\delta}_B \mathrm{tr}_{G/H} \left(\frac{1}{2} \mathcal{A}_\mu(x)\mathcal{A}_\mu(x) + iC(x)\bar{C}(x) \right). \tag{4}$$

Using the decomposition,

$$\mathcal{A}_\mu(x) = \Omega_\mu(x) + Q_\mu(x) = \frac{i}{g} U(x)\partial_\mu U(x)^\dagger + U(x)\mathcal{V}_\mu(x)U^\dagger(x), \tag{5}$$

the action of the TQFT is given by

$$S_{TQFT} := \int d^D x \; i\delta_B \bar{\delta}_B \; \mathrm{tr}_{G/H} \left(\frac{1}{2} \Omega_\mu(x)\Omega_\mu(x) + iC(x)\bar{C}(x) \right), \tag{6}$$

and the partition function is given by

$$Z_{TQFT} = \int [dU][dC][d\bar{C}][d\phi] \exp\{-S_{TQFT}[\Omega_\mu]\}, \tag{7}$$

where U is an element of the gauge group G.

4. Strategy of a proof

We consider D-dim. QCD (with a gauge group G) for $D > 2$.

4.1. Step 1:[7]

In MAG, D-dim. QCD (QCD_D) is reformulated as a perturbative deformation of $TQFT_D$. MAG is a partial gauge fixing which fixes the coset part G/H and leaves the maximal torus group H invariant.

4.2. Step2:[7]

It is shown that $TQFT_D$ is equivalent to the coset G/H nonlinear sigma model (NLSM) in (D-2) dimensions. This is a consequence of Parisi-Sourlas dimensional reduction due to the hidden supersymmetry in TQFT (6).

As extensively discussed more than 20 years ago, QCD_4 and $NLSM_2$ have common properties: renormalizability, asymptotic freedom ($\beta(g) < 0$), dynamical mass generation, existence of instanton solution, no phase transition for any value of coupling constant (one phase), etc. This similarity between two theories can be understood from this above correspondence,

$$QCD_4 \supset TQFT_4 \Longleftrightarrow G/H\ NLSM_2 \tag{8}$$

For $G = SU(2)$, G/H NLSM is equivalent to O(3) NLSM. Existence of 2D instanton is guaranteed for any N, because $\Pi_2(SU(N)/U(1)^{N-1}) = \Pi_1(U(1)^{N-1}) = \mathbf{Z}^{N-1}$.

4.3. Step3:[9]

For the non-Abelian Wilson loop,

$$W^C[\mathcal{A}] := \mathrm{tr}\left[\mathcal{P}\exp\left(iq\oint_C \mathcal{A}_\mu^A(x)T^A dx^\mu\right)\right], \tag{9}$$

we define the full string tension σ by

$$\sigma := -\lim_{A(C)\to\infty}\frac{1}{A(C)}\ln\langle W^C[\mathcal{A}]\rangle_{YM_4}. \tag{10}$$

On the other hand, the diagonal string tension σ_{Abel} is defined by

$$\sigma_{Abel} := -\lim_{A(C)\to\infty}\frac{1}{A(C)}\ln\left\langle W^C[a^\Omega]\right\rangle_{TFT_4}, \tag{11}$$

using the diagonal Wilson loop,

$$W^C[a^\Omega] = \exp\left(iq\oint_C dx^\mu a_\mu^\Omega(x)\right), \quad a_\mu^\Omega(x) := \Omega_\mu^3(x) := \mathrm{tr}(T^3\Omega_\mu(x)). \tag{12}$$

From the dimensional reduction, we find

$$\langle W^C[a^\Omega]\rangle_{TFT_4} = \left\langle W^C[a^\Omega]\right\rangle_{NLSM_2}. \tag{13}$$

Then it is shown that in the large Wilson loop limit

$$|\sigma - \sigma_{Abel}| \downarrow 0 \quad (A(C)\uparrow\infty). \tag{14}$$

This is derived from the non-Abelian Stokes theorem. For the large (non-intersection planar) Wilson loop, the full string tension σ is saturated by the diagonal string tension σ_{Abel}. This explains the Abelian dominance and magnetic monopole dominance.

4.4. Step4:[7]

The whole problem is reduced to calculating

$$\left\langle W^C[a^\Omega]\right\rangle_{NLSM_2} = \left\langle e^{i\frac{q}{g}8\pi Q_S}\right\rangle_{NLSM_2}, \quad Q_S = \frac{1}{8\pi}\int_S d^2z\, \epsilon_{\mu\nu}\mathbf{n}\cdot(\partial_\mu\mathbf{n}\times\partial_\nu\mathbf{n}). \tag{15}$$

Note that the integrand of Q_S is the instanton density in NLSM_2. So, Q_S counts the number of instantons minus anti-instantons inside the area S bounded by the Wilson loop C. In order to perform the actual calculation, we adopt the the naive instanton calculus (dilute gas approximation). This leads to the area law of the diagonal Wilson loop and the non-zero diagonal string tension σ_{Abel} for the fractional charge q (i.e., when q/g is not an integer). More systematic instanton calculations enable us to identify the instanton solution with the Coulomb gas of vortices.

Note that all the above steps are exact except for the instanton calculus of the Wilson loop in NLSM. The above results imply that the non-zero string tension σ in QCD_4 follows from the non-zero diagonal string tension σ_A in NLSM_2. The problem of proving area law in QCD_4 is reduced to the equivalent problem in NLSM_2.

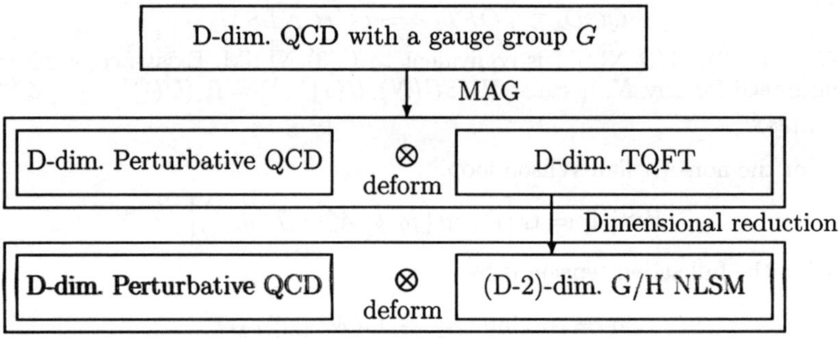

Similar strategy is also applied to QED_4 to prove the existence of strong coupling confinement phase.[8] This follows from the existence of Berezinski-Kosterlitz-Thouless transition of the O(2) NLSM_2.

References

1. A.M. Polyakov, Phys. Lett. B **59**, 82 (1975).
2. A.M. Polyakov, Nucl. Phys. B **120**, 429 (1977).
3. N. Seiberg and E. Witten, Nucl. Phys. B **426**, 19 (1994); **431**, 484 (1994).
4. G. 't Hooft, Nucl. Phys. B **190** [FS3], 455 (1981).
5. E. Witten, Commun. Math. Phys. **117**, 353 (1988).
6. K.-I. Kondo, hep-th/9709109 (revised), Phys. Rev. D **57**, 7467 (1998).
   K.-I. Kondo, hep-th/9803063, Prog. Theor. Phys. Suppl., No. 130, in press.
7. K.-I. Kondo, hep-th/9801024 (revised), Phys. Rev. D (1998), in press.
8. K.-I. Kondo, hep-th/9803133 (revised), Phys. Rev. D (1998), in press.
9. K.-I. Kondo, hep-th/9805153 (revised), Phys. Rev. D (1998), in press.

QUANTUM AND STRING SHAPE FLUCTUATIONS IN THE DUAL MONOPOLE NAMBU-JONA-LASINIO MODEL WITH DUAL DIRAC STRINGS

Alexandra Müller, Manfried Faber, Andrei N. Ivanov
Institut für Kernphysik, Technische Universität Wien, A-1040 Vienna, Austria

ABSTRACT

The Monopole Nambu-Jona-Lasinio (MNJL) model with dual Dirac strings realizes a superconducting mechanism of quark confinement. Integrating over quantum and string shape fluctuations the magnetic monopole condensate is calculated as a functional of the dual Dirac string shape. We find substantial contributions of the fluctuations to the monopole condensate near the string. It is shown, that in accordance with the results obtained in Compact Quantum Electrodynamics (CQED), the MNJL model can provide a suppression of the monopole condensate in the vicinity of a dual Dirac string.

1. Introduction

A dual superconductor seems to be a very promising model for the nonperturbative QCD vacuum leading to color confinement. It has been shown[1] that Compact Quantum Electrodynamics (CQED), an Abelian gauge theory formulated on lattice and possessing all nonperturbative phenomena of QCD, behaves like an effective dual superconductor with magnetic monopoles.

The Nambu-Jona-Lasinio (NJL) model[2] can be regarded as a field theoretical generalization of the BCS theory of superconductivity. It also possesses a nonperturbative vacuum similar to the ground state of a superconductor in the superconducting (condensed) phase. The latter has been the promoting idea to put the NJL-model on the foundation of a continuum space-time model which realizes the nonperturbative properties of CQED.

The Monopole Nambu-Jona-Lasinio (MNJL) model with dual Dirac strings, where in the non-condensed phase magnetic monopoles are described by a massless fermion field $\chi(x)$, is suggested as a continuum analogy of CQED. In the MNJL-model massless magnetic monopoles interact via local four-fermion interaction given by[3]

$$\mathcal{L}(x) = \bar{\chi}(x) i\gamma^{\mu}\partial_{\mu}\chi(x) + G[\bar{\chi}(x)\chi(x)]^2 - G_1[\bar{\chi}(x)\gamma^{\mu}\chi(x)][\bar{\chi}(x)\gamma_{\mu}\chi(x)] \quad (1)$$

where G and G_1 are positive phenomenological constants. Lagrangian (1) is invariant under magnetic $U(1)$ gauge transformation. Calculating the energy of the ground state one can see that it is more advantageous for the system to be in the condensed (superconducting) phase, characterized by a non-zero value of the magnetic monopole condensate $< \overline{\chi}\chi > \neq 0$, which plays the role of an order parameter. The wave function of the ground state in the condensed phase $|0\rangle^{(M_\chi)}$ is similar to the BCS-wave function, and reads

$$|0\rangle^{(M_\chi)} = \prod_{\vec{p},\lambda=\pm 1} \left[\sqrt{\frac{1+\beta_{\vec{p}}}{2}} + \lambda \sqrt{\frac{1-\beta_{\vec{p}}}{2}} a^{(0)\dagger}(\vec{p},\lambda)\, b^{(0)\dagger}(-\vec{p},\lambda) \right] |0\rangle , \tag{2}$$

where $\beta_{\vec{p}} = \vec{p}/\sqrt{\vec{p}^2 + M_\chi^2}$ is the velocity of massive monopoles with mass M_χ, $a^{(0)\dagger}$ ($b^{(0)\dagger}$) denotes the creation operator of a massless monopole (antimonopole) with momentum $\vec{p}$ and helicity λ, $|0\rangle \equiv |0\rangle^{(0)}$ is the wave function of the ground state in the non-condensed phase. Unlike the BCS-wave function, which violates electromagnetic $U(1)$ gauge invariance, $|0\rangle^{(M_\chi)}$ is invariant under magnetic $U(1)$ gauge transformation.

The transition of the system to the condensed phase, i.e. the magnetic monopole condensation, accompanies the creation of massive magnetic monopoles $\chi_M(x)$ with mass M_χ and $\overline{\chi}\chi$ collective excitations with quantum numbers of a scalar meson $\sigma(x)$ with mass $M_\sigma = 2M_\chi$ and of a dual vector field $C^\mu(x)$ with mass M_C . The monopole mass is related to the condensate via the gap equation $M_\chi = -2G\,\langle\overline{\chi}\chi\rangle$ derived from the effective Lagrangian of the scalar field and the dual vector field by virtue of the suppression of transitions $\sigma \leftrightarrow$ vacuum[3] *.

Integrating out the magnetic monopole fields $\chi(x)$ we arrive at an effective Lagrangian for the collective excitations $\sigma(x)$ and $C^\mu(x)$.[3]

$$\begin{aligned} \mathcal{L}_{eff}(x) &= \frac{1}{4} F_{\mu\nu}(x) F^{\mu\nu}(x) + \frac{1}{2} M_C^2 C_\mu(x) C^\mu(x) + \frac{1}{2} \partial_\mu \sigma(x)\, \partial^\mu \sigma(x) \\ &\quad - \frac{1}{2} M_\sigma^2 \sigma^2(x) \left[1 + \kappa \frac{\sigma(x)}{M_\sigma} \right]^2 + \mathcal{L}_{free\,quark}(x), \end{aligned} \tag{3}$$

where the field strength tensor $F^{\mu\nu}(x)$ is defined as $F^{\mu\nu}(x) = \mathcal{E}^{\mu\nu}(x) - \epsilon^{\mu\nu\alpha\beta}\partial_\alpha C_\beta(x)$, and $\mathcal{E}_{\mu\nu}(x)$ is the dual electric field related to the dual Dirac string. A massive dual-vector field coupled to a dual Dirac string acquires the shape of a dual Abrikosov flux line and obeys the equation of motion

$$\left(\Box + M_C^2\right) C^\nu\left[\mathcal{E}(x)\right] = -\partial_\mu {}^*\mathcal{E}^{\mu\nu}(x) . \tag{4}$$

In the plane perpendicular to the dual Dirac string the vector field $C^\nu\left[\mathcal{E}(x)\right]$ is localized in the region of the order $\mathcal{O}(1/M_C)$ around the string and decreases as $e^{-M_C|\vec{x}_\perp|}$ for $|\vec{x}_\perp| \to \infty$.

*When matching the gap equation $M_\chi = -2G\,\langle\overline{\chi}\chi\rangle$ with $M_\chi = -2(0.75G + G_1)\,\langle\overline{\chi}\chi\rangle$ derived as one-loop corrections to the mass of the monopole field by using Eq.(1) we fix G_1 in terms of G: $G_1 = 0.25G$.

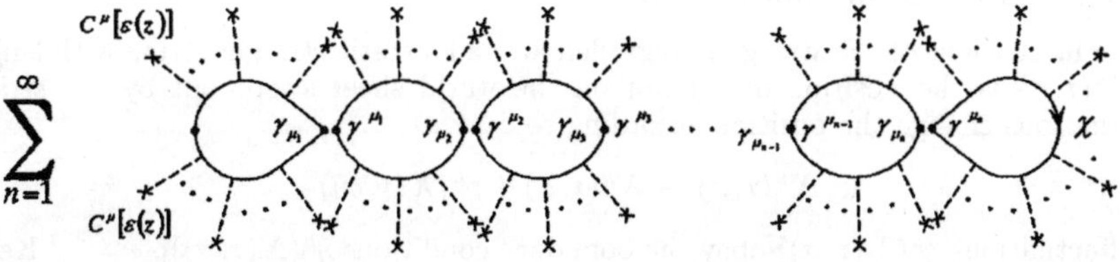

Fig. 1. Magnetic monopole loops describing the main contributions to the magnetic monopole condensate caused by quantum fluctuations of the C_μ -field.

2. Dual vector field fluctuations

Physical observables can be expressed in terms of Green functions of the monopole fields,[3]

$$G(x_1,...) = \langle 0|\, T\left(\overline{\chi}(x_1)\Gamma_1\chi(x_1)...\right)|0\rangle_{conn.} =^{(M_\chi)} \langle 0|\, T\left(\overline{\chi}_M(x_1)\Gamma_1\chi_M(x_1)...\right) \times e^{i\int d^4x\,\{-g\overline{\chi}_M(x)\gamma^\mu\chi_M(x)C_\mu(x) - \kappa\overline{\chi}_M(x)\chi_M(x)\sigma(x)\}}|0\rangle^{(M_\chi)}_{conn.}. \tag{5}$$

The computation of the r.h.s of Eq. (5) we perform in the tree approximation of the σ-field exchange valid at $M_\chi \gg M_C$. We keep only leading terms in the large M_χ expansion. This corresponds to the London limit in the dual Higgs model $M_\sigma \to \infty$.

The monopole condensate $\langle\overline{\chi}(x)\chi(x),\mathcal{E}\rangle$ is calculated as depending on the shape of the dual Dirac string,

$$\langle\overline{\chi}(x)\chi(x),\mathcal{E}\rangle - \langle\overline{\chi}\chi\rangle_0 =^{(M_\chi)} \langle 0|\, T(\overline{\chi}_M(x)\chi_M(x) \times e^{i\int d^4z\,\{-g\overline{\chi}_M(z)\gamma^\mu\chi_M(z)C_\mu(z) - \kappa\overline{\chi}_M(z)\chi_M(z)\sigma(z)\}}|0\rangle^{(M_\chi)}_{conn.}. \tag{6}$$

By using path integral representation, we integrate over dual vector field fluctuations around the dual Abrikosov flux line: $C^\mu(x) = C^\mu[\mathcal{E}(x)] + c^\mu(x)$. As a result we arrive at an expression containing only massive monopole fields $\chi_M(x)$. The vacuum expectation value of this expression can be given by the magnetic monopole diagrams depicted in Fig. 1 :

$$\langle\overline{\chi}(x)\chi(x),\mathcal{E}\rangle = \langle\overline{\chi}\chi\rangle_0\{1 + \frac{M_C^2 + \frac{g^2}{2G_1}}{M_C^2 - \frac{g^2}{6G_1}}\frac{\kappa^2}{96\pi^2}\frac{1}{M_\chi^2}g^2 C_\mu[\mathcal{E}(x)]\,C^\mu[\mathcal{E}(x)]\}. \tag{7}$$

This gives the explicit dependence of the monopole condensate on the Abrikosov flux line.

3. Dual Dirac string fluctuations

Fluctuations of the string are regarded around a static straight string with length L. $X^{\mu}(\tau,\sigma)$ is the position of a point on the world sheet swept out by the string. Fluctuations change the position according to:

$$X^{\mu}(\tau,\sigma) \to X^{\mu}(\tau,\sigma) + \eta^{\mu}(X(\tau,\sigma)) \tag{8}$$

The fluctuations $\eta^{\mu}(X(\tau,\sigma))$ obey the boundary conditions $\eta^{\mu}(X(\tau,\sigma))|_{\partial S} = 0$.[3] Keeping only Gaussian fluctuations and integrating them out we get

$$\langle\overline{\chi}(x)\chi(x),\mathcal{E}\rangle = \langle\overline{\chi}\chi\rangle_0 \left[1 - \frac{M_C^2 + \frac{g^2}{2G_1}}{M_C^2 - \frac{g^2}{6G_1}} \frac{\kappa^2}{96\pi^2} \frac{M_C^2}{M_\chi^2} K_1^2 (M_C |\vec{x}_\perp|)\right], \tag{9}$$

where $K_1(M_C|\vec{x}_\perp|)$ is a McDonalds function.
Due to CQED[4] we know that the monopole condensate should vanish in the vicinity of the dual Dirac string. In the MNJL model such a vanishing can be obtained only if the mass of the dual vector field M_C obeys the constraint

$$M_C^2 \geq \frac{g^2}{6G_1} \frac{1 + \frac{g^2}{12\pi^2}}{1 - \frac{g^2}{36\pi^2}} \tag{10}$$

At distances $r >> 1/2M_C$ the magnetic monopole condensate Eq.(9) tends to the value of the order parameter $\langle\overline{\chi}\chi\rangle_0$.

4. Conclusions

We have calculated the contributions to the magnetic monopole condensate caused by quantum and string shape fluctuations around the Abrikosov flux line. It is shown that in accordance with CQED the obtained contributions provide the vanishing of the monopole condensate at distances close to the dual Dirac string. In turn, at distances far from the string the contributions vanish exponentially as $e^{-2M_C|\vec{x}_\perp|}$. This is a much steeper fall off than the one of the Abrikosov flux line which vanishes as $e^{-M_C|\vec{x}_\perp|}$ at large distances. Thus, in the spatial region $1/2M_C \leq |\vec{x}_\perp| \leq 1/M_C$ the magnetic condensate is of the order of $\langle\overline{\chi}\chi\rangle_0$ and also the Abrikosov flux line has a perceptible value. The existence of this region plays an important role for realization of confinement in the MNJL model.

References

1. P.Becher, H.Joos, Z. Phys. C 15 (1982) 343; V. Singh, D. Browne, R. Haymaker, Phys. Rev. D47 (1990) 1715.
2. Y. Nambu, G. Jona -Lasinio, Phys. Rev.122 (1961) 345; ibid. 124 (1961) 246.
3. M. Faber, A.N. Ivanov, W. Kainz, N.I. Troitskaya, Z. Phys. C 74 (1997) 721; Phys. Lett. B386 (1996) 198; M. Faber, A.N. Ivanov, A. Müller, N.I. Troitskaya, M. Zach, hep-th/9805166, submitted to Europ. Phys. Jour. C.
4. M. Zach, M. Faber, W. Kainz, P. Skala, Phys. Lett. B358 (1995) 325, hep-lat/ 9508017.

Intersection between Microscopic and Macroscopic Abelian Dominance in the Confinement Physics of QCD

Hiroko Ichie*
Department of Physics, Tokyo Institute of Technology, Tokyo, 152-8551, Japan

Hideo Suganuma
Research Center for Nuclear Physics, Osaka University, Osaka, 567-0047, Japan

ABSTRACT

We study abelian dominance for confinement in terms of the local gluon properties in the maximally abelian (MA) gauge, where the diagonal component of the gluon is maximized by the gauge transformation. We find microscopic abelian dominance on the link-variable for the whole region of β in the lattice QCD in the MA gauge. The off-diagonal angle variable, which is not constrained by the MA-gauge fixing condition, tends to be random besides the residual gauge degrees of freedom. Within the random-variable approximation for the off-diagonal angle variable, we analytically prove that off-diagonal gluon contribution $W^{\rm off}$ to the Wilson loop obeys the perimeter law in the MA gauge. The perimeter-law behavior of $W^{\rm off}$ is also confirmed using the lattice QCD simulation. This indicates macroscopic abelian dominance for the string tension.

1. Introduction

In the low-energy region of QCD, there appear interesting phenomena such as color confinement and chiral symmetry breaking reflecting the strong gauge-coupling. However, because of nonperturbative and nonabelian nature, these phenomena are difficult to treat analytically, and it is desired to extract the relevant degrees of freedom for description of infrared phenomena.

In 1974, Nambu proposed an idea that quark confinement can be interpreted using the dual version of the superconductivity.[1] In the superconductor, Cooper-pair condensation leads to the Meissner effect, and the magnetic flux is squeezed like a quasi-one-dimensional tube as the Abrikosov vortex. In this dual-superconductor picture for the QCD vacuum, the squeezing of the color-electric flux between quarks is realized by the dual Meissner effect as the result of condensation of color-magnetic monopoles. However, there are two following large gaps between QCD and the dual superconductor picture; 1) This picture is based on the abelian gauge theory, while QCD is a nonabelian gauge theory. 2) The dual-superconductor scenario requires condensation of magnetic monopoles as the key concept, while QCD does not have such a

*E-mail : ichie@th.phys.titech.ac.jp

monopole as the elementary degrees of freedom. As the connection between QCD and the dual superconductor scenario, 't Hooft proposed the concept of the abelian gauge fixing[2] with assumption of abelian dominance for the infrared QCD. The abelian gauge fixing is defined so as to diagonalize a suitable gauge-dependent variable $\Phi[A_\mu(x)]$ and reduces QCD into an abelian gauge theory, where the off-diagonal element of the gluon field behaves as a charged matter field. Moreover, in the abelian gauge, color-magnetic monopoles appear as topological objects corresponding to the nontrivial homotopy group $\Pi_2(\mathrm{SU(N_c)/U(1)^{N_c-1}}) = \mathbf{Z}_\infty^{\mathrm{N_c}-1}$. If monopole condenses, the scenario of color confinement by the dual Meissner effect would be realized in QCD. In this paper, with the help of the lattice QCD simulation, we study intersection between abelian dominance of the gluon field (microscopic variable) and confinement force (macroscopic variable) as the theoretical basis of dual superconductor picture.

2. Microscopic Abelian Dominance in the Maximally Abelian Gauge

Abelian dominance on the confinement force have been investigated using the lattice QCD simulation in the maximally abelian (MA) gauge.[4] In terms of the link variable $U_\mu(s) \equiv U_\mu^0(s) + i\tau^a U_\mu^a(s)$, the MA gauge fixing is defined by maximizing $R \equiv \sum_{s,\mu} \mathrm{tr}\{U_\mu(s)\tau_3 U_\mu^\dagger(s)\tau_3\} = \sum_{s,\mu}\{(U_\mu^0(s))^2+(U_\mu^3(s))^2-(U_\mu^1(s))^2-(U_\mu^2(s))^2\}$ through the gauge transformation. In the MA gauge, the off-diagonal components, U_μ^1 and U_μ^2, are forced to be small, and therefore the QCD system seems describable only by U(1)-like variables approximately. The MA gauge is a sort of the abelian gauge, because the MA gauge fixing diagonalizes $\Phi(s) \equiv \sum_{\mu,\pm} U_{\pm\mu}(s)\tau_3 U_{\pm\mu}^\dagger(s)$ with $U_{-\mu}(s) \equiv U_\mu^\dagger(s-\mu)$. In this section, we study abelian dominance on the link variable $U_\mu(s)$.

In the lattice formalism, the SU(2) link variable U_μ(s) is factorized as

$$U_\mu(s) = \begin{pmatrix} \cos\theta_\mu(s) & -\sin\theta_\mu(s)e^{-i\chi_\mu(s)} \\ \sin\theta_\mu(s)e^{i\chi_\mu(s)} & \cos\theta_\mu(s) \end{pmatrix} \begin{pmatrix} e^{i\theta_\mu^3(s)} & 0 \\ 0 & e^{-i\theta_\mu^3(s)} \end{pmatrix} \equiv M_\mu(s)u_\mu(s).$$

Here, the $\mathrm{U(1)_3}$ link-variable $u_\mu(s)$ corresponds to the diagonal gluon part and behaves as the abelian gauge filed in the MA gauge, while $M_\mu(s)$ corresponds to the off-diagonal gluon part.

In order to investigate abelian dominance on the link variable in the MA gauge, we define "abelian projection rate"[4] as $R_{\mathrm{Abel}} = \cos\theta_\mu(s) \in [0,1]$ with $0 \le \theta_\mu \le \frac{\pi}{2}$. For instance, the SU(2) link variable becomes completely diagonal if $\cos\theta = 1$, while it becomes off-diagonal if $\cos\theta = 0$. In Fig.1, we show local abelian projection rate R_{Abel} expressed by the arrow $(\sin\theta, \cos\theta)$ in a typical configuration of the lattice QCD. In the MA gauge, most of all SU(2) link variables become U(1)-like. For the quantitative argument, we show in Fig.2 the probability distribution $P(R_{\mathrm{Abel}})$ of the abelian projection rate R_{Abel}. Without gauge fixing, one finds the average $\langle R_{\mathrm{Abel}}\rangle = \frac{2}{3}$. In the MA gauge, the off-diagonal component of the SU(2) link variable is forced to be reduced, and R_{Abel} approaches to unity; one obtains $\langle R_{\mathrm{Abel}}\rangle_{\mathrm{MA}} \simeq 0.93$ on 16^4 lattice with $\beta = 2.4$. Thus, we find *microscopic abelian dominance* on the link variable.

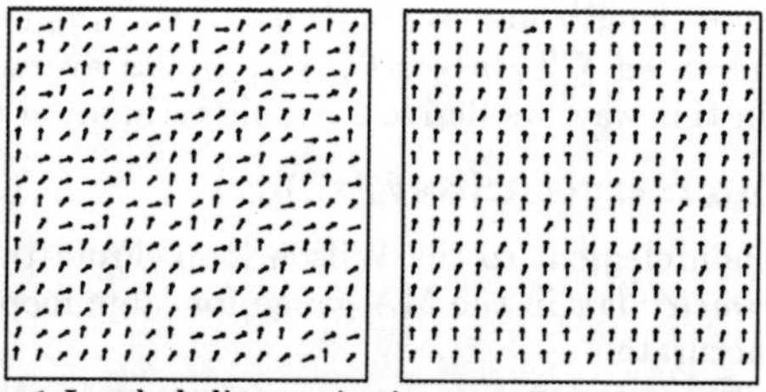

Fig.1 Local abelian projection rate $R_{\rm Abel} \equiv \cos\theta$ ($0 \le \theta_\mu \le \frac{\pi}{2}$) at β= 2.4 on 16^4 lattice without gauge fixing (left) and in MA gauge fixing (right). The arrow expresses $(\sin\theta, \cos\theta)$.

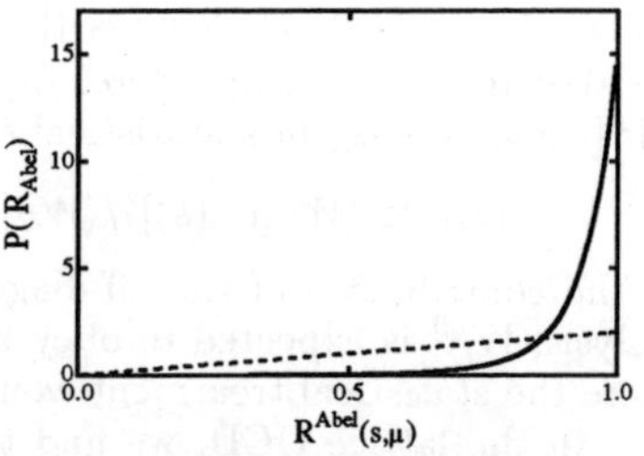

Fig.2 The probability distribution $P(R_{\rm Abel})$ of abelian projection rate $R_{\rm Abel}$ at $\beta = 2.4$ on 16^4 lattice in the MA gauge (solid curve) and without gauge fixing (dashed curve).

3. Semi-analytical Proof of Abelian Dominance for Confinement

In the MA gauge, the diagonal element $\cos\theta_\mu(s)$ in $M_\mu(s)$ is maximized by the gauge transformation. Then, the off-diagonal element $e^{i\chi_\mu(s)}\sin\theta_\mu(s)$ is forced to take a small value in the MA gauge, and therefore the approximate treatment on the off-diagonal element would be allowed in the MA gauge. Moreover, the angle variable $\chi_\mu(s)$ is not constrained by the MA gauge-fixing condition at all, and tends to take a random value[4] besides the residual $U(1)_3$ gauge degrees of freedom. Hence, $\chi_\mu(s)$ in the MA gauge can be regarded as a *random angle variable* in a good approximation. In this section, we first investigate properties of $\chi_\mu(s)$ using the lattice QCD simulation, and then study the origin of abelian dominance on confinement.

We examine the randomness of $\chi_\mu(s)$ using the lattice QCD simulation in the MA gauge with $U(1)_3$ Landau-gauge fixing.[4] We show in Fig.3 the probability distribution $P(\Delta\chi)$ of the correlation $\Delta\chi(s) \equiv {\rm mod}_\pi|\chi_\mu(s) - \chi_\mu(s+\hat\nu)| \in [0,\pi]$, which is the difference between two neighboring angle variables, at β=0, 1.0, 2.4, 3.0. In the strong-region as $\beta \le 1$, $\chi_\mu(s)$ behaves as a random variable, and there is no correlation between neighboring χ_μ. On the other hand, in the weak-coupling region, the smallness of $\sin\theta_\mu$ makes off-diagonal components more irrelevant in the MA gauge, which permits the approximate treatment on $\chi_\mu(s)$. Thus, we can take the random-variable approximation for $\chi_\mu(s)$ as a good approximation in the whole region of β in the MA gauge.

Next, let us consider the Wilson loop $\langle W_C[U_\mu(s)]\rangle \equiv \langle {\rm tr}\Pi_C U_\mu(s)\rangle$ in the MA gauge. In calculating $\langle W_C[U_\mu(s)]\rangle$, the expectation value of $e^{i\chi_\mu(s)}$ in $M_\mu(s)$ vanishes as $\langle e^{i\chi_\mu(s)}\rangle \simeq \int_0^{2\pi} d\chi_\mu(s) \exp\{i\chi_\mu(s)\} = 0$ within the random-variable approximation on $\chi_\mu(s)$. Then, the off-diagonal factor $M_\mu(s)$ appearing in the Wilson loop $W_C[U_\mu(s)]$ becomes a diagonal matrix, $U_\mu(s) \equiv M_\mu(s)u_\mu(s) \to \cos\theta_\mu(s)u_\mu(s)$.

Then, for the $I \times J$ rectangular C, the Wilson loop $W_C[U_\mu(s)] \equiv \langle {\rm tr}\Pi_{i=1}^L U_{\mu_i}(s_i)\rangle$ in the MA gauge is estimated as

$$\begin{aligned}\langle W_C[U_\mu(s)]\rangle &\simeq \langle {\rm tr}\Pi_{i=1}^L \cos\theta_{\mu_i}(s_i)u_{\mu_i}(s_i)\rangle_{\rm MA} = \langle \Pi_{i=1}^L \cos\theta_{\mu_i}(s_i)\cdot {\rm tr}\Pi_{j=1}^L u_{\mu_j}(s_j)\rangle_{\rm MA} \\ &\simeq \exp\{L\langle \ln(\cos\theta_\mu(s))\rangle_{\rm MA}\}\ \langle W_C[u_\mu(s)]\rangle_{\rm MA}, \qquad (1)\end{aligned}$$

where $L \equiv 2(I+J)$ denotes the perimeter length and $W_C[u_\mu(s)] \equiv \mathrm{tr}\Pi_{i=1}^L u_{\mu_i}(s_i)$ the abelian Wilson loop. Here, we have replaced $\sum_{i=1}^L \ln\{\cos(\theta_{\mu_i}(s_i)\}$ by its average $L\langle \ln\{\cos\theta_\mu(s)\}\rangle_{\rm MA}$ in a statistical sense. In this way, we derive a simple estimation as

$$W_C^{\rm off} \equiv \langle W_C[U_\mu(s)]\rangle / \langle W_C[u_\mu(s)]\rangle_{\rm MA} \simeq \exp\{L\langle \ln(\cos\theta_\mu(s))\rangle_{\rm MA}\} \tag{2}$$

for the contribution of the off-diagonal gluon element to the Wilson loop. From this analysis, $W_C^{\rm off}$ is expected to obey the *perimeter law* in the MA gauge for large loops, where the statistical treatment would be accurate.

In the lattice QCD, we find that $W_C^{\rm off}$ seems to obey the perimeter law for the Wilson loop with $I, J \geq 2$ in the MA gauge (Fig.4). We find also that the lattice data of $W_C^{\rm off}$ as the function of L are well reproduced by the above analytical estimation with microscopic information on $\cos\theta_\mu(s)$ as $\langle \ln\{\cos\theta_\mu(s)\}\rangle_{\rm MA} \simeq -0.082$ for $\beta = 2.4$.

Thus, the off-diagonal contribution $W_C^{\rm off}$ to the Wilson loop obeys the perimeter law in the MA gauge, and therefore the SU(2) string-tension $\sigma_{\rm SU(2)} \equiv -\lim_{I,J\to\infty} \frac{1}{IJ}\ln\langle W_{I\times J}[U_\mu(s)]\rangle$ coincides with to the abelian string-tension $\sigma_{\rm Abel}$,

$$\sigma_{\rm SU(2)} = -2\langle \ln\{\cos\theta_\mu(s)\}\rangle_{\rm MA}\frac{I+J}{IJ} + \sigma_{\rm Abel} \quad \stackrel{I,J\to\infty}{\longrightarrow} \quad \sigma_{\rm Abel}. \tag{3}$$

Thus, abelian dominance for the string tension, $\sigma_{\rm SU(2)} = \sigma_{\rm Abel}$, can be proved in the MA gauge by approximating the off-diagonal angle variable $\chi_\mu(s)$ as a random variable.

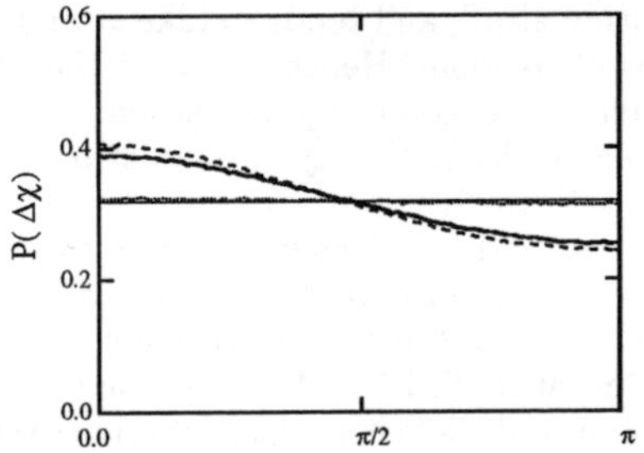

Fig.3 The probability distribution $P(\Delta\chi)$ of the correlation $\Delta\chi \equiv {\rm mod}_\pi(|\chi_\mu(s) - \chi_\mu(s+\hat\nu)|)$ in the MA gauge with $U(1)_3$ Landau-gauge fixing at $\beta = 0$ (thin line), 1.0 (dotted curve), 2.4 (solid curve), 3.0 (dashed curve).

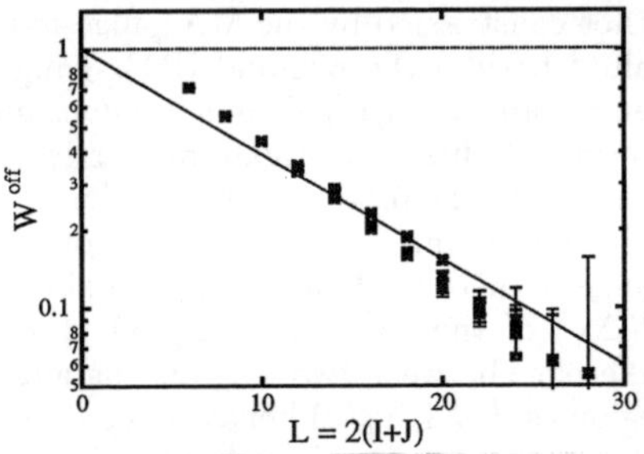

Fig.4 The comparison between the analytical estimation (straight line) and the lattice data ($\times$) of the off-diagonal gluon contribution $W_C^{\rm off}$ for the Wilson loop as the function of $L \equiv 2(I+J)$ in the MA gauge at $\beta = 2.4$.

References

1. Y. Nambu, Phys. Rev. **D10** (1974) 4262.
2. G. 't Hooft, Nucl. Phys. **B190** (1981) 455.
3. H. Ichie and H. Suganuma, Proc. of Int. Workshop on "Future Directions in Quark Nuclear Physics", 1998, (World Scientific): hep-lat/9807006.
4. H. Ichie and H. Suganuma, preprint, hep-lat/9807025,
5. H. Suganuma, H. Ichie, A. Tanaka and K. Amemiya, Prog. Theor. Phys. Suppl. **131** (1998): hep-lat/9804027, and references.

From the Lattice QCD to the Dual Ginzburg-Landau Theory

H. Suganuma, K. Amemiya and A. Tanaka
Research Center for Nuclear Physics, Osaka University, Osaka 567-0047, Japan

ABSTRACT

We study the numerical derivation of the dual Ginzburg-Landau (DGL) theory from the lattice QCD in the maximally abelian (MA) gauge on the basis of abelian dominance and monopole condensation at the infrared scale.

1. QCD and Dual Superconductor Picture for Confinement

The fundamental theory of the strong interaction is established as Quantum chromodynamics (QCD), however, it is still hard to understand the nonperturbative QCD (NP-QCD) phenomena like color confinement and dynamical chiral-symmetry breaking in the infrared strong-coupling region.

As for the confinement mechanism, Nambu proposed the *dual superconductor picture* for quark confinement, based on the electro-magnetic duality.[1] In this picture, there occurs the one-dimensional squeezing of the color-electric flux between quarks by the dual Meissner effect due to condensation of bosonic color-magnetic monopoles. However, there are *two large gaps* between QCD and the dual-superconductor picture.

(1) This picture is based on the *abelian gauge theory* subject to the Maxwell-type equations, where electro-magnetic duality is manifest, while QCD is a non-abelian gauge theory.

(2) The dual-superconductor scenario requires condensation of (color-) magnetic monopoles as the key concept, while QCD does not have such a monopole as the elementary degrees of freedom.

These gaps can be simultaneously fulfilled by taking the *'t Hooft abelian gauge fixing*,[2] the partial gauge fixing on $SU(N_c)/U(1)^{N_c-1}$. In the abelian gauge, QCD is reduced into an abelian gauge theory, where the off-diagonal gluon behaves as a charged matter field similar to $W_\mu^\pm$ in the Standard Model (SM) and provides a color-electric current in terms of the residual abelian gauge symmetry. As a remarkable fact in the abelian gauge, color-magnetic monopoles appear as topological objects reflecting the nontrivial homotopy group $\Pi_2(SU(N_c)/U(1)^{N_c-1}) = Z_\infty^{N_c-1}$ like the GUT monopole.

Thus, in the abelian gauge, QCD is reduced into an abelian gauge theory including both the electric current j_μ and the magnetic current k_μ, which is expected to provide the theoretical basis of the dual-superconductor picture for confinement.

2. Infrared Abelian Dominance in the Maximally Abelian Gauge

We summarize abelian and monopole dominances for NP-QCD (confinement, DχSB, instantons) observed in the lattice QCD in the maximally abelian (MA) gauge.

(a) Without gauge fixing, it is difficult to extract relevant degrees of freedom for NP-QCD, and all the gluon components equally contribute to NP-QCD.

(b) In the MA gauge, QCD is reduced into an abelian gauge theory including the electric current j_μ and the magnetic current k_μ. In the MA gauge, the lattice QCD shows *abelian dominance* for NP-QCD: only the diagonal gluon is relevant for NP-QCD, while off-diagonal gluons do not contribute to NP-QCD. In the lattice QCD, there appears the *global network of the monopole world-line covering the whole system in the MA gauge.*

(c) The diagonal gluon can be decomposed into the 'photon part' and the 'monopole part', corresponding to the separation of j_μ and k_μ. In the MA gauge, the lattice QCD shows *monopole dominance* for NP-QCD: the monopole part ($k_\mu \neq 0$, $j_\mu = 0$) leads to NP-QCD, while the photon part ($j_\mu \neq 0$, $k_\mu = 0$) seems trivial like QED and does not contribute to NP-QCD.

Recently, we study the gluon propagator in the MA gauge using the SU(2) lattice QCD simulation [3]. As for the residual $U(1)_3$ gauge symmetry, the $U(1)_3$ Landau gauge fixing is imposed to extract most continuous gauge configuration and to compare with the continuum theory. We find that the off-diagonal charged gluon $A_\mu^\pm(x)$ behaves as a massive field with the effective mass $m_{\rm ch} \simeq 0.9$ GeV in the MA gauge.

Then, the origin of infrared abelian dominance in the MA gauge can be interpreted in terms of the generation of the effective mass $m_{\rm ch}$ of the charged gluon induced by the MA gauge fixing. Due to the effective mass $m_{\rm ch}$, the charged gluon propagation is limited within the short range as $r \lesssim m_{\rm ch}^{-1}$ in the MA gauge, and hence charged gluons $A_\mu^\pm$ cannot contribute to the long-distance physics, which leads to abelian dominance for NP-QCD.

Here, $m_{\rm ch} \simeq 0.9$ GeV provides the *critical scale on abelian dominance* in the MA gauge. At the long distance as $r \gtrsim m_{\rm ch}^{-1}$, the charged-gluon contribution can be neglected like the weak bosons in SM and infrared abelian dominance holds. On the other hand, at the short distance as $r \lesssim m_{\rm ch}^{-1}$, charged gluons $A_\mu^\pm$ contribute as well as the abelian gluon and abelian dominance is to be broken.

3. The Dual Wilson Loop and Infrared Monopole Condensation

We next study the dual Higgs mechanism by monopole condensation in the NP-QCD vacuum in the MA gauge. Since QCD is described by the 'electric variable' as quarks and gluons, the 'electric sector' of QCD has been well studied with the Wilson loop or the inter-quark potential, however, the 'magnetic sector' of QCD is hidden and still unclear. To investigate the magnetic sector directly, it is useful to introduce the 'dual (magnetic) variable' as the *dual gluon* B_μ, which is the dual partner of the diagonal gluon and directly couples with the magnetic current k_μ.

In the MA gauge, the monopole part holds essence of NP-QCD, so that we analyze the monopole part here. Because of the absence of the electric current j_μ in the monopole part, the dual gluon B_μ can be introduced as the regular field satisfying $(\partial \wedge B)_{\mu\nu} = {}^*F_{\mu\nu}$ and the dual Bianchi identity, $\partial^\mu{}^*(\partial \wedge B)_{\mu\nu} = j_\nu = 0$. Since the dual Higgs mechanism is the screening effect on the magnetic flux, the inter-monopole potential is expected to be short-range Yukawa-type, and the dual gluon B_μ becomes massive in the monopole-condensed vacuum. To examine the inter-monopole potential, we define *the dual Wilson loop* W_D as the loop-integral of the dual gluon,

$$W_D(C) \equiv \exp\{i\frac{e}{2}\oint_C dx_\mu B^\mu\} = \exp\{i\frac{e}{2}\iint d\sigma_{\mu\nu}{}^*F^{\mu\nu}\},$$

which is the *dual version of the abelian Wilson loop* $W_{\rm Abel}(C) \equiv \exp\{i\frac{e}{2}\oint_C dx_\mu A^\mu\} = \exp\{i\frac{e}{2}\iint d\sigma_{\mu\nu}F^{\mu\nu}\}$. Here, we have set the test monopole charge as $e/2$. The potential between the monopole and the anti-monopole is derived from the dual Wilson loop as $V_M(R) = -\lim_{T\to\infty}\frac{1}{T}\ln\langle W_D(R,T)\rangle$. Using the SU(2) lattice QCD in the MA gauge, we study the dual Wilson loop and the inter-monopole potential in the monopole part.[3] Except for the short distance, the inter-monopole potential can be almost fitted by the Yukawa potential $V_M(r) = -\frac{(e/2)^2}{4\pi}\frac{e^{-m_B r}}{r}$ after removing the finite-size effect of the dual Wilson loop. In the MA gauge, the dual gluon mass is estimated as $m_B \simeq 0.5$GeV, which can be regarded as the direct evidence of the dual Higgs mechanism by monopole condensation at the infrared scale.

4. The Dual Ginzburg-Landau Theory from QCD in the MA Gauge

To summarize, the lattice QCD in the MA gauge exhibits abelian dominance and monopole condensation in the infrared region, and is described as the dual Ginzburg-Landau (DGL) theory [4] as the infrared effective theory directly based on QCD as shown in Fig.1. The DGL lagrangian without quarks is expressed as

$$\mathcal{L}_{\rm DGL} = \frac{1}{2}{\rm tr}(\partial_\mu B_\nu - \partial_\nu B_\mu)^2 + {\rm tr}[\hat{\mathcal{D}}_\mu, \chi]^\dagger[\hat{\mathcal{D}}^\mu, \chi] - \lambda {\rm tr}(\chi^\dagger\chi - v^2)^2],$$

where the last term is introduced reflecting infrared monopole condensation. Here, $\hat{\mathcal{D}}_\mu \equiv \hat{\partial}_\mu + igB_\mu$ is the dual covariant derivative with g the dual gauge coupling obeying the Dirac condition $eg = 4\pi$. The dual gauge field $B^\mu \equiv B_3^\mu T_3^{\rm dual} + B_8^\mu T_8^{\rm dual}$ is defined on the dual Cartan sub-algebra $\vec{H}^{\rm dual} = (T_3^{\rm dual}, T_8^{\rm dual})$. The QCD-monopole field is defined on the nontrivial dual root vectors $E_\alpha^{\rm dual}$: $\chi \equiv \sqrt{2}\sum_{\alpha=1}^3 \chi_\alpha E_\alpha^{\rm dual}$.

Obviously, there are *two cautions* on and the applicable scale and fake breaking of the gauge group in the DGL theory. (For typical misunderstanding of the DGL theory based on the ignorance of these important notions, see Shuryak's summary.[5])

(1) The DGL theory is the effective theory at the infrared scale as $r \gtrsim m_{\rm ch}^{-1} \simeq 0.2$fm. For the short-distance physics, off-diagonal gluons should be taken into account.

(2) The DGL theory can be constructed from QCD in the MA gauge, and only the abelian gauge group remains in a manifest form due to the MA gauge fixing, which leads to the fake breaking of ${\rm SU}(N_c)/{\rm U}(1)^{N_c-1}$.

Finally, we mention the useful criterion on the SU(N_c) gauge dependence in the DGL theory: if the variable $O^{\rm Abel}[A_\mu(x)]$ defined in the MA gauge is invariant on the residual symmetry $H \equiv {\rm U}(1)_{\rm local}^{N_c-1} \times {\rm Weyl}_{\rm global}^{N_c} \subset {\rm SU}(N_c)_{\rm local}$, $O^{\rm Abel}[A_\mu(x)]$ is proved to be SU(N_c) gauge invariant.[6] For instance, the dual gluon B_μ is Weyl-variant, so that B_μ is not color-singlet. On the other hand, the QCD-monopole fluctuation $\tilde{\chi}(x) \equiv \sum_{\alpha=1}^3\{\chi_\alpha(x) - v\}$ in the monopole-condensed vacuum in the DGL theory is H-invariant, so that $\tilde{\chi}(x)$ is color-singlet and physically appears as a scalar glueball with 0^{++}, like the Higgs scalar fluctuation in the Standard Model.

References

1. Y. Nambu, Phys. Rev. **D10** (1974) 4262.
2. G. 't Hooft, Nucl. Phys. **B190** (1981) 455.
3. H. Suganuma, H. Ichie, A. Tanaka and K. Amemiya, Prog. Theor. Phys. Suppl. **131** (1998) in press : hep-lat/9804027.
4. H. Suganuma, S. Sasaki and H. Toki, Nucl. Phys. **B435** (1995) 207.
5. For a typical amateur criticism on the DGL theory, E. Shuryak, this proceedings.
6. H. Suganuma, M. Fukushima, H. Ichie and A. Tanaka, Nucl. Phys. **B** (Proc. Suppl.) **65** (1998) 29.

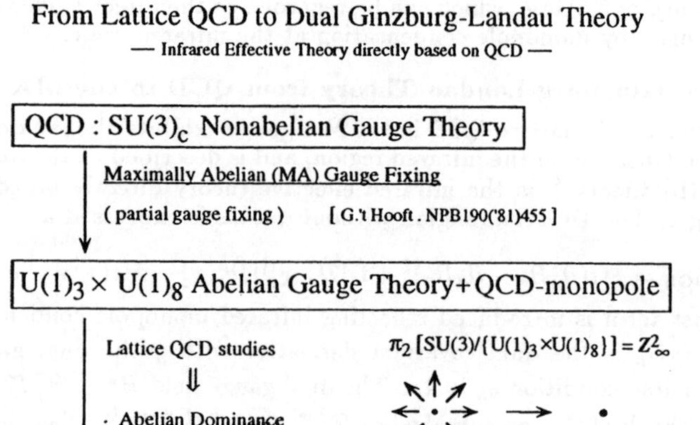

Figure 1: Construction of the dual Ginzburg-Landau (DGL) theory from the lattice QCD in the maximally abelian (MA) gauge.

HYBRID MESONS AND RELATIVISTIC STRINGS

Theodore J. Allen* and M.G. Olsson
Physics Department, University of Wisconsin-Madison
1150 University Avenue, Madison, Wisconsin, USA

and

Siniša Veseli
Theory Group, Fermi National Accelerator Laboratory
P.O. Box 500, Batavia, IL 60510

ABSTRACT

We use the relativistic string to model a QCD flux tube with fixed ends. We demonstrate that the energy of the string well approximates the lattice energies of states with the same quantum numbers recently found by Morningstar et al.

Increasingly sophisticated lattice QCD simulations provide strong evidence that QCD predicts the formation of flux tubes between quarks and antiquarks that are separated by more than a fraction of a fermi. In this talk we will present a relativistic formulation of the effective dynamics of these flux tubes.

The leading order relativistic action for the center of the flux tube should simply be the string tension times the area swept out by the world-sheet of the string.

Hybrid mesons are quark-antiquark boundstates in which the glue is in an excited state. These states are represented by the excited states of the flux tube. The first description of these states by Isgur and Paton[1] was essentially non-relativistic. More recently, Chris Michael[2] showed that his lattice simulations were consistent with the energy formula for a fixed relativistic string.

Lattice QCD allows us to ignore the quark contributions to the action and examine the dynamics of the glue alone. We give a short explanation of the string theory describing the dynamics and quantum numbers of a flux tube with fixed ends.

We start with the Nambu-Goto-Polyakov action, which is the string tension a times the area of the string world-sheet, with fixed end boundary conditions,

$$S_{\mathrm{NGP}} = -\frac{a}{2}\int d\tau \int_0^\pi d\sigma \sqrt{-h}\, h^{ab}\partial_a X^\mu \partial_b X^\nu \eta_{\mu\nu}, \quad \mathbf{X}(0,\tau) = \mathbf{x}_1, \quad \mathbf{X}(\pi,\tau) = \mathbf{x}_2.$$

Classically, this action is equivalent to the square root Nambu-Goto action. The equations of motion that follow from the action are

$$\Box X^\mu = \partial_a(\sqrt{-h}h^{ab}\partial_b X^\mu) = 0, \qquad \partial_a X^\mu \partial_b X_\mu = \frac{1}{2}h_{ab}\, h^{cd}\partial_c X^\mu \partial_d X_\mu. \tag{1}$$

---

*Presenter

There is a reparametrization invariance of the theory that must be gauge fixed. We choose the orthonormal gauge condition, $\sqrt{-h}\, h^{ab} = \eta^{ab}$, and use the residual symmetry of the action to choose our time coordinate τ in terms of the time in the rest frame of the glue, $X^0 = \frac{P^0}{\pi a}\tau$. We may solve the equations of motion (1) by making a mode expansion of the string position.

$$\mathbf{X}(\sigma,\tau) \;=\; \frac{\sigma}{\pi}\mathbf{R} - \frac{1}{\sqrt{\pi a}}\sum_{n\neq 0}\frac{\hat{\alpha}_n^k}{n}\sin(n\sigma)e^{-in\tau}, \tag{2}$$

where $\mathbf{R} = \mathbf{x}_2 - \mathbf{x}_1$. Upon quantization, the Fourier coefficients $\hat{\alpha}_n^i$ become independent oscillators obeying the commutation relations $[\hat{\alpha}_n^i, \hat{\alpha}_m^j] = m\,\delta_{n+m,0}\,\delta^{ij}$. The second equation of motion in (1) may be expressed in terms of its Fourier coefficients,

$$L_n \;=\; \sum_m \hat{\alpha}_m\cdot\hat{\alpha}_{n-m} = 0, \quad \text{with} \quad \hat{\alpha}_0 \equiv -\mathbf{R}\sqrt{\frac{a}{\pi}}. \tag{3}$$

This equation of motion is actually a constraint in the Dirac sense and will be used to define physical states. We use the Gupta-Bleuler method, in which we require that the "annihilation" parts of the constraints annihilate the physical states, to implement the constraints. The constraint conditions $L_n = 0$ are in fact second class because even classically they do not form a closed algebra.

We select the physical states from the Fock space of the oscillators. The states of the Fock space are fully described by the number of phonons in each of the modes. Thus we will have $\mathbf{N}_1, \mathbf{N}_2, \ldots$ phonons in the $n = 1, 2, \ldots$ modes, etc.

$$\begin{aligned}\text{Fock space} \;&=\; |\mathbf{R}, \mathbf{N}_1, \mathbf{N}_2, \ldots\rangle, \\ L_{-n}\,|\psi_{\text{physical}}\rangle \;&=\; 0, \quad n \geq 0.\end{aligned} \tag{4}$$

The Virasoro conditions (4) for non-zero n eliminate the "longitudinal" modes. To see this, we imagine taking $R \to \infty$ and ignoring the piece quadratic in the oscillators.

$$L_n \;\sim\; R\,\hat{\alpha}_n^L \equiv R\,(\hat{\mathbf{R}}\cdot\hat{\alpha}_n) \tag{5}$$

For ease in describing the physical states, we will assume that we can treat them as strictly transverse for any fixed separation R. The same analysis can be applied to the exact states with more work, yielding the same results. It is convenient to work in the helicity basis for the transverse phonons.

The mass-shell relation $(P^0)^2 = (aR)^2 + 2\pi a(N + c)$ follows from the zero-mode Virasoro condition.

$$\begin{aligned}L_0 \;&=\; \frac{1}{2\pi a}\left\{-(P^0)^2 + (aR)^2 + 2\pi a(N+c)\right\}, \\ N \;&=\; \sum_{n=0}^{\infty}\hat{\alpha}_{-n}\cdot\hat{\alpha}_n \;=\; \sum_{n=1}^{\infty} n(N_n^+ + N_n^-),\end{aligned} \tag{6}$$

where N_n^+ and N_n^- are respectively the number of phonons in the mode n with positive and negative helicity.

The first quantum number we examine is the angular momentum, Λ, of the glue along the quark axis. We find

$$\Lambda \;=\; \hat{\mathbf{R}}\cdot\mathbf{L} \;=\; \frac{i}{2}\epsilon_{ijk}\hat{R}^i\sum_{n=1}^{\infty}\frac{1}{n}\alpha_n^{[j}\alpha_{-n}^{k]} \;=\; \sum_{n=1}^{\infty}(N_n^+ - N_n^-). \qquad (7)$$

The second quantum number of the glue is its CP value. Charge conjugation turns a quark into an antiquark and reverses the flux. The effect is to take $\sigma \to \pi - \sigma$. Parity then simply takes a position vector to its negative. We find

$$\mathbf{X}(\sigma) \overset{CP}{\to} -\mathbf{X}(\pi-\sigma) \quad\Rightarrow\quad \alpha_n \overset{CP}{\to} (-1)^n\alpha_n \quad\Rightarrow\quad CP = (-1)^N. \qquad (8)$$

The last quantum number is the parity under reflection in a plane containing the quark axis. This is a symmetry of the system and implies a degeneracy in the $+\Lambda$ and $-\Lambda$ states. For states with $\Lambda = 0$, the states may be classified by their parity under this reflection.

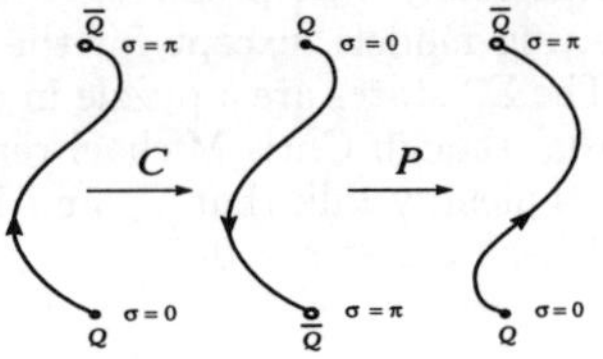

Fig. 1. The action of CP.

We use the standard molecular notation to label the various string excitation states. States with $\Lambda = 0, 1, 2, \ldots$ are denoted by $\Sigma, \Pi, \Delta, \ldots$, which is just a fancy way of saying $S, P, D, \ldots$. The state's value of CP is denoted by a subscript g or u depending on whether it is even or odd. Finally, the Σ states are superscripted with the value of the reflection parity through a plane containing the quark axis.

Colin Morningstar graciously sent us the results[3] of one of his lattice simulations for the energy of two fixed quarks. We analyzed these data by fitting the lowest state with $N = 0$ string energy plus a one gluon exchange potential,

$$V_{\Sigma_g^+} \;=\; -\frac{4\alpha_s}{3r} + \sqrt{(ar)^2 + 2\pi ac + C}. \qquad (9)$$

to obtain the best fit values

$$\begin{aligned} \alpha_s &= \;\; 0.234, & ar_0^2 &= 1.34, \\ Cr_0 &= -2.51, & c &= 0. \end{aligned}$$

We then compare the data to the now parameter-free curves

$$\begin{aligned} V &= \sqrt{(ar)^2 + 2\pi a(N+c) + C} \quad \text{(solid lines)}, \\ V &= \frac{\alpha_s}{6r} + \sqrt{(ar)^2 + 2\pi a(N+c) + C} \quad \text{(dashed lines)}. \end{aligned} \qquad (10)$$

The first potential is just a shifted string energy (lattice data do not determine absolute energies) while the second has an octet one gluon exchange potential added in. The argument for this is that as $r \to 0$ the string may continue on to a single constituent gluon plus quark-antiquark state, in which case the quarks are in a color octet state. It is important to note that these are not "fits," but are "predictions" as we have determined all the parameters from the lowest energy Σ_g^+ state. The data either fit or do not fit. Notice that there is general agreement for the states Σ^+, Π, and Δ, with the Π states being especially striking. On the other hand, the Σ^- states do not agree well.

We have shown by construction that single relativistic strings can be consistently constructed in $D = 4$ to model flux tubes. Though this theory is only an effective theory, it can be pushed quite far. We found that recent lattice data were reasonably well predicted by relativistic string models, except for the Σ^- states. The Σ^- states are a puzzle in this formalism, though Chris Michael reminds us in his plenary talk that Σ_u and Π_u must be degenerate at $r = 0$.

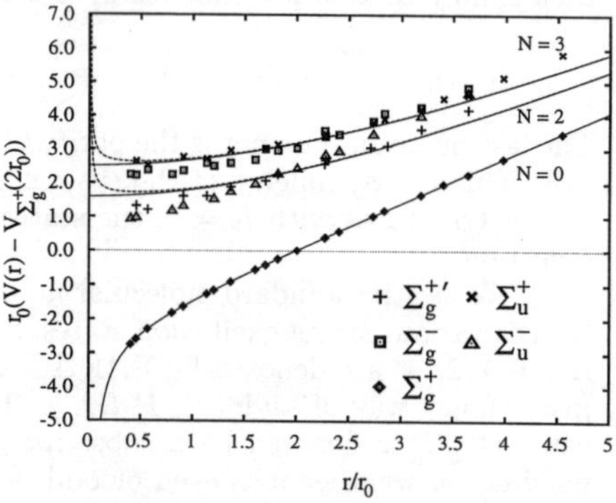

Fig. 2. The Σ states. ($\Lambda = 0$)

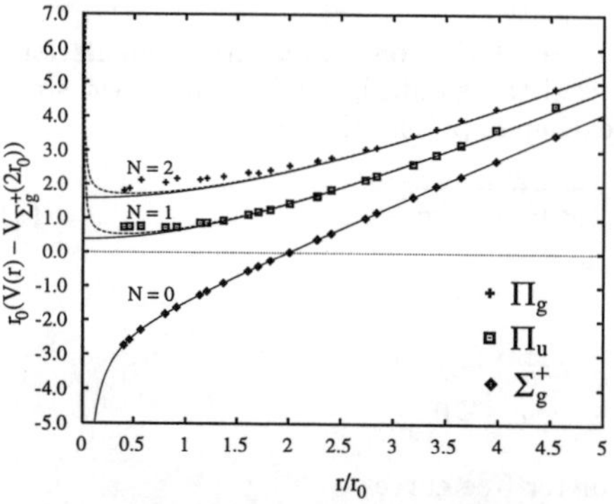

Fig. 3. The Π states. ($\Lambda = 1$)

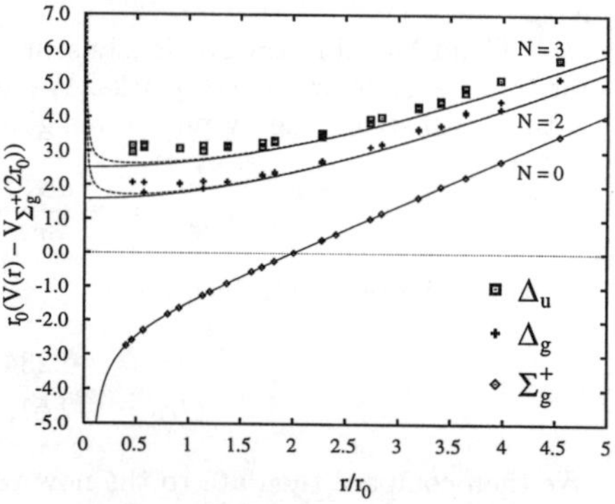

Fig. 4. The Δ states. ($\Lambda = 2$)

References

1. N. Isgur and J. Paton, Phys. Rev. **D31**, 2910 (1985).
2. S. Perantonis and C. Michael, Nucl. Phys. **B347**, 854 (1990).
3. K.J. Juge, J. Kuti, and C.J. Morningstar, Proceedings of the XVth International Symposium on Lattice Field Theory, Nucl. Phys. B (Proc. Suppl.) **63**, 326 (1998); hep-lat/9709131.

OPTIMIZED δ-EXPANSION IN QCD; A CHALLENGE

Dieter Gromes
Institut für Theoretische Physik, Philosophenweg 16,
D-69120 Heidelberg, Germany.
e-mail: d.gromes@thphys.uni-heidelberg.de

ABSTRACT

After briefly recalling the general idea of the optimized δ-expansion, we formulate an approach for QCD. The Lagrangian is split into a free and an interacting part in such a way that an arbitrary free propagator shows up which has to be fixed later. Manifest gauge invariance is maintained through the introduction of a string. The gluon vacuum polarization in one loop order consists of 20 contributions now, it is strictly transversal as it should. Several problems arise. The renormalized propagator is not automatically finite, which, however, appears not as a serious problem. A calculation of the β-function is much simpler than that of the propagator. The application of the principle of minimal sensitivity is problematic, a two loop calculation appears necessary to obtain non trivial results. We did not get a break through up to now, but hope that the hitherto extremely successful method of the optimized δ-expansion will also lead to interesting non perturbative results in gauge theories.

1. A Toy Model

The general method of the optimized δ-expansion is best illustrated for a simple example like the integral

$$z(g) = \frac{1}{\sqrt{2\pi}} \int_{\infty}^{\infty} \exp(-\frac{x^2}{2} - gx^4) dx. \quad (1)$$

It is well known that naive perturbation theory with respect to the "coupling constant" g gives a series with convergence radius zero. A more clever expansion is obtained by writing the exponent as

$$-\lambda \frac{x^2}{2} + \delta[(\lambda - 1)\frac{x^2}{2} - gx^4]. \quad (2)$$

This agrees with the original one for δ=1. The expansion is now performed around the "free term" $-\lambda x^2/2$, where λ is an arbitrary parameter for the moment, the formal expansion parameter is δ, which is put equal to one at the end.

The essential point is the following: While the exact result is, of course, independent of the artificially introduced parameter λ, any truncated approximation will depend on it. For instance, in first order in δ, after putting $\delta = 1$, one immediately obtains the approximation

$$z_1(g;\lambda) = \frac{3}{2}\lambda^{-1/2} - \frac{1}{2}\lambda^{-3/2} - 3g\lambda^{-5/2}. \tag{3}$$

Two principles are at our disposal to find the "optimal" value of λ to be used. The "principle of minimal sensitivity" (PMS) requires that the result depends as little upon λ as possible, i.e. one should look for an extremum of $z_n(g;\lambda)$ with respect to λ. In our example this leads to $\lambda_1 = [1 + \sqrt{1+40g}]/2$. The principle of "fastes apparent convergence" (FAC) postulates that the approximation z_n coincides with the lower order result z_{n-1}, which, in our case, is $1/\sqrt{\lambda}$. This would lead to $\lambda_1 = [1 + \sqrt{1+24g}]/2$. Obviously λ becomes a non trivial function of g. So perturbation theory plus a little bit of school mathematics leads to a non perturbative result! If one goes to higher orders in δ (even orders have no extremum here, the simplest thing is to restrict to odd orders) one has to determine the extremum for λ_n again. The fact that λ_n depends on the order leads to a sequence of approximations which, for simple models like the present one or the anharmonic oscillator, converges uniformly and very rapidly to the exact value.[1] In fig. 1 one may see the impressive result already obtained in first order.

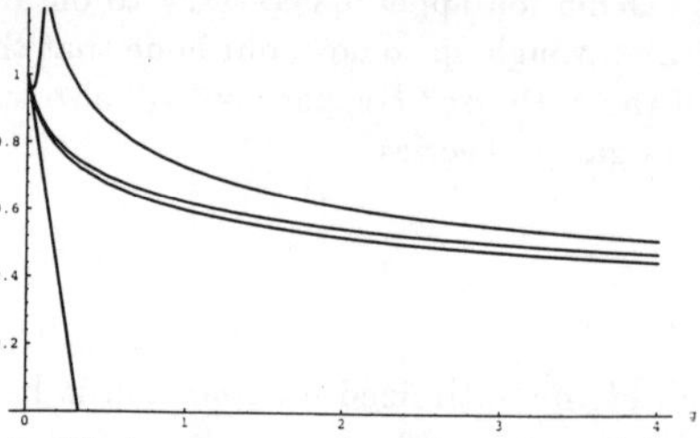

Figure 1: Various approximations for $z(g)$ in (1):
From bottom to top: First order weak coupling expansion, first order optimized δ-expansion, exact result, second order weak coupling expansion, first order strong coupling expansion.

2. Yang-Mills-Theory

The applications in field theory up to now were rather modest with respect to the splitting of the Lagrangian into a free and an interaction part. Only the mass parameter or the whole free Lagrangian was scaled (see e.g.[2] for references). The full power of the method can, however, be expected when starting with the most general free action. In Φ^4-theory, e.g. this would mean to use the non local expression $S_0 = (1/2)\int\int \Phi(x)K(x-y)\Phi(y)dxdy$, with $K(x-y)$ an essentially arbitrary form factor. In gauge theories the situation is clearly more complicated. To keep manifest gauge invariance for every δ, one has to connect the fields by strings when taking the arguments apart. We use the string in the octet representation,

$$U(x, y; \delta) = P[\exp i\delta g_0 \int_y^x T^a A_\mu^a(z) dz^\mu], \tag{4}$$

and generally replace g_0 by δg_0, not only here but also in the definition of the field strength tensor. We then write the classical YM action (which is recoverd for $\delta = 1$) as

$$S^{(cl)} = -\frac{1}{4}\int [(1-\delta^2)K(x-y) + \delta^2 \delta(x-y)] F_{\mu\nu}^a(x;\delta) U^{ab}(x,y;\delta) F^{\mu\nu b}(y;\delta) dx dy. \tag{5}$$

The free action is the part which survives for $\delta = 0$. Gauge fixing can be done essentially in the usual way. The free propagator derived from (5) is

$$D_{\mu\nu}^{ab}(q) = \delta^{ab} D(q)(g_{\mu\nu} - \xi \frac{q_\mu q_\nu}{q^2}), \quad \text{with} \quad D(q) = \frac{1}{q^2 K(q)}, \tag{6}$$

and $K(q)$ the Fourier transform of $K(x)$ in (5). Expanding the action $S = S_0 + \delta S^{(1)} + \delta^2 S^{(2)} + \cdots$, one obtains an infinite series originating from the expansion of the string $U(x, y; \delta)$. The various terms are best visualized in graphical form refering to momentum space (fig. 2).

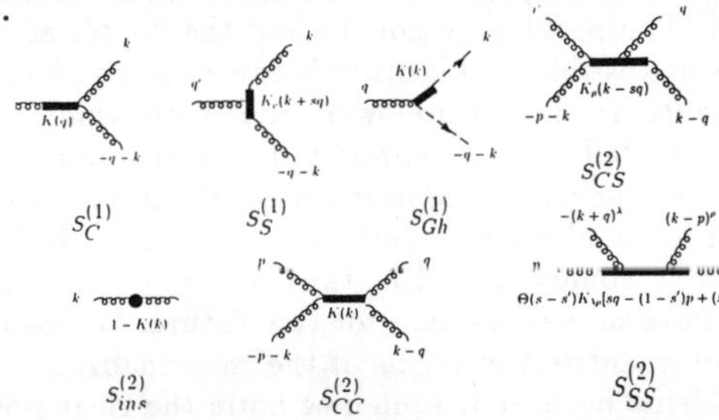

Figure 2: Interaction terms in momentum space

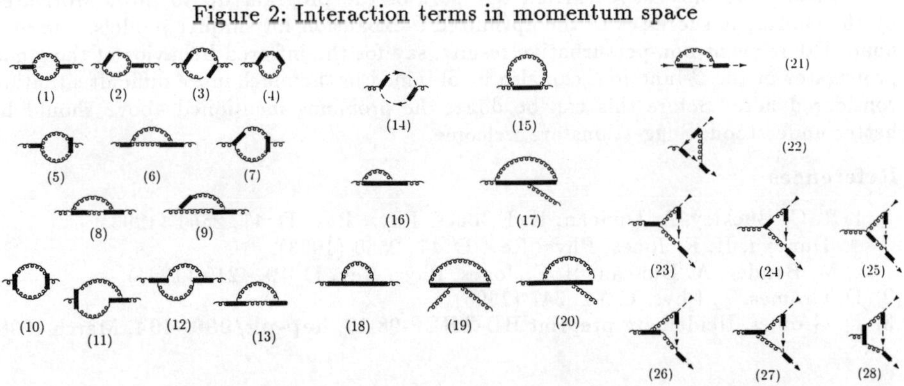

Figure 3: Gluon vacuum polarization, ghost self-energy, and ghost-gluon vertex

A thick line denotes the presence of a form factor $K(q)$, thick lines with one or more gluons attached at the interior of the line arise from the expansion of the string and involve an additional integration over one or more string parameters s, s' from 0 to 1. One may next calculate the gluon vacuum polarization, the ghost self-energy and the ghost-gluon vertex in one-loop order. The graphical representations are shown in fig. 3, for details and the analytical expressions we have to refer to.[3]

As a consistency test one can check the transversality of the vacuum polarization which arises in a rather tricky way; the string contributions are essential.

3. Problems and Perspectives

One can next make an ansatz for $K(q)$ or $D(q)$ (which are related by (6)), e.g. in terms of a dispersion relation, and calculate whatever one is interested in. Generally, renormalized quantities are now not necessarily finite. We don't consider this as a serious problem because we know from ordinary perturbation theory that divergences finally have to cancel, therefore one may drop divergent contributions by some definite description. An explicit calculation of the gluon propagator will, however, become rather complicated.

It is much easier to calculate the β-function, because only divergent contributions are relevant. This was carried through in.[3] The result is independent of the gauge parameter ξ as it should. The problem is now to find the "optimal" form factor $K(q)$. An application of FAC is impossible in this order because the one-loop calculation is the lowest non-trivial contribution; there is no lower order with which one could compare. An application of PMS also fails. In the lowest non-trivial order considered here, the extremum should appear for the usual free propagator, the next (two-loop) order would then lead to a modification and to non perturbative results. But the extremum of the one loop calculation is somewhere else, thus not reproducing the lowest order perturbative result. We have no explanation for this failure. We also mention that the rewriting of renormalized quantities in terms of the renormalized coupling constant g causes conceptional problems because it somehow blurs the clear power counting in δ.

Clearly we only have carried out part of the program up to now. Motivated by the enormous successes of the optimized δ-expansion for simpler models, one may hope that relevant non-perturbative results, say for the infrared behavior of the gluon propagator or the β-function, can also be obtained in the much more difficult situation considered here. Before this can be done, the problems mentioned above should be better understood. Suggestions are welcome.

References

1. I. R. C. Buckley, A. Duncan, H. F.Jones, Phys. Rev. **D 47**, 2554 (1993);
   A. Duncan, H. F. Jones, Phys. Rev. **D 47**, 2560 (1993),
   C.M. Bender, A. Duncan, H. F. Jones, Phys. Rev. **D 49**, 4219 (1994).
2. D. Gromes, Z. Phys. **C 71**, 347 (1996).
3. D. Gromes, Heidelberg preprint HD-THEP-98-13, **hep-ph/9803494**, March 1998.

GLUEBALL SPECTRUM FROM A QCD–MOTIVATED MODEL HAMILTONIAN*

P.O. Hess, A. Weber, C.R. Stephens, S.A. Lerma H. and J.C. López V.†
Instituto de Ciencias Nucleares, UNAM,
Circuito Exterior C.U., A.P. 70–543, 04510 México D.F., Mexico

ABSTRACT

We approximate the dynamics of low–energy pure QCD by an effective Hamiltonian depending only on the constant modes. Its eigenstates can be classified according to the irreducible representations of an $O(8)$ group. The glueball spectrum as measured on the lattice is well reproduced, and the level ordering with respect to spin is naturally explained.

In the last few years pure QCD lattice calculations have improved to the point where a substantial part of the glueball spectrum has become visible.[1,2] Our theoretical understanding of glueballs is, however, still quite rudimentary. One would like to have a simple model which at least explains the qualitative features of the spectrum, in particular the apparently random ordering with respect to angular momentum. In this contribution we will propose such a simple model which relies heavily on a group–theoretical analysis of many–gluon systems. A much more detailed account of the model will be presented elsewhere.[3]

The first step towards the effective description we have in mind is to restrict the dynamics (where we consider pure gluonic QCD throughout) to the spatially constant modes of the gluonic field. All other modes are considered as "integrated out", which of course is impossible to do in any strict sense with the presently available methods. Our approach is motivated by the work of Lüscher and van Baal and Koller in finite volumes.[4] In principle, the energies of the low–lying states in the effective description can be shown to tend towards the energies of the zero–momentum states in the infinite–volume limit.

As our goal is a Hamiltonian description, we will use the temporal gauge $A^0_a = 0$ (a the color index) and impose Gauss' law, $G_a \Psi(A) = 0$, on the physical states. Here G_a are the generators of *global* (time–independent) gauge transformations. Using an adiabatic approximation we arrive at an effective Hamiltonian of the form

$$H_{eff} = -\frac{1}{2B} \frac{\partial^2}{(\partial A^i_a)^2} + V_{eff}(A^i_a) \, , \tag{1}$$

*This work, as well as A.W.'s participation in the conference, was supported by CONACyT grant 3298P–E9608.

†e–mail addresses: hess@theo.physik.uni–giessen.de, axel@nuclecu.unam.mx, stephens@nuclecu.unam.mx, alerma@nuclecu.unam.mx, vieyra@pythia.nuclecu.unam.mx

where the constant B will be determined by fitting to the lattice glueball spectrum.

The next important step consists in introducing appropriate "collective" coordinates that reflect the symmetries of the theory. We choose the parametrization

$$A_a^i = \sum_{k=1}^{3} \varrho_k D_{ki}(\theta)\Delta_{k+5,a}(\phi)\,, \tag{2}$$

where D is a 3×3 rotation matrix in 3-space, while Δ is an 8×8 rotation matrix in color space, of which only the last three rows appear in (2). Consequently, the angles θ describe spatial rotations and the angles ϕ rotations in color space, where the parametrization of Δ may be chosen in such a way that only 18 different angles ϕ appear. Out of these, 8 are associated with (global) gauge transformations.

The intensity distribution of the gluon field is described by the gauge invariant tensor $q^{ij} = A_a^i A_a^j$. While the spatial orientation of the distribution is given by the angles θ, its moments are characterized by the variables ϱ_k, $k = 1,2,3$. Writing

$$\varrho_k^2 = \frac{\varrho^2}{3}\left[1 + 2b\cos\left(c - \frac{2\pi}{3}k\right)\right]\,, \tag{3}$$

ϱ^2 can be identified with the monopole moment, or total intensity of the distribution, while b describes the absolute deformation and c the deviation from axial symmetry associated with the quadrupole moment.

We furthermore *assume* that the potential V_{eff} can be approximated by a function of ϱ^2 only. We parametrize it in the vicinity of its minimum by a linear combination of the functions $1/\varrho^2$ and ϱ^2, where the former essentially serves to shift the position of the minimum. The coefficients of the linear combination will again be determined by a fit to the lattice results. In terms of the new variables, the effective Hamiltonian can be written as

$$H_{eff} = -\frac{1}{2B}\frac{\partial^2}{\partial\varrho^2} + \frac{\mathbf{R}^2}{2\varrho^2} + \frac{C}{2}\varrho^2\,, \tag{4}$$

where $\mathbf{R}^2$ represents an operator in the variables b, c, θ and ϕ.

In order to diagonalize H_{eff} we follow closely a procedure used in the context of nuclear collective models.[5] It basically consists of expanding H_{eff} in powers of b and $1/\sigma^2$, where $\sigma^2 = \omega_1 + \omega_2 + \omega_3 + 6$ (see below). The corrections to the lowest-order contributions will be hidden in the redefinition ("renormalization") of the parameters appearing in these contributions. As a result of this procedure the new effective Hamiltonian can be diagonalized analytically.

It is, however, far from trivial to implement Bose statistics and the color zero condition (Gauss' law) for physical states. It turns out to be most convenient to use a classification scheme for many-gluon states based on the following group chain:

$$\begin{array}{ccccc} U(24) & \supset & U(8) & \times & U(3) \\ & & \cup & & \cup \\ U(3)' & \approx & O(8) & & SU(3)_J \\ \cup & & \cup & & \cup \\ SO(3)' & & SU(3)_C & & SO(3)_J \end{array} \tag{5}$$

The group $U(24)$ arises as a consequence of the formally $24 = 8 \times 3$ (color $\times$ space) degrees of freedom of the gluon field. Bose statistics is enforced by choosing a totally symmetric irreducible representation (irrep) of $U(24)$. We denote this irrep by its Young tableau $[N]$, where N is the number of constituent gluons, which turns out to be a good quantum number within our model. Due to the symmetry the irreps of the subgroups $U(8)$ and $U(3)$ possess the *same* Young tableaux $[h_1 h_2 h_3]$, $h_1 + h_2 + h_3 = N$. The reduction to $SO(3)_J$ then determines the different possible values of J, the angular momentum of the state. The possible $O(8)$ irreps are parametrized by $(\omega_1 \omega_2 \omega_3 0)$, where $\omega_k \le h_k$, $k = 1, 2, 3$. They are constrained by the condition that they contain a color singlet, i.e. a $(0, 0)$ irrep of $SU(3)_C$. The irreps of $O(8)$ play a central rôle in the classification as they determine the parity and charge conjugation of the states.[3] The approximate equality of $O(8)$ and a $U(3)'$ group refers to the equality of the matrix elements of the last three generators of $O(8)$, between states of the above type, and the matrix elements of the generators of the $SO(3)'$ subgroup of $U(3)'$, between states of the $[\omega_1 \omega_2 \omega_3]$ irrep of $U(3)'$ (with the same values of ω_k), in the limit $\sigma^2 \to \infty$.[5]

The $O(8)$ irreps, together with the quantum number $\mathcal{L}$ of the $SO(3)'$ group, determine the ϕ–dependence of the eigenstates of H_{eff}. We can interpret this part of a state as $\omega_1 + \omega_2 + \omega_3$ ($\neq 1, 2$) constituent gluons coupled to color zero and angular momentum $\mathcal{L}$. The rest of the constituent gluons are coupled to pairs of color zero and angular momentum zero or two, corresponding to excitations of the monopole or quadrupole moment of the intensity distribution, respectively. Both kinds of excitations are described by harmonic oscillators and characterized by the eigenvalues, n_ϱ and N_2, of the respective number operators. The number of constituent gluons can then be expressed as $N = \omega_1 + \omega_2 + \omega_3 + 2(n_\varrho + N_2)$, and the angular momentum of the state is obtained as the (angular momentum) sum of $\mathcal{L}$ and L, the total angular momentum of the spin–two gluon pairs.

If we assume that the energy of the states rises with the number of constituent gluons, our simple model already gives us a *qualitative* explanation of the level ordering of the glueball spectrum as presently observed. This fact is somewhat remarkable as it relies only on a group theoretical analysis of the problem, and hence on the symmetries of the effective Hamiltonian, but not on the details of the latter. On a more *quantitative* level, we can use the glueball masses as measured on the lattice[1,2] to fit the six parameters appearing in H_{eff}. The resulting spectrum of H_{eff} is presented in Table 1. Note that the masses of all excited states (marked with an asterisk in Table 1) are *predicted* in the sense that the corresponding lattice values have not been used for the parameter fit. We furthermore predict glueball states 3^{+-*} at 4.69 GeV and 2^{--*} at 4.44 GeV, which should be accessible to lattice measurements in the near future.

The agreement of our model spectrum with the lattice calculations is very good considering the simplicity of the model. However, it must be said that our spectrum is in general a lot *denser* than that measured on the lattice, especially in the $PC = ++$ sector. There is however, at this moment, no discrepancy with the lattice data due to the fact that for a given J^{PC} only the lowest–lying states and a few corresponding first excitations have been measured.

$(\omega_1\omega_2\omega_3 0)$	$\mathcal{L}$ (λ,μ)	n_ϱ	N_2	J^{PC}	mass[GeV]	Ref. 1	Ref. 2
(0000)	0 (0,0)	1	0	0^{++}	1.61	1.61±.03	1.63±.03
(0000)	0 (0,0)	0	1	2^{++}	2.13	2.23±.22	2.39±.01
(0000)	0 (0,0)	0	2	0^{++*}	3.02	—	2.84±.04
(0000)	0 (0,0)	0	2	2^{++*}	3.02	—	3.29±.02
(0000)	0 (0,0)	0	3	3^{++}	3.69	3.92±.48	3.69±.04
(2200)	2 (0,2)	0	1	1^{++}	4.50	3.96±.31	4.12±.05
(1110)	0 (0,0)	0	0	0^{-+}	2.19	2.23±.37	2.59±.03
(1110)	0 (0,0)	0	1	2^{-+}	3.41	3.01±.18	3.07±.02
(1110)	0 (0,0)	1	0	0^{-+*}	3.80	—	3.64±.04
(3220)	1 (1,0)	0	0	1^{-+}	4.07	× 3.71±.39	4.18±.03
(1110)	0 (0,0)	0	2	2^{-+*}	4.30	—	3.93±.02
(1110)	0 (0,0)	0	3	3^{-+}	5.03	× 5.83±.66	4.67±.05
(2110)	1 (1,0)	0	0	1^{+-}	3.03	2.90±.26	2.94±.02
(2110)	1 (1,0)	0	1	2^{+-}	4.09	3.89±.66	4.10±.04
(2110)	1 (1,0)	0	1	3^{+-}	4.09	× 6.18±.89	3.54±.02
(4220)	0 (2,0)	0	0	0^{+-}	4.77	× 2.99±.75	4.74±.05
(2210)	1 (0,1)	0	0	1^{--}	3.38	4.36±.48	3.85±.04
(3110)	0 (2,0)	0	0	0^{--}	3.84	3.93±.48	4.94±.05
(3110)	2 (2,0)	0	0	2^{--}	3.86	3.94±.35	3.93±.02
(3000)	3 (3,0)	0	0	3^{--}	3.91	× 5.74±.89	4.13±.08

Table 1. The spectrum of our model Hamiltonian compared to the different lattice measurements. The first columns display our classification of the states — (λ,μ) refers to the irreps of the $SU(3)'$ subgroup of $U(3)'$ — and the usual J^{PC} classification. Mass values marked with a cross give upper limits only.

References

1. G.S. Bali, K. Schilling, A. Hulsebos, A.C. Irving, C. Michael and P.W. Stephenson, Phys. Lett. **B309**, 378 (1993); X.Q. Luo, Q. Chen, S. Guo, X. Fang and J. Liu, Nucl. Phys. **B** (Proc. Suppl.) **53**, 243 (1997); C. Michael, in *Confinement, Duality, and Nonperturbative Aspects of QCD*, ed. by P. van Baal, NATO ASI Series B 368 (Plenum Press, New York, 1998).
2. C. Morningstar and M. Peardon, Phys. Rev. **D56**, 4043 (1997); M. Peardon, Nucl. Phys. **B** (Proc. Suppl.) **63**, 22 (1998).
3. P.O. Hess, A. Weber, C.R. Stephens, S.A. Lerma H. and J.C. López V., *Glueball Spectrum from an Effective Hamiltonian*, in preparation.
4. M. Lüscher, Phys. Lett. **B118**, 391 (1982); Nucl. Phys. **B219**, 233 (1983); J. Koller and P. van Baal, Nucl. Phys. **B302**, 1 (1988).
5. O. Castaños, A. Frank, E. Chacón, P.O. Hess and M. Moshinsky, J. Math. Phys. **23**, 2537 (1982).

EXPLICIT QUARK-HADRON DUALITY IN 1+1 DIMENSIONS*

Richard F. Lebed
Jefferson Lab, Newport News, Virginia 23606, USA

ABSTRACT

Explicit quark-hadron duality in the limit of heavy quark mass is studied using the 't Hooft model, where both partonic and hadronic amplitudes may be computed exactly. Results for weak decays of heavy mesons are presented for both standard spectator decays, where the duality limit is convincingly approached, and annihilation decays of the valence quark-antiquark pair, where the approach to asymptotic duality is much less precise.

1. Introduction

One of the central issues in the study of confinement is quark-hadron duality, the manner in which partial widths for exclusive hadronic processes sum to the corresponding inclusive total at the partonic level. In this talk I present a brief summary of recent advances[1,2] in the study of this issue, obtained in collaboration with Benjamín Grinstein.

It is widely believed that the weak decay of a meson ("$\bar{B}$") containing a very heavy quark ("b") should be well approximated by the partonic decay of a free b. From the physical point of view, the heavy b quark is indifferent to the dynamics of the light spectator antiquark, and this reasoning is built into nonrelativistic quark models, as well as heavy quark effective theory and the "practical version"[3] of the operator product expansion (OPE), which expands in powers of $1/m_b$. Exactly how this "duality limit" is achieved, or how closely it is satisfied, are open questions.

2. Method of Calculation

We address these issues in terms of a soluble toy theory chosen to resemble real QCD as closely as possible. The calculations are performed in the context of the 't Hooft model,[4] which is the soluble theory of QCD in one space and one time dimension with an infinite number N_c of color charges. Here "soluble" means that each hadronic Green function (for example, meson wavefunctions and transition amplitudes) may be calculated exactly (albeit numerically) in terms of quark masses. In particular, the nth

*Work supported by the Department of Energy under contract No. DE-AC05-84ER40150.

meson eigenfunction ϕ_n of mass eigenvalue μ_n for quarks of masses M and m is found to satisfy

$$\mu_n^2 \phi_n^{M\overline{m}}(x) = \left(\frac{M_R^2}{x} + \frac{m_R^2}{1-x} \right) \phi_n^{M\overline{m}}(x) - \int_0^1 dy \, \phi_n^{M\overline{m}}(y) \, \mathrm{Pr} \, \frac{1}{(y-x)^2}, \tag{1}$$

where masses are written in units of the gauge coupling constant $g\sqrt{N_c/2\pi}$, since in 1+1 dimensions g has dimensions of mass, while in the large N_c limit[5] it scales as $1/\sqrt{N_c}$. The renormalized quark mass is then $m_R^2 = m^2 - 1$, while x, the sole remaining kinematic invariant, represents the fraction of the meson momentum carried by the quark in light-cone coordinates.

Meson transition amplitudes, which are needed for decay widths, may be written in terms of integral double and triple overlaps of 't Hooft wavefunctions.[6] The nonleptonic decays represented in these calculations are of the type $\bar{B} \to \pi_k \pi_m$, where, *e.g.*, π_k represents the kth radial excitation of the pion.

The partonic width, on the other hand, may be obtained through a straightforward calculation leading to a closed-form expression in terms of elliptic integrals with arguments that are functions of the quark masses. Results[1] of the hadronic and partonic calculations are compared in Fig. 1*a*.

Physics in 1+1 dimensions possesses a number of peculiar features, most notably the lack of spin (since spatial rotations are absent), as well as singularities in two-body phase space at threshold. These singularities appear as spikes in Fig. 1*a*, but remarkably do not affect the speed at which approximate duality is achieved.

That the exceptional agreement between hadronic and partonic calculations is not accidental may be verified by studying the partial width arising from each exclusive channel. In particular, in Fig.1*b* we exhibit the partial width for the decay of $\bar{B}$ into two ground-state π channels, which clearly does not saturate the result of Fig. 1*a*.

The small discrepancy between the two curves at large M has received attention in the literature. Numerically, it is well-fit by a relative correction of order $1/M$, which should not be present in expansions of the Ref.[3] type. Recent work[7] claims that such a correction is not present, partly on the basis of rigorous analytic studies of 't Hooft model solutions, but also partly on estimates of scaling behavior in hadronic threshold regions. The issue, for the moment, appears to be unresolved; however, even if Ref.[7] proves to be absolutely correct, the presence of a numerical effect *simulating* a forbidden correction may provide valuable insight into processes in 3+1 dimensions that appear to violate duality at $O(1/M)$, such as the lifetime difference between Λ_b and $\bar{B}$.

3. Annihilation Decays

In the decays studied in the previous section, the light spectator quark of the $\bar{B}$ is unaffected by the decay (except for binding effects). It is natural to consider also how well duality holds for decays in which it weakly annihilates with the b quark.[2]

In this case, one finds a new complication in that powers of $N_c \to \infty$ no longer appear homogeneously in the amplitude. The reason for this behavior is clear: There are special values of quark masses for which the $\bar{B}$ can annihilate resonantly into a single, highly excited π meson. Since the presence of each additional meson in large N_c costs

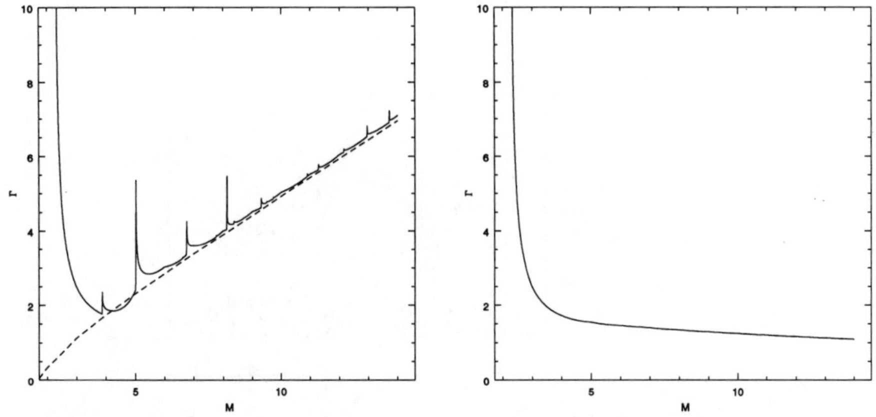

Fig. 1. (a) The full decay width for the sum of nonleptonic exclusive modes in the decay of $\bar{B}$ as a function of heavy quark mass M, with light quark mass $m = 0.56$. The overall coefficient in the width is suppressed for convenience. The dashed line is the tree-level parton result. (b) The partial hadronic width from (a) in lowest exclusive channel.

an extra factor of $\sqrt{N_c}$ in the amplitude, one-meson decay widths should dominate the expected two-meson widths by a factor of N_c. However, this enhancement appears only very near the special values of b quark mass for which such resonances are allowed, since large N_c mesons have widths $\Gamma_{\text{res}} \propto 1/N_c$.

In a physically meaningful picture, these resonances must be "integrated out" into the two-meson channel, where they appear as intermediate mesons with Breit-Wigner widths in the form

$$\frac{i}{\mu_B^2 - \mu_{\text{res}}^2 + i\mu_{\text{res}}\Gamma_{\text{res}}}. \tag{2}$$

The propagator, like meson masses, is $O(N_c^0)$ except in the immediate vicinity of $\mu_B = \mu_{\text{res}}$, where it is promoted to $O(N_c^1)$. It turns out that this is exactly the factor needed for large N_c counting consistency. On the other hand, one finds that different powers of N_c are necessarily mixed up in obtaining a physically reasonable picture of these decays, and a particular value of N_c must be chosen in order to exhibit numerical results. Since the 't Hooft model is exact for $N_c = \infty$, N_c is chosen as large as numerically practical.

The results for $N_c = 20$ are shown in Fig. 2, where the prominent cusps in Γ appear at values of M for which single-resonance decays can occur. In order to compare to the corresponding partonic rate, obviously some form of averaging, or "smearing," process is required. This is achieved here by weighting the exact result by a series of Gaussian functions whose widths are chosen as small as possible, but that still produce a smooth result. Since larger N_c means sharper spikes, while the range in M over which to smear is finite, numerically allowed choices of N_c are bounded above. One then finds

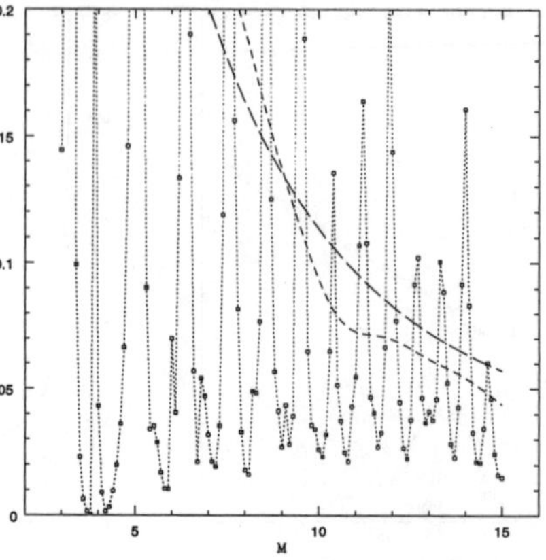

Fig. 2. Total width of $\bar{B}$ meson through annihilation decays (dotted line) as a function of heavy quark mass M, with $N_c = 20$. The width smeared by Gaussian averaging is in short dashes, while the partonic result is in long dashes.

reasonable agreement between the two curves, to about 20%, but it is not nearly so startling as that presented in Fig. 1*a*. Also, the partonic and smeared hadronic curves do not yet even appear to have achieved the same asymptotic power law behavior.

4. Conclusions

The 't Hooft model provides a unique simplified environment in which to study many of the currently intractable issues of the strong interaction. Even the issue of quark-hadron duality raises a number of interesting subtleties, whose resolution may help in fathoming the complexities of the 3+1 dimensional problem.

References

1. B. Grinstein and R. F. Lebed, Phys. Rev. D **57**, 1366 (1998).
2. B. Grinstein and R. F. Lebed, Report No. UCSD/PTH 98-14 and JLAB-THY-98-18, hep-ph/9805404 (unpublished).
3. M. A. Shifman, A. I. Voloshin, and V. I. Zakharov, Nucl. Phys. **B147**, 385 and 448 (1979); V. Novikov, M. Shifman, A. Vainshtein, and V. Zakharov, Nucl. Phys. **B249**, 445 (1985).
4. G. 't Hooft, Nucl. Phys. **B75**, 461 (1974).
5. G. 't Hooft, Nucl. Phys. **B72**, 461 (1974).
6. M. B. Einhorn, Phys. Rev. D **14**, 3451 (1976); B. Grinstein and P. F. Mende, Nucl. Phys. B **425**, 451 (1994).
7. I. Bigi, M. Shifman, N. Uraltsev and A. Vainshtein, Minnesota U. Report No. TPI-MINN-97-29-T, hep-ph/9805241 (unpublished).

FIRST PRINCIPLE $q\bar{q}$ SALPETER EQUATION, SPECTRUM AND REGGE TRAJECTORIES.

M. Baldicchi and G.M. Prosperi
Dipartimento di Fisica dell'Università di Milano and I.N.F.N.
Via Celoria,16 20133 Milano, Italy

ABSTRACT

Starting from a first principle Salpeter equation we computed the heavy-heavy, light-light and heavy-light quarkonium spectrum and the ground light-light Regge trajectories. We neglect spin-orbit structures and exclude from our treatment the light pseudoscalar states which in principle would require the use of the full Bethe-Salpeter equation due to the chiral symmetry breaking problem. We found a good agreement with the experimental meson masses and straight Regge trajectories with the right slope and intercepts. Since the parameter are completely specified by the heavy quarkonium spectrum and by high energy scattering, for what concerns light-light and light-heavy quarkonium systems our calculation is essentially parameter free.

In preceding papers[1] $q\bar{q}$ Bethe-Salpeter and q Schwinger-Dyson equations were obtained essentially from first principle in a Wilson loop context. Only the assumption $i\ln W = i(\ln W)_{\rm pert} + \sigma S_{\rm min}$ on the evaluation of the logarithm of the Wilson loop correlator as the sum of a perturbative and an area term was used. Unfortunately a direct treatment of the full B-S equation in its four dimensional form seems to be beyond the present possibilities. From it, however, by a standard method a Salpeter equation was derived with the Hamiltonian of the form

$$H = \sqrt{m_1^2 + \mathbf{k}^2} + \sqrt{m_2^2 + \mathbf{k}^2} + V, \tag{1}$$

with

$$\begin{aligned}\langle \mathbf{k}|V|\mathbf{k}'\rangle = \frac{1}{2\sqrt{w_1 w_2 w_1' w_2'}}\Big\{ &\frac{4}{3}\frac{\alpha_s}{\pi^2}\Big[-\frac{1}{\mathbf{Q}^2}\left(q_{10}q_{20} + \mathbf{q}^2 - \frac{(\mathbf{Q}\cdot\mathbf{q})^2}{\mathbf{Q}^2}\right) + \\ &+\frac{i}{2\mathbf{Q}^2}\mathbf{k}'\times\mathbf{k}\cdot(\boldsymbol{\sigma}_1+\boldsymbol{\sigma}_2) + \frac{1}{2\mathbf{Q}^2}\Big[q_{20}(\boldsymbol{\alpha}_1\cdot\mathbf{Q}) - q_{10}(\boldsymbol{\alpha}_2\cdot\mathbf{Q})\Big] + \\ &+\frac{1}{6}\boldsymbol{\sigma}_1\cdot\boldsymbol{\sigma}_2 + \frac{1}{4}\left(\frac{1}{3}\boldsymbol{\sigma}_1\cdot\boldsymbol{\sigma}_2 - \frac{(\mathbf{Q}\cdot\boldsymbol{\sigma}_1)(\mathbf{Q}\cdot\boldsymbol{\sigma}_2)}{\mathbf{Q}^2}\right) + \frac{1}{4\mathbf{Q}^2}(\boldsymbol{\alpha}_1\cdot\mathbf{Q})(\boldsymbol{\alpha}_2\cdot\mathbf{Q})\Big] + \\ &+\frac{1}{(2\pi)^3}\int d^3\mathbf{r}\, e^{i\mathbf{Q}\cdot\mathbf{r}} J^{\rm inst}(\mathbf{r},\mathbf{q},q_{10},q_{20})\Big\}\end{aligned} \tag{2}$$

and

$$J^{\rm inst}(\mathbf{r},\mathbf{q},q_{10},q_{20}) = \frac{\sigma r}{q_{10}+q_{20}}\Big[q_{20}^2\sqrt{q_{10}^2-\mathbf{q}_T^2}+q_{10}^2\sqrt{q_{20}^2-\mathbf{q}_T^2}+$$
$$+\frac{q_{10}^2q_{20}^2}{|\mathbf{q}_T|}\Big(\arcsin\frac{|\mathbf{q}_T|}{q_{10}}+\arcsin\frac{|\mathbf{q}_T|}{q_{20}}\Big)\Big]-\frac{\sigma}{r}\Big[\frac{q_{20}}{\sqrt{q_{10}^2-\mathbf{q}_T^2}}\Big(\mathbf{r}\times\mathbf{q}\cdot\boldsymbol{\sigma}_1+iq_{10}(\mathbf{r}\cdot\boldsymbol{\alpha}_1)\Big)+$$
$$+\frac{q_{10}}{\sqrt{q_{20}^2-\mathbf{q}_T^2}}\Big(\mathbf{r}\times\mathbf{q}\cdot\boldsymbol{\sigma}_2-iq_{20}(\mathbf{r}\cdot\boldsymbol{\alpha}_2)\Big)\Big]. \tag{3}$$

In eq.(1-3) the perturbative term has been evaluated only at the first order in the coupling constant α_s, the indices 1 and 2 denote the quark and the antiquark, $\mathbf{k}'$ and $\mathbf{k}$ denote the final and the initial center of mass momentum of the quark, $w_j = \sqrt{m_j^2+\mathbf{k}^2}, w_j' = \sqrt{m_j^2+\mathbf{k}'^2}, j=1,2$. σ is the string tension, $\mathbf{q} = \frac{\mathbf{k}+\mathbf{k}'}{2}$; $\mathbf{Q} = \mathbf{k}'-\mathbf{k}$; $q_{j0} = \frac{w_j+w_j'}{2}$, $q_T^h = (\delta^{hk}-\hat{r}^h\hat{r}^k)q^k$ is the transverse momentum, while α_j^k are the usual Dirac matrices $\gamma_j^0\gamma_j^k$, and σ_j^k the 4×4 Dirac spin matrices $i/4\,\varepsilon^{knm}[\gamma_j^n,\gamma_j^m]$. In the heavy masses limit this potential reduces to that discussed in ref.[2] that is known to reproduce well the spin averaged multiplets[3] in the heavy quarkonium cases and gives the correct fine and hyperfine splittings[4] if even the second order contributions in α_s are included. On the other hand if we consider only the confinement part of the Hamiltonian (1-3) in the $m_1 = m_2 = 0$ limit, setting $\mathbf{k} = \mathbf{k}'$ in the cinematic factors, we obtain $H_{\rm hqt} = 2|\mathbf{q}| + \frac{\pi}{4}\sigma r$ that gives straight Regge trajectories with slope

$$\alpha' = \frac{1}{8\frac{\pi}{4}\sigma} = \frac{1}{2\pi\sigma} \tag{4}$$

instead of the slope $\alpha' = 1/8\sigma$ given by the more commonly used linear rising potential $V = \sigma r$. The slope (4) is the same obtained in the Nambu-Goto string model and gives the experimental slope $\alpha' = 0.88$ GeV^{-2} for $\sigma \simeq 0.18$ GeV^2, which is the value used e.g. in ref.[3] All these observations make us hope that the Hamiltonian (1-3) will reproduce well both the heavy-heavy, light-light and heavy-light quarkonium spectrum and the light-light Regge trajectories. We have shown it with a direct numerical computation. Our results have been reported in ref,[5] where some details on the way followed to overcome the numerical difficulties were explained.[6,7]

We have adopted the following parameters: $\alpha_s = 0.363$, $\sigma = 0.175$ GeV^2, $m_c = 1.405$ GeV, $m_b = 4.81$ GeV, $m_s = 200$ MeV, $m_u = 10$ MeV; no *ad hoc* constant C has been added to the potential. The first four values have to be compared with those obtained from heavy quarkonium fits (e.g.[3,8]). The light quark masses are taken as the average values of the current masses reported from the Particle Data Group.[9] No attempt of optimizing the parameter was made.

As an example for the heavy-heavy systems we report in fig. 1 the $c\bar{c}$ spectrum and for the light-light systems we report in fig. 2 the $u\bar{u}$ Regge trajectories. However we obtained similar good results also for the other treated systems.[5] The results for heavy-light systems are reported in table 2 and compared with the experimental spin

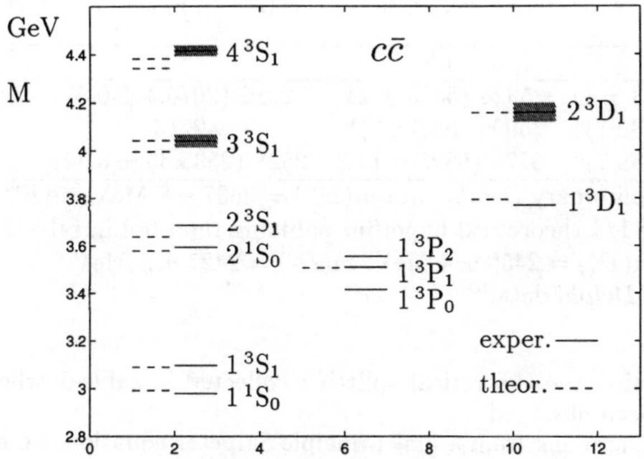

Fig. 1. $c\bar{c}$ quarkonium spectrum.

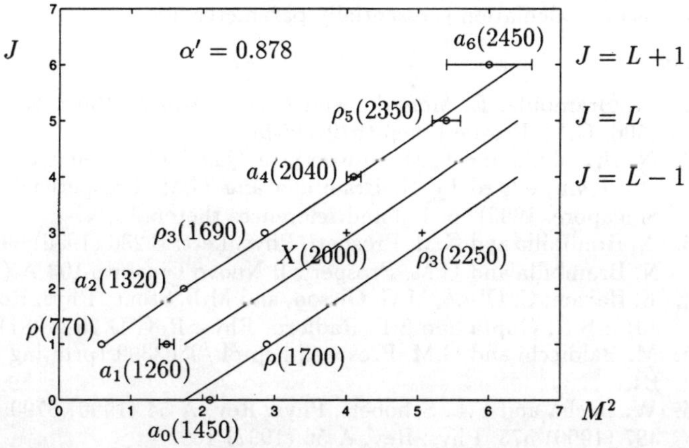

Fig. 2. Ground triplet $u\bar{u}$ Regge trajectories. Theoretical results (full line) compared with experimental data (circlet). Cross denote less established masses.

Table 1. Theoretical results for $q\bar{q}$ hyperfine splitting (MeV). Experimental data are enclosed in brackets.

State	$u\bar{c}$	ub	$c\bar{c}$	bb	$s\bar{c}$	sb
1S	111 (141 ± 1)	59 (46 ± 3)	97 (117 ± 2)	102	108 (144)	60 (47 ± 4)
2S	59	38	59 (92 ± 5)	42	62	40

Table 2. Theoretical results for $u\bar{c}$, $u\bar{b}$, $s\bar{c}$, $s\bar{b}$ systems (MeV). Experimental data are enclosed in brackets.

State	$u\bar{c}$	ub	$s\bar{c}$	sb
1S	1973 (1973 ± 1)	5326 (5313 ± 2)	2080 (2076.4 ± 0.5)	5418 (5404.6 ± 2.5)
2S	2600 (2623±?)[a]	5906 (5897±?)[a]	2713	6004
1P	2442 (2438±?)[b]	5777 (5825 ± 14)[c]	2528 (2535.35 ± 0.34)	5848 (5853 ± 15)

[a]Obtained from preliminary *Delphi* data $m(D^{*\prime}) = 2637 \pm 8$ MeV, $m(B^{*\prime}) = 5906 \pm 14$ MeV[10] subtracting 1/4 theoretical hyperfine splitting reported in table 1.
[b]Estimated from $m(D_2^*) = 2459 \pm 4$ MeV, $m(D_1) = 2427 \pm 5$ MeV.
[c]From preliminary *Delphi* data.[10]

averaged masses using the theoretical splitting collected in table 1 when the singlet state has not yet been observed.

In conclusion, starting from a first principle Salpeter equation, we have obtained a good reproduction of the meson spectrum involving heavy and light quarks with the exception of the light pseudoscalar states. Also the light-light Regge trajectories are well reproduced. Since the parameter are completely specified by the heavy quarkonium spectrum and by high energy scattering, for what concerns light-light and light-heavy quark systems our calculation is essentially parameter free.

References

1. N. Brambilla, E. Montaldi and G.M. Prosperi, Phys. Rev. D 54 (1996) 3506; G.M. Prosperi, hep-th/9709046.
2. N Brambilla and G.M. Prosperi, in *Quark Confinement and the Hadron Spectrum*, edited by N. Brambilla and G.M. Prosperi (World Scientific, Singapore, 1995), p. 113 and references therein.
3. N. Brambilla and G.M. Prosperi, Phys. Lett. B 236 (1990) 69; A. Barchielli, N. Brambilla and G.M. Prosperi, Il Nuovo Cimento 103 A (1990) 59.
4. F. Halzen, C. Olson, M.G. Olsson, and M.L. Stong, Phys. Rev. D 47 (1993) 3013; S.N. Gupta and S.F. Radford, Phys. Rev. D 24 (1981) 2309.
5. M. Baldicchi and G.M. Prosperi, hep-ph/9803390 (printing on Phys. Lett. B).
6. W. Lucha and F.F. Schöberl, Phys. Rev. A 54 (1996) 3790; Phys. Lett. B 387 (1996) 573; Phys. Rev. A 56 (1997) 139.
7. J.W. Norbury, D.E. Kahana and K.Maung Maung, Can. J.Phys. 70 (1992) 86; K.Maung Maung, D.E. Kahana and J.W. Norbury, Phys. Rev. D 47 (1993) 1182; R. Chen, L. Sorrillo and K. Maung Maung, preprint.
8. M.G. Olsson, in *Quark Confinement and the Hadron Spectrum*, edited by N. Brambilla and G.M. Prosperi (World Scientific, Singapore, 1994), p. 76 and references therein; A.Yu. Dubin, A.B. Kaidalov and Yu.A. Simonov, Phys. Lett. B 323 (1994) 41.
9. Particle Data Group, R.M. Barnett *et al.*, Phys. Rev. D 54 (1996).
10. A. Pullia (private communication).

RENORMALIZED EFFECTIVE HAMILTONIAN APPROACH TO QCD

D. G. Robertson,[1] E. S. Swanson,[1] A. P. Szczepaniak,[2] C.-R. Ji[1] and S. R. Cotanch[1]
[1]*Department of Physics, North Carolina State University, Raleigh, NC 27695, USA*
[2]*Department of Physics and Nuclear Theory Center, Indiana University, Bloomington, IN 47405, USA*

ABSTRACT

Continuing our previous QCD Hamiltonian studies in the gluonic and quark sectors, we describe a new renormalization procedure which generates an effective Hamiltonian. The formulation, which is in the Coulomb gauge, provides an improved framework for investigating hadron structure.

1. Introduction

For the past few years the NCSU group has focused upon the development of a comprehensive relativistic, many-body description of hadron structure[1,2]. This approach has many virtues, particularly the explicit incorporation of gluonic degrees of freedom and a rigorous connection to QCD at short distances. Calculations[2] of the glueball spectrum have been quite encouraging, reproducing the available lattice data.

This talk details recent results from a new renormalization procedure for the Coulomb gauge QCD Hamiltonian. There are several reasons for selecting the Coulomb gauge. First, it is completely physical; there are no spurious gauge degrees of freedom or ghosts. Second, both $q\bar{q}$ and gluon-gluon interactions appear explicitly in the canonical Hamiltonian which facilitates contact with quark model phenomenology. Finally, there appears to be a natural scenario for confinement in the Coulomb gauge, as originally suggested by Gribov[3] and recently formalized by Zwanziger[4].

This Hamiltonian must be supplemented with a regulator, which is applicable nonperturbatively and is consistent with the many-body techniques we eventually wish to employ. Next the renormalized Hamiltonian is constructed, which includes the correct cut-off dependent counterterms such that physical quantities (e.g., eigenvalues) are cut-off independent. It is inevitable that the regulators at our disposal will violate subsets of Lorentz covariance and gauge invariance, so that the structure of the counterterm Hamiltonian may be expected to be quite complicated. Due to asymptotic freedom, however, it should be reasonable to construct these operators using perturbation theory, provided the energy cut-off is not too low.

The result is a Hamiltonian formulation of QCD which is fully covariant and gauge invariant through some order in perturbation theory. To effectively apply this Hamiltonian, further approximations and phenomenological considerations are necessary. An approach we have initiated involves making a variational *Ansatz* for the physical vacuum state, based on the BCS "pairing" form. Quasi-particle degrees of freedom emerge naturally with constituent-scale masses and may be identified with constituent quarks and gluons. One then can construct approximate bound states of these constituents utilizing standard many-body techniques such as Tamm-Dancoff and RPA.

2. Regularization and Renormalization

Our starting point is the canonical QCD Hamiltonian in the Coulomb gauge $\nabla \cdot \mathbf{A} = 0$. The dynamical degrees of freedom are the transverse gauge fields $\mathbf{A}$, their conjugate momenta $\mathbf{\Pi}$, and the quark field ψ. The canonical Hamiltonian may be written as

$$H_{\rm can} = H_0 + H_{gg} + H_{qg} + H_C \; , \tag{1}$$

where H_0 is the free Hamiltonian for quarks and transverse gluons, H_{gg} contains the self-interactions between transverse gluons, and H_{qg} contains quark-gluon interactions. H_C is the Coulomb term,

$$H_C = \frac{1}{2} g^2 \int d^3x d^3y \, \mathcal{J}^{-1} \rho^a(\mathbf{x}) K^{ab}(\mathbf{x},\mathbf{y}) \mathcal{J} \rho^b(\mathbf{y}) \; , \tag{2}$$

where $\mathcal{J} = \det[\nabla \cdot \mathbf{D}]$ is the Fadeev-Popov determinant, and the kernel K is given by

$$K^{ab}(\mathbf{x},\mathbf{y}) = \langle \mathbf{x}, a | (\nabla \cdot \mathbf{D})^{-1} (-\nabla^2) (\nabla \cdot \mathbf{D})^{-1} | \mathbf{y}, b \rangle \; . \tag{3}$$

The color charge density ρ^a includes both matter and gluonic contributions, $\rho^a(\mathbf{x}) = \psi^\dagger(\mathbf{x}) T^a \psi(\mathbf{x}) + f^{abc} \mathbf{A}^b(\mathbf{x}) \cdot \mathbf{\Pi}^c(\mathbf{x})$, so that H_C represents a complicated set of instantaneous quark-quark, quark-gluon, and gluon-gluon interactions.

Perturbative schemes for renormalizing Hamiltonians typically suffer from technical problems related to the occurrence of vanishing energy denominators. One way to avoiding this difficulty is the cut-off method of Głazek and Wilson[5]. In this approach one considers the basis formed by eigenstates of the free Hamiltonian H_0, with eigenvalues E_n. The theory is then regulated by suppressing matrix elements of H between states for which the absolute value of the free energy difference $E_{nm} \equiv E_n - E_m$ is large. Specifically, we define matrix elements of the regulated Hamiltonian $H(\Lambda)$ by

$$\langle n | H(\Lambda) | m \rangle \equiv E_n \delta_{nm} + f_{mn}(\Lambda) \langle n | H_I | m \rangle \; , \tag{4}$$

where $H_I = H - H_0$ is the interaction. Here $f_{nm}(\Lambda)$ is some convenient function which is unity for $|E_{nm}| \ll \Lambda$ and vanishes for $|E_{nm}| \gg \Lambda$. In Ref.[1] we chose

$$f_{nm}(\Lambda) = e^{-E_{nm}^2/\Lambda^2} \; . \tag{5}$$

Note that, since the cut-off is defined in terms of *free* energy differences, it is not fully Lorentz covariant (though it is of course rotationally invariant). It also violates gauge invariance, though this issue is difficult to address directly in the present context since we work in a fixed gauge. The counterterm Hamiltonian will therefore also contain Lorentz- and gauge-noninvariant terms, which are necessary to correct for the violations induced by the regulator. These complications are in practice unavoidable if one wishes to work with Hamiltonians in a relativistic field theory.

At this point we have a fully regulated formulation of QCD in the Coulomb gauge. Our next task is to remove the dependence on the cut-off parameter Λ by adding counterterms to the Hamiltonian. We thus obtain the renormalized effective Hamiltonian, $H_{\rm eff}(\Lambda)$, which may then be analyzed nonperturbatively using many-body techniques.

An elegant way of determining the counterterms involves designing a similarity transformation which changes the cut-off in all matrix elements of H. Because the transformed Hamiltonian is equivalent to the original one, physical predictions (e.g., eigenvalues) are unchanged. One then searches for Hamiltonians which are form invariant, or "coherent," under this transformation. Such self-similar Hamiltonians will automatically yield cut-off independent predictions and thus represent renormalized Hamiltonians. Specifically, we write

$$H(\Lambda') = U(\Lambda';\Lambda)H(\Lambda)U^{\dagger}(\Lambda';\Lambda)\,, \tag{6}$$

where U is a unitary matrix constructed so that the matrix elements $\langle m|H(\Lambda')|n\rangle$ are all proportional to $\exp(-E_{nm}^2/\Lambda'^2)$. Construction of a suitable U is may be accomplished following the formulation of Wegner[6]. We define the operator T via

$$\frac{dU(\Lambda;\Lambda')}{d(\Lambda^{-2})} \equiv T(\Lambda)U(\Lambda;\Lambda')\,. \tag{7}$$

The evolution of H under a change in Λ is then given by

$$\frac{dH(\Lambda)}{d\Lambda^{-2}} = [T(\Lambda), H(\Lambda)]\,, \tag{8}$$

from which T is seen to be a generator of infinitesimal scale transformations. It is straightforward to show that the choice

$$T(\Lambda) = [H_0, H(\Lambda)] \tag{9}$$

results in the transformation properly changing the value of the cut-off parameter in the regulating functions used for all matrix elements of the Hamiltonian[1].

Eqn. (8) can then be analyzed order-by-order in perturbation theory by expanding the interaction $H_I = \sum_p g^p H_p$. At second order, for example, we find

$$\langle n|\delta V_2|m\rangle = \frac{1}{2}\sum_l \left(\frac{1}{E_{nl}} + \frac{1}{E_{ml}}\right)\left(e^{-2E_{nl}E_{ml}/\Lambda^2} - e^{-2E_{nl}E_{ml}/\Lambda'^2}\right)\langle n|V_1(\Lambda)|l\rangle\langle l|V_1(\Lambda)|m\rangle\,, \tag{10}$$

where $\langle n|H_p(\Lambda)|m\rangle \equiv \exp(-E_{nm}^2/\Lambda^2)\langle n|V_p(\Lambda)|m\rangle$ defines the "reduced" interaction V_p, and $\langle n|\delta V_2|m\rangle \equiv \langle n|V_2(\Lambda')|m\rangle - \langle n|V_2(\Lambda)|m\rangle$. This shift in the interaction explicitly accounts for the physics removed by changing the cut-off from Λ to Λ'.

As discussed above, requiring form-invariance of the Hamiltonian under the similarity transformation determines the counterterm structure uniquely. We have implemented this constraint and determined the complete effective Hamiltonian through second order. The operators that appear in the counterterm Hamiltonian include, as expected, all possible one-body operators which are rotationally invariant (and respect, e.g., time-reversal invariance, charge conjugation, etc.). These are a gluon "mass" term, $\mathbf{A}^2$, the operators $\mathbf{\Pi}^2$ and $(\nabla_i A_j)^2$, and, for fermions, $\bar{\psi}\psi$. Each has a calculable coefficient, except for the particular combination of $\mathbf{\Pi}^2$ and $(\nabla_i A_j)^2$ which has the structure of a gluon wavefunction renormalization. For this combined operator, the *variation* of the coefficient under changes in the cut-off is calculable, and gives the gauge field anomalous dimension: $\gamma_A = -(g^2/16\pi^2)(4C_A/3)$, where C_A is the Casimir invariant of the adjoint representation. See Ref.[1] for full details.

3. Work in Progress and Future Directions

Having obtained the renormalized Hamiltonian, we can now proceed to construct a model for the ground state and analyze excitations such as glueballs, mesons and hybrids. In Ref.[1], we focused attention on the gluonic sector and made a variational BCS vacuum *Ansatz*. This leads to a gap equation which may be solved numerically to give the spectral function of the quasi-particle constituent gluons.

Work in progress includes extending the similarity analysis to higher orders, applying this smooth cut-off treatment to the quark sector and designing improved models for the vacuum. At third order, one should see the running of the coupling, as well as the appearance of many new operators in the counterterm Hamiltonian. Going beyond BCS is expected to be particularly important in a combined treatment of the quark and gluon sectors. In addition, other composite operators, such as currents, must be properly renormalized which is also possible using the formalism presented in Ref.[1].

Finally, there are a host of phenomenological applications to be explored. Not only meson spectra but also form factors, electro-weak and hadronic decays, etc. can be predicted and confronted with available data. This comprehensive analysis should provide significant insight into hadron structure as well as a key linkage between fundamental QCD and the successful phenomenological quark model.

Acknowledgments

Financial support is provided by U. S. Department of Energy.

References

1. D. G. Robertson, E. S. Swanson, A. P. Szczepaniak, C.-R. Ji, and S. R. Cotanch, submitted to Phys. Rev. D.
2. A. Szczepaniak, E. S. Swanson, C.-R. Ji, and S. R. Cotanch, Phys. Rev. Lett. **76**, 2011 (1996); A. Szczepaniak and E. S. Swanson, Phys. Rev. D **55**, 1578 (1997); E. S. Swanson and A. Szczepaniak, hep-ph/9804219.
3. V. N. Gribov, Nucl. Phys. B **139**, 1 (1978).
4. D. Zwanziger, Nucl. Phys. B **485**, 185 (1997); A. Cucchieri and D. Zwanziger, Phys. Rev. Lett. **78**, 3814 (1997).
5. St. D. Głazek and K. G. Wilson, Phys. Rev. D **48**, 5863 (1993); *ibid.* **49**, 4214 (1994).
6. F. Wegner, Ann. Physik **3**, 77 (1994).

QUANTUM MECHANICS OF CONFINEMENT AND CHIRAL SYMMETRY BREAKING IN TWO-DIMENSIONAL QCD

Alexei Nefediev

Institute of Theoretical and Experimental Physics, Moscow 117259, Russia

ABSTRACT

The system of light quark and heavy anti-quark source is studied in 1+1 QCD in the large N_C limit. Making use of the modified Fock–Schwinger gauge allows to consider simultaneously the spectroscopical problem of the $q\bar{Q}$ bound states and the problem of the light quark Green function. The Dirac-type equation for the spectrum of the system is proved to be equivalent to the well-known 't Hooft one in the one body limit. The unitary transformation from the Dirac–Pauli representation to the Foldy–Wouthuysen one is carried out explicitly, and it is shown that the equation in the Foldy–Wouthuysen representation can be treated as a gap equation which defines the light quark self-energy in the modified Fock–Schwinger gauge. The Foldy–Wouthuysen angle is found to play the role of the Bogoliubov–Valatin one and to give the standard value of the chiral condensate. Connections of the given formalism to the standard four-dimensional QCD are outlined and discussed.

For the first time the two-dimensional model of QCD in the limit of infinite number of colours N_C was considered in 1974 by 't Hooft[1] and the celebrated equation of the same name was derived in the light-cone gauge. Four years later, in 1978, this equation was re-derived in the axial gauge.[2] So the model seems to have been studied well enough. Still it attracts considerable attention as a problem with features very much similar to those of standard four-dimensional QCD.

Besides the usual assumption $N_C \to \infty$ limit that allows to sum up only planar diagrammes, we make use of the so-called modified Fock-Schwinger or Balitsky gauge[3]

$$A_1^a(x_0, x) = 0 \quad A_0^a(x_0, 0) = 0 \tag{1}$$

As soon as gauge (1) is a kind of radial one, the gluonic field can be expressed in terms of the field strength tensor that yields the gluon propagator in the form

$$K_{00}^{ab}(x_0 - y_0, x, y) = \delta^{ab}\frac{g^2}{2}\delta(x_0 - y_0)(|x - y| - |x| - |y|) \equiv \delta^{ab}K(x, y), \tag{2}$$

and other components equal to zero.

Note that K can be naturally broken into local ($K^{(1)} \sim |x-y|\delta(x_0-y_0)$) and non-local ($K^{(2)} \sim (|x|+|y|)\delta(x_0-y_0)$) parts.

Green function for the $q-\bar{Q}$ system under consideration has the form *

$$S_{q\bar{Q}}(x,y) = \frac{1}{N_C}\int D\psi D\bar{\psi} DA_\mu \exp\left\{-\frac{1}{4}\int d^2x F_{\mu\nu}^{a2} - \int d^2x \bar{\psi}(i\hat{\partial}-m-\hat{A})\psi\right\}\times$$

$$\times\bar{\psi}(x) S_{\bar{Q}}(x,y|A)\psi(y), \tag{3}$$

where anti-quark Green function $S_{\bar{Q}}$ is introduced. The main advantage of our peculiar gauge choice is the fact that the anti-quark is decoupled completely so that $S_{\bar{Q}}$ can be substituted in a very simple form:

$$S_{\bar{Q}}(x,y|A) = S_{\bar{Q}}(x-y);\ S_{\bar{Q}} = -i\left(\frac{1+\gamma_0}{2}\theta(t)e^{-iMt} + \frac{1-\gamma_0}{2}\theta(-t)e^{iMt}\right)\delta(x). \tag{4}$$

On integrating gluon degrees of freedom in (3), we arrive at the effective Lagrangian for the light quark which leads in turn to the Schwinger–Dyson equation

$$(i\hat{\partial}_x - m)S(x,y) + \frac{iN_C}{2}\int d^2z \gamma_0 S(x,z)\gamma_0 K(x,z) S(z,y) = \delta^{(2)}(x-y), \tag{5}$$

where

$$S(x,y) = \frac{1}{N_C} S^\alpha_\alpha(x,y). \tag{6}$$

It is very instructive to note here that the role played by Green function (6) is twofold. By construction S is the Green function of the light quark, but due to a very passive part of the static anti-quark it plays the role of the Green function of the whole $q\bar{Q}$ system as well, so that the problem of the light quark Green function and the spectroscopical problem for the $q\bar{Q}$ system can be considered simultaneously. We shall get back to this statement later on while discussing the chiral properties of the model.

Approach based on spectral decomposition of the Green function (6) turns out very useful †, so one has

$$S(q_{10},q_1,q_{20},q_2) = 2\pi\delta(q_{10}-q_{20})\left(\sum_{\varepsilon_n>0}\frac{\varphi_n^{(+)}(q_1)\bar{\varphi}_n^{(+)}(q_2)}{q_{10}-\varepsilon_n} + \sum_{\varepsilon_n<0}\frac{\varphi_n^{(-)}(q_1)\bar{\varphi}_n^{(-)}(q_2)}{q_{10}+\varepsilon_n}\right). \tag{7}$$

To proceed further we assume that the Foldy-Wouthuysen operator $T(p) = e^{\frac{1}{2}\theta_F(p)\gamma_1}$ diagonalizing equation (5) exists and that angle θ_F is the same for all n.

With such an assumption applied Schwinger–Dyson equation (5) reduces to the Dirac-type equation in the Hamiltonian form ($f = \frac{g^2 N_C}{4\pi}, \alpha = \gamma_0\gamma_1, \beta = \gamma_0,$):

$$(\alpha p + \beta m)\varphi_n^0(p) - \pi f\int dq dk(\beta\cos\theta_F(q) + \alpha\sin\theta_F(q))K(p-q,k-q)\varphi_n^0(k) = E_n\varphi_n^0, \tag{8}$$

*We adopt the following γ-matrix convention[2]: $\gamma_0 = \sigma_3$, $\gamma_1 = i\sigma_2$, $\gamma_5 = \sigma_1$

†The approach based on diagrammatic technics leads to the same results[4]

where $\varphi_n^0(p)$ being scalar wave function

$$\varphi_n^{(+)}(p) = \varphi_n^0(p)T^+(p)\begin{pmatrix}1\\0\end{pmatrix}, \ \varphi_n^{(-)}(p) = \varphi_n^0(p)T^+(p)\begin{pmatrix}0\\1\end{pmatrix}. \tag{9}$$

Let us consider only local (*i.e.* generated by $K^{(1)}$) part of interaction which reduces to a mass operator Σ and can be naturally parametrized via two scalar functions $E(p)$ and $\theta(p)$ in the convenient form

$$\Sigma(p) \equiv [E(p)cos\theta(p) - m] + \gamma_1 [E(p)sin\theta(p) - p]. \tag{10}$$

Self-consistency condition for such a parametrization makes $E(p)$ and $\theta(p)$ satisfy a system of coupled equations[2]:

$$\begin{cases} E(p)cos\theta(p) = m + \frac{f}{2}\int \frac{dk}{(p-k)^2}cos\theta(k) \\ E(p)sin\theta(p) = p + \frac{f}{2}\int \frac{dk}{(p-k)^2}sin\theta(k), \end{cases} \tag{11}$$

It is easy to verify that if we identify the Foldy-Wouthuysen angle θ_F with the Bogoliubov–Valatin one θ[‡] then the non-local interaction diagonalizes as well, so that the Foldy-Wouthuysen representation of equation (8) takes the Schrödinger-type form

$$\varepsilon_n\varphi_n^0(p) = E(p)\varphi_n^0(p) - f\int \frac{dk}{(p-k)^2}cos\frac{\theta(p)-\theta(k)}{2}\varphi_n^0(k). \tag{12}$$

Equation (12) is nothing but the one-body limit of the well-known 't Hooft equation.[2]

As mentioned above, Green function of the $q\bar{Q}$ system constructed from the solutions of equation (12) is the one of the light quark as well, so that the chiral condensate can be easily calculated

$$< \bar{q}q >= -i \underset{x\to y+}{Tr}\ S(x,y) = -\frac{N_C}{\pi}\int_0^\infty dp\ cos\theta(p). \tag{13}$$

Condensate (13) does not vanish in the chiral limit $m \to \infty$ and coincides with the standard value[7,8]

$$< \bar{q}q >_{m=0} = -0.29N_C\sqrt{2f}. \tag{14}$$

A reasonable question may arise, whether it is worth reproducing old results with a new complicated method. Still the answer is positive, since the given method can be easily generalized to describe the four-dimensional QCD. The Schwinger–Dyson equation similar to (5) takes the following form in Euclidean space[5]

$$(-i\hat{\partial}_x - im)S(x,y) + \int d^4zK_{\mu\nu}(x,z)\gamma_\mu S(x,z)\gamma_\nu S(z,y) = \delta^{(4)}(x-y). \tag{15}$$

‡ $\theta(p)$ obviously plays the role of the Bogoliubov–Valatin angle as it describes the rotation from the bare particle with the free dispersion law $\sqrt{p^2+m^2}$ to the dressed "physical" particle with dispersion $E(p)$.

The main object governing the quark dynamics is the correlation function D:

$$< F^a_{\mu\nu}(x) F^b_{\lambda\rho}(y) >= \frac{\delta^{ab}}{N_C^2 - 1} D(x-y)(\delta_{\mu\lambda}\delta_{\nu\rho} - \delta_{\mu\rho}\delta_{\nu\lambda}) \tag{16}$$

with the kernel of equation (15) being proportional to D.

Function D rapidly decreases at large Euclidean distances and this decrease is governed by the gluonic correlation length T_g. In the string picture T_g defines the radius of the string formed between q and $\bar{Q}$. The two limiting cases, $mT_g \gg 1$ and $mT_g \ll 1$, should be treated separetely, as it is clearly seen from the non-relativistic expansion of the interaction in (15).

The case of large mT_g gives quite a natural result for interaction[6]

$$V(r) = \left(\frac{5}{6} + \frac{1}{6}\gamma_0\right)\sigma r + corrections, \quad V_{FW}(r) = \sigma r - \frac{\vec{\sigma}\vec{l}}{4m^2 r} + O\left(\frac{\sigma r}{mT_g}\right), \tag{17}$$

which is in agreement with the Eichten–Feinberg–Gromes results,[9,10] whereas the opposite limit of small mT_g leads to a self-inconsistency,[11] as the corrections $\sim O((mT_g)^{-2})$ and diverge in the given limit.

A natural interpretation of such results is offered by the string picture of confinement. In the case of "thick" string ($T_g \gg \frac{1}{m}$) the quark interacts with the gluonic field rather than with a formed string, so the quark dynamics is local and potential.

The opposite limit of "thin" string ($T_g \ll \frac{1}{m}$) is just the case realized in the two-dimensional 't Hooft model, where strings are infinitely thin (the system just lacks of extra transverse dimensions to allow the string to swell). The interaction is sufficiently non-local (see the integral term in the r.h.s. of equation (12)) and the same behaviour is expected in the string limit ($T_g \to 0$) of the four-dimensional QCD. Regge picture of the rotating string should follow from equation (5) in this case. Such a picture helps to resolve the well-known problem with local confinement: incorrect trajectory slope for scalar interaction *vs* Klein paradox for vector one.

References

1. G. 't Hooft, Nucl. Phys. **B72**, 461 (1974);
2. I.Bars, M.B.Green, Phys.Rev. **D17**, 537 (1978);
3. I.I.Balitsky, Nucl.Phys. **B254**, 166 (1985);
4. Yu.S.Kalashnikova, A.V.Nefediev hep-ph/9711347, Phys.At.Nucl. in press;
5. Yu.A.Simonov, Phys.At.Nucl. **60**, 2069 (1997);
6. N.Brambilla and A.Vairo, Phys.Lett. **B407**, 167 (1997);
7. Ming Li, L.Wilets and M.C.Birse, J.Phys.G: Nucl.Phys. **13**, 915 (1987);
8. A.R.Zhitnitsky, Phys.Lett. **B165**, 405 (1985);
9. E.Eichten, F.L.Feinberg, Phys.Rev **D23**, 2724 (1981);
10. D.Gromes, Z.Phys. **C26**, 401 (1984);
11. Yu.S.Kalashnikova, A.V.Nefediev, Phys.Lett. **B414**,149 (1997)

SECTION "B"

LATTICE THEORY AND NUMERICAL SIMULATIONS

GLUON EXCITATIONS OF THE STATIC-QUARK POTENTIAL

K.J. Juge, J. Kuti, and C. Morningstar*

University of California at San Diego, La Jolla, California 92093-0319

ABSTRACT

The spectrum of gluon excitations in the presence of a static quark-antiquark pair is presented. Our results are obtained from computer simulations of gluons on anisotropic space-time lattices using an improved gauge-field action. Measurements for quark-antiquark separations r ranging from 0.1 fm to 4 fm and for various orientations on the lattice are made. Discretization errors and finite volume effects are taken into account. Surprisingly, the spectrum does not exhibit the expected onset of the universal π/r Goldstone excitations of the effective QCD string, even for r as large as 4 fm.

1. Introduction

Accurate knowledge of the properties of the stationary states of glue in the presence of the simplest of color sources, that of a static quark and antiquark separated by some distance r, is an important stepping stone on the way to understanding confinement. It is generally believed that at large r, the linearly-growing ground-state energy of the glue is the manifestation of the confining flux whose fluctuations can be described in terms of an effective string theory. The lowest-lying excitations are then the Goldstone modes associated with spontaneously-broken transverse translational symmetry. Expectations are less clear for small r. The determination of the energies of glue in the presence of a static quark-antiquark pair is also the first step in the Born-Oppenheimer treatment of conventional and hybrid heavy-quark mesons.[1] A better understanding of hybrid quarkonium should provide valuable insight into the nature of light hybrid mesons, which are currently of great experimental and theoretical interest.

Even the simplest property, the energy spectrum, of the stationary states of glue interacting with a static quark-antiquark pair is not accurately known. The main goal of this work is to remedy this. Here, we present, for the first time, a comprehensive determination of the low-lying spectrum of gluonic excitations in the presence of a static quark-antiquark pair. In this initial study, the effects of light quark-antiquark pair creation are ignored. A few of the energy levels for r less than 1 fm have been studied before.[2] Our results for these quantities have significantly improved precision, and we have extended the range in r to 4 fm. Most of the energy levels presented here have never been studied before. Some of our results were previously reported.[3]

*Talk presented by C. Morningstar

2. Computation of the energies

We adopt the standard notation from the physics of diatomic molecules and use Λ to denote the magnitude of the eigenvalue of the projection $\vec{J}_g \cdot \hat{\mathbf{r}}$ of the total angular momentum $\vec{J}_g$ of the gluons onto the molecular axis $\hat{\mathbf{r}}$. The capital Greek letters $\Sigma, \Pi, \Delta, \Phi, \ldots$ are used to indicate states with $\Lambda = 0, 1, 2, 3, \ldots$, respectively. The combined operations of charge conjugation and spatial inversion about the midpoint between the quark and the antiquark is also a symmetry and its eigenvalue is denoted by η_{CP}. States with $\eta_{CP} = 1(-1)$ are denoted by the subscripts g (u). There is an additional label for the Σ states; Σ states which are even (odd) under a reflection in a plane containing the molecular axis are denoted by a superscript $+$ $(-)$. Hence, the low-lying levels are labelled Σ_g^+, Σ_g^-, Σ_u^+, Σ_u^-, Π_g, Π_u, Δ_g, Δ_u, and so on. For convenience, we use Γ to denote these labels in general.

The glue energies $E_\Gamma(\vec{r})$ were extracted from Monte Carlo estimates of generalized Wilson loops. The expectation value of the path-ordered exponential of the gauge field along a closed loop is known as a Wilson loop. The familiar static-quark potential $E_{\Sigma_g^+}(r)$ can be obtained from the large-t behaviour $\exp[-tE_{\Sigma_g^+}(r)]$ of the Wilson loop for a rectangle of spatial length r and temporal extent t. The two spatial segments of such a Wilson loop can be replaced by a complicated *sum* of spatial paths, all sharing the same starting and terminating sites, in order to extract the $E_\Gamma(\vec{r})$ energies, as long as the sum of paths transforms as Γ under all symmetry operations. Using several different sums of paths (we typically use 3 to 22 such sums) then produces a matrix of Wilson loop correlators $W_\Gamma^{ij}(r,t)$.

Monte Carlo estimates of the $W_\Gamma^{ij}(r,t)$ matrices were obtained in eight simulations using an improved gauge-field action.[4] Our use of anisotropic lattices in which the temporal lattice spacing a_t was much smaller than the spatial spacing a_s was crucial for resolving the glue spectrum, particularly for large r. To hasten the onset of asymptotic behaviour, iteratively-smeared spatial links[4] were used in the generalized Wilson loops. The temporal segments in the Wilson loops were constructed from thermally-averaged links, whenever possible, to reduce statistical noise. To facilitate the removal of finite-spacing errors by extrapolating to the continuum limit, results for several values of the lattice spacing were obtained; simulations were done using $a_s \approx 0.29, 0.27, 0.22, 0.19$, and 0.12 fm. Two runs corresponding to the same a_s but different a_t were done to provide a measure of the a_t^2 errors in our results (our action has $O(a_t^2, a_s^4)$ errors). Such information is important for carrying out the $a_s \to 0$ extrapolations. Agreement of energies obtained using different quark-antiquark orientations on the lattice was used to check the smallness of finite-spacing errors and to help identify the continuum Λ value corresponding to each level (there are only discrete symmetries on the lattice). For this reason, results were obtained not only for cases in which the molecular axis coincided with an axis of the lattice, but also for separation vectors such as $(r,r,r)/\sqrt{3}$ and $(r,r,0)/\sqrt{2}$.

The matrices $W_\Gamma^{ij}(r,t)$ were reduced in the data fitting phase to single correlators and 2×2 correlation matrices using the variational method. The lowest-lying glue energies were then extracted from these reduced correlators by fitting a single exponential and a sum of two exponentials, the expected asymptotic forms, in various ranges $t_{\min}$

to t_{max} of the source-sink separation. The two-exponential fits were used to check for consistency with the single-exponential fits, and in cases of favourable statistics, to extract the first-excited state energy in a given channel.

Three additional runs on small lattices were done to verify that finite-volume errors in our results were negligible. We confirmed the smallness of the a_s/a_t renormalization for two values of the QCD coupling by extracting the ground state potential from Wilson loops in different orientations. The hadronic scale parameter $r_0 \approx 0.5$ fm was used to determine the lattice spacing.[4] The additive ultraviolet-divergent self-energies of the static sources were removed by expressing all of our results with respect to $\Sigma_g^+(r_0)$. Finite-lattice spacing errors were removed by extrapolating our simulation results for $r_0[E_\Gamma(r) - E_{\Sigma_g^+}(r_0)]$ to the $a_s \to 0$ continuum limit. These extrapolations were carried out by fitting all of our simulation results to an ansatz $F_{\rm cont}(r) + a_s^4\, F_{\rm latt}(r)$: a ratio of a polynomial of degree $p+1$ over a polynomial of degree p, where $p = 1$ or 2, was found to work well for the continuum limit form $F_{\rm cont}(r)$, and $F_{\rm latt}(r)$ was chosen empirically to be a sum of three terms $1/\sqrt{r}$, $1/r$, and $1/r^2$. All fits yielded χ^2/dof near unity. Continuum Λ values were easily identified in all cases but one: we were unable to distinguish between a Π_u' and Φ_u interpretation for the on-axis E_u' level.

3. Results

Our continuum-limit extrapolations are shown in Fig. 1. The ground-state Σ_g^+ is the familiar static-quark potential. A linearly-rising behaviour dominates the Σ_g^+ potential once r exceeds about 0.5 fm and we find no deviations from the linear form up to 4 fm. The lowest-lying excitation is the Π_u. There is definite evidence of a band structure at large r: the Σ_g', Π_g, and Δ_g form the first band above the Π_u; the Σ_u^+, Σ_u^-, Π_u'/Φ_u, and Δ_u form another band. The Σ_g^- is the highest level at large r. This band structure breaks down as r decreases below 2 fm. In particular, two levels, the Σ_g^- and Σ_u^-, drop far below their large-r partners as r becomes small. Note that for r above 0.5 fm, all of the excitations shown are stable with respect to glueball decay. As r decreases below 0.5 fm, the excited levels eventually become unstable as their gaps above the ground state Σ_g^+ exceed the mass of the lightest glueball.

A universal feature of any low-energy description of a thin fluctuating flux tube is the presence of Goldstone excitations associated with the spontaneously-broken transverse translational symmetry. These transverse modes have energy separations above the ground state given by multiples of π/r. The level orderings and approximate degeneracies of the gluon energies at large r match, without exception, those expected of the Goldstone modes. However, the precise $m\pi/r$ gap behaviour is not observed (see Fig. 1). For separations less than 2 fm, the gluon energies lie well below the Goldstone energies and the Goldstone degeneracies are no longer observed. The two Σ^- states are in violent disagreement with expectations from a fluctuating string. The gluon energies also cannot be explained in terms of a Nambu-Goto string (whose quantization in four dimensions is problematical, besides).

These results are rather surprising. Our results cast serious doubts on the validity of treating glue in terms of a fluctuating string for quark-antiquark separations less than 2 fm. Note that such a conclusion does not contradict the fact that the $\Sigma_g^+(r)$ energy rises linearly for r as small as 0.5 fm. A linearly-rising term is not necessarily indicative

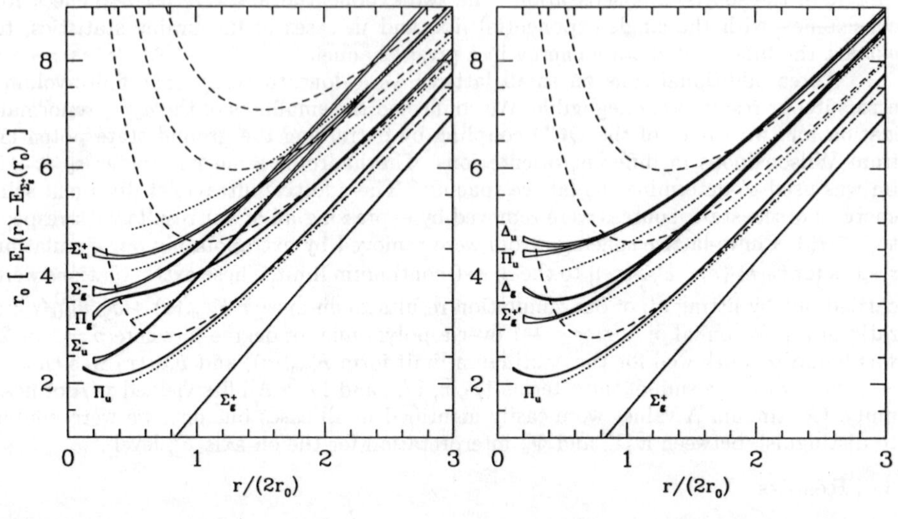

Fig. 1. Plot of the continuum-limit extrapolations (with uncertainties as indicated) for $r_0[E_\Gamma(r) - E_{\Sigma_g^+}(r_0)]$ against $r/(2r_0)$ for various Γ. The dashed lines indicate the locations of the $m\pi/r$ gaps above the Σ_g^+ curve for $m = 1$, 2, 3, and 4. The dotted curves are the naive Nambu-Goto energies in four-dimensions. Note that we cannot rule out a Φ_u interpretation for the curve labelled Π'_u.

of a string: for example, the adiabatic bag model predicts a linearly-rising ground state much before the onset of string-like behaviour, even in the spherical approximation.[1] For r greater than 2 fm, there are some tantalizing signatures of Goldstone mode formation, yet significant disagreements still remain. To what degree these discrepancies can be explained in terms of a distortion of the Goldstone mode spectrum arising from the spatial fixation of the quark and antiquark sources is currently under investigation. This work was supported by the U.S. DOE, Grant No. DE-FG03-97ER40546.

References

1. P. Hasenfratz, R. Horgan, J. Kuti, J. Richard, Phys. Lett. B **95**, 299 (1980).
2. S. Perantonis and C. Michael, Nucl. Phys. **B 347**, 854 (1990); I.J. Ford, R.H. Dalitz, and J. Hoek, Phys. Lett. B **208**, 286 (1988); N. Campbell *et al.*, Phys. Lett. B **142**, 291 (1984).
3. K.J. Juge, J. Kuti, and C. Morningstar, Nucl. Phys. B (Proc. Suppl.) **63**, 326 (1998).
4. C. Morningstar and M. Peardon, Phys. Rev. D **56**, 4043 (1997).

The $Q\bar{Q}$ Potential in the Color-Dielectric Formulation of the Transverse Lattice

Bob Klindworth and Matthias Burkardt
Department of Physics, New Mexico State University, Las Cruces, NM 88003, U.S.A.

ABSTRACT

The $Q\bar{Q}$ potential is calculated in the Color-Dielectric Formulation of Transverse Lattice QCD. In such a formulation an effective potential is used to enforce the $SU(N_c)$ symmetry of the link fields. This effective potential is truncated at fourth order. Surprisingly, there is striking rotational invariance in the ground state $Q\bar{Q}$ potential as well as in the first hybrid mode.

1. Introduction

Light front field theories have many advantages. First, high energy experiments often involve large momentum transfers. Therefore, it seems plausible that such experiments could be well-described in a frame moving at the speed of light. Second, light front wavefunctions can be regarded as correlation functions at equal light front time. From these wavefunctions it is comparatively easy to extract important physical observables like parton distribution functions, and these observables have a direct physical interpretation in terms of light-front coordinates.

The transverse lattice is a marriage of light front field theory with the very successful lattice gauge theory. One first defines the light front variables $x^{\pm} = \frac{1}{\sqrt{2}}(x^0 \pm x^3)$ and keeps them continuous. The remaining two directions (x^1, x^2) are discretized. Like all light front theories, the transverse lattice lacks manifest rotational invariance. For this reason calculating the $Q\bar{Q}$ potential on the transverse lattice is a good way of testing the dynamics of our model.

In previous work we calculated the $Q\bar{Q}$ potential using 2+1 dimensional Transverse Lattice QCD.[2] In such a theory the lattice gauge fields are matrices associated with the links between lattice sites. Ideally, we would like to work with an effective potential that has minima in $SU(N_c)$. However, as explained in Ref. [3], there are difficulties with the Light Front formulation of such a theory. Instead we use the Color-Dielectric formulation in which the link variables are assumed to be smeared variables. This framework allows us to cover a wider range of physical distances using relatively few degrees of freedom.

The results of the original 2+1 calculation exhibited surprising rotational invariance. This was quite surprising because we used a very crude effective potential, accepting only terms up to second order in the link fields. In the current work we repeat the analysis from Ref. [2], including terms up to fourth order in the effective potential and extending the calculation to 3+1 dimensions.

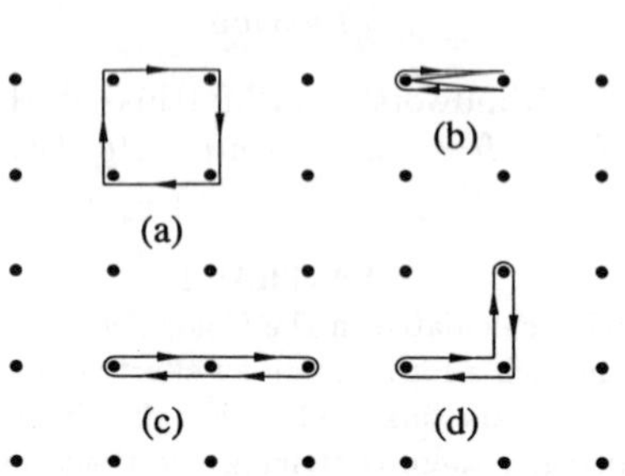

Fig. 1. Relevant fourth order interactions in the effective potential for the link fields. (a) canonical plaquette interaction, (b) local interaction with coupling c_1, (c) longitudinal nonlocal interaction with coupling c_L, (d) transverse nonlocal interaction with coupling $c_\perp$. In $2+1$ dimensions only (b) and (c) contribute.

2. The Effective Potential

The purpose of the effective potential is to constrain the link fields to be members of $SU(N_c)$. The terms in the effective potential up to fourth order are[3]

$$\begin{aligned} V_{eff}(U) &= \mu^2 Tr(U_i U_i^\dagger) + \frac{\lambda_1}{aN_c} Tr(U_i U_i^\dagger U_i U_i^\dagger) + \frac{\lambda_2}{aN_c} Tr(U_{i+1} U_i^\dagger U_{i+1} U_i^\dagger) \\ &+ \frac{\lambda_3}{aN_c^2} Tr(U_i U_i^\dagger) Tr(U_i U_i^\dagger) \end{aligned} \quad (1)$$

Our initial calculation truncated V_{eff} at second order. The second and the fourth terms in this expansion do not contribute due to our Fock Space truncation. We know that our truncation is valid in the case of the ground state, so neglecting these terms seems plausible. It is hoped that this is also valid for the lowest excited states. Thus, the only fourth order term which contributes to our calculation is the nonlocal interaction.

$$V_{eff}(U) = \mu^2 Tr(U_i U_i^\dagger) + \frac{\lambda_2}{aN_c} Tr(U_{i+1} U_i^\dagger U_{i+1} U_i^\dagger)$$

It was found in Ref. [3] that the glueball spectrum could be fit to reliable Euclidean Lattice Monte Carlo data as long as μ^2 and λ_2 lie along a scaling trajectory. In our calculation, we verified that points off of the scaling trajectory give rise to potentials which are not rotationally invariant.

3. Results in 2+1 Dimensions

With the effective potential fixed we calculated the $Q\bar{Q}$ Potential as before using the 2+1 dimensional Transverse Lattice. We used Discretized Light Cone Quantization.[8]

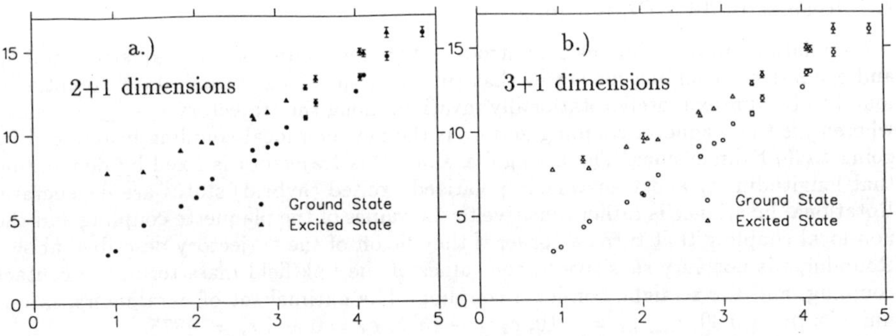

Fig. 2. a.) Static $Q\bar{Q}$ potential in the color dielectric formulation of the $\perp$ lattice in 2+1 dimensions versus the distance $r \equiv \sqrt{x_\perp^2 + x_L^2}$ between the Q and the $\bar{Q}$. Circles: ground state, triangles: first adiabatic hybrid state. The points represent calculations where the $Q\bar{Q}$ pair was separated by up to 4 lattice units in the transverse direction as well as by various separations in the longitudinal direction. In 3+1 dimensions, as long as the $\perp$ separation between the Q and the $\bar{Q}$ is along one of the $\perp$ axis, the results are identical. b.) Same as a.), but for a $3+1$ dimensional $\perp$ lattice. The transverse separations between the $Q\bar{Q}$ pair are $(n_x, n_y) = (1,0), (2,0), (3,0), (1,1)$ and $(2,1)$,.

The main differences when compared to the calculation in Ref. [2] are the addition of a "mass" term to the link fields and the inclusion of an attractive (Lorentz-) scalar interaction between adjacent links. The resulting potential has an anisotropy in the points with zero transverse separation resulting from the fact that the non-local coupling does not act on states whose transverse separation is less than two. Rotational invariance is restored by introducing a new term in the effective potential.[9] In momentum space this interaction has the form,

$$V_{contact} = \frac{c_{contact}}{v^+} \frac{1}{\sqrt{k^+ k'^+}} \tag{2}$$

where k^+ and k'^+ are the incoming/outgoing momenta of the link fields which interact with the external charges. Note that this interaction acts only on the first and last link field in the "chain" connecting the $Q\bar{Q}$ pair since it involves the external charges. For the same reason, it does not conserve gluon momentum. This interaction has a Lorentz structure similar to the 4-point interaction between fermions and bosons that arises when the constrained component of the spinor field is eliminated.

In addition to studying the ground state potential we can also study the rotational invariance and qualitative shape of the first excited state. Surprisingly, this state also exhibits striking rotational invariance. This is surprising because higher Fock components, which we omit, could contribute significantly to the excited states.

4. Results in 3+1 Dimensions

Calculations in 3+1 dimensions introduce two new couplings: a plaquette coupling and a non-local coupling for states that turn a corner. The ground state potential is found to be approximately rotationally invariant along the trajectory $c_p - c_\perp = const.$, where c_p is the plaquette coupling and $c_\perp$ is the new non-local coupling introduced by going to 3+1 dimensions. The ambiguity along this trajectory is fixed by demanding that longitudinally and transversely polarized excited (hybrid) states are degenerate. Rotational invariance is rather sensitive to the values of the plaquette coupling and the non-local coupling that turns a corner if they lie off of the trajectory described above. Roundness is not very sensitive to the values of the link field mass term, the contact coupling, and the straight non-local coupling. The optimal set of parameters, in our units is: $m^2 = 0.30$, $c_{contact} = -.10$, $c_2 = -.1375$, $c_\perp = 0$ and $c_p = .4875$.

The first excited state of the 3+1 dimensional potential exhibits a slight anisotropy. This is an expected result of our Fock Space truncation. However, the shape at intermediate to long distances is encouraging and it does agree rather well with results obtained on a Euclidean lattice.[5]

5. Summary and Outlook

The static $Q\bar{Q}$ potential provides a test of the dynamics of our model. In all, the rotational invariance of the ground state potential is striking. However, the fact that rotational invariance is not extremely sensitive to all of the parameters in the effective potential suggests that other observables are needed in order to pin down these couplings unambiguously.[9]

Alternatively, it is conceivable that one could determine the coefficients of the effective potential on the Euclidean lattice where one can make the transverse lattice spacing small. Not only could one be rather confident of the calculations done from such a theory, but one could also systematically determine the dependence of observables on high orders in the effective potential. Such work is currently underway.[10]

References

1. K. G. Wilson et al., Phys. Rev. D **49**, 6720 (1994); M. Burkardt, Advances Nucl. Phys. **23**, 1 (1996); S. J. Brodsky et al., Phys. Rept. **301**, 299 (1998).
2. M. Burkardt and B. Klindworth, Phys. Rev. D **55**, 1001 (1997) (and references therein)
3. S. Dalley and B. van de Sande, Phys. Rev. D **56**, 7917 (1997)
4. Nathan Isgur and Jack Paton, Phys. Rev. D **31**, 2910 (1985)
5. K.J. Juge, J. Kuti and C.J. Morningstar, Nucl. Phys. Proc. Suppl. **63**, (1998).
6. W. A. Bardeen and R. B. Pearson, Phys. Rev. D **14**, 547 (1976); W. A. Bardeen, R. B. Pearson and E. Rabinovici, *ibid* **21**, 1037 (1980).
7. M. Burkardt, in "Theory of Hadrons and Light-Front QCD", edited by S. Glazek, (World Scientific, Singapore, 1995).
8. H.-C. Pauli and S. J. Brodsky, Phys. Rev. D **32**, 1993 (1985); 2001 (1985).
9. S. Dalley and B. van de Sande, hep-th/9806231.
10. B. Klindworth and M. Burkardt, *work in progress.*

DUAL LATTICE SIMULATIONS OF FLUX TUBES

Martin Zach, Manfried Faber, Alexandra Müller, Peter Skala
Institut für Kernphysik, Technische Universität Wien, A-1040 Vienna, Austria

ABSTRACT

Abelian gauge theories formulated on a space-time lattice can be used as a prototype for investigating the confinement mechanism. In $U(1)$ lattice gauge theory it is possible to perform a dual transformation of the path integral. Simulating the obtained dual theory (which corresponds to a certain limit of a dual Higgs model) including external sources, we perform a very accurate analysis of flux tubes with respect to the dual superconductor picture. Dual flux tube simulations are also performed in the full Abelian Higgs model, in order to obtain non-perturbative control over quantum and string fluctuations, and for a comparison to the results of dual QCD.

1. Introduction

The formation of colour-electric flux tubes provides an intuitive physical picture for permanent confinement of quarks within hadrons. Moreover, this picture seems to be realized within QCD, as has widely been observed in lattice simulations. The mechanism leading to flux tube formation, however, is still subject to analytical and numerical investigations.

The shortcoming of the corresponding lattice calculations, however, is the fact that regarding sufficiently long flux tubes, i. e. large lattice sizes, becomes a numerically very difficult task. A simpler gauge theory which also exhibits confinement is $U(1)$; it can be used as a prototype for investigating the formation of flux tubes.

Furthermore, it was realized many years ago[1] that one can perform a duality transformation of the path integral in compact Abelian gauge theories. In this way a new partition function is obtained which can be regarded as a limit of the dual non-compact Abelian Higgs model.[2] Besides this fact, the dually transformed theory can be used as a very efficient tool for the calculation of expectation values in the presence of external charges.[3,4] The numerical advantages of such dual flux tube simulations also hold for the more general case of dual Higgs models.

In this talk we review the results obtained in dually transformed $U(1)$ lattice gauge theory and present some first calculations in the dual Abelian Higgs model. We will focus on one of the key questions within the dual superconductor picture, namely whether the confinement vacuum corresponds to a type-I or a type-II superconductor, and we will investigate the role of quantum fluctuations of the dual degrees of freedom.

2. Dually transformed $U(1)$ lattice gauge theory

The expectation value of a physical observable like the electric field or the magnetic current in the presence of external sources (represented by Polyakov loops) can be rewritten as averaging over the dual degrees of freedom, as has been shown in previous work.[3] The dual representation of the Polyakov loops can be interpreted as a dual Dirac sheet connecting the electric charges.

This dual formulation opens the possibility for very accurate numerical calculations for several reasons: Contrary to standard lattice simulations, the confinement phase is the weakly coupled one in the dual theory, therefore there are less quantum fluctuations. Even more important is the fact that it is not necessary to project the charge–anticharge state out of the vacuum: On a lattice of same size charge pairs with arbitrary distance can be simulated with equal accuracy, and increasing the time extent of the loops (i. e. of the lattice) does not influence the quality of the result, either.

Further advantages are the implementation of doubly charged flux tubes[4] by double Dirac strings which also yields results of equal accuracy, and the extension to periodically closed flux tubes. These "torelons" are free of end effects and allow for a more reliable calculation of pure flux tube properties like the string tension.

An interesting physical task is the testing of the validity of a dual London equation for $U(1)$ flux tubes.[5] In the dual formulation we were able to investigate rather long flux tube lengths up to 20 lattice spacings. Our results[3] showed that the agreement between the $U(1)$ data and the predictions of a classical model of Maxwell and London equations are very good for small charge distances, but there is absolutely no agreement at large distances ($>$ 10 lattice spacings). A possible interpretation within the dual superconductor picture is that the confining $U(1)$ behaves as an effective type-I superconductor, rather than an extreme type-II as an exact validity of the dual London equation would suggest. In fact we got support for this conjecture also from a completely independent investigation: simulating doubly charged flux tubes. We found[4] that in four-dimensional $U(1)$ there is an attractive interaction between flux tubes for the coupling β approaching the phase transition.

Thus one might look at $U(1)$ lattice gauge theory from two points of view: From a "microscopical" it can be regarded as a double limit of a dual Higgs model with both vanishing London penetration length and vanishing coherence length. From a "macroscopical" point of view it looks like a classical dual type-I superconductor. This peculiarity is related to the fluctuations in the dual theory: The observed attraction between flux tubes was shown to be a purely "quantum-mechanical" effect of the dually transformed $U(1)$ theory.[4]

3. The dual Abelian Higgs model

The action of the full Abelian Higgs model including a dual Dirac string (denoted with ${}^*n$) can be written in the form

$$S = \beta \sum_{plaqu.} (d^*\theta + 2\pi^*n)^2 \; - \; 2\gamma \sum_{links} |\overline{{}^*\Phi}|^2 \cos\left(d^*\chi - {}^*\theta\right) \; + \; \kappa \sum_{sites} V({}^*\Phi), \qquad (1)$$

where ${}^*\theta$ is the dual gauge field and ${}^*\Phi$ the dual Higgs field, located on the sites of the dual lattice. $V({}^*\Phi)$ is the symmetry-breaking Higgs potential with its minimum at $^*\Phi = 1$. For the squared covariant derivative of the Higgs current the compact formulation has been chosen, because this allows for fluctuations of the fluxoid string.[3] In the classical model β/γ is the squared London penetration depth, and γ/κ the squared coherence length. The double limit $\gamma \to \infty$, $\kappa \to \infty$ exactly corresponds to the dually transformed $U(1)$ theory discussed in the last section.

If we compare this model to the parameters of dual QCD[6] after an Abelian ansatz, we find the correspondence

$$\beta = \frac{4}{3}\frac{1}{g^2}, \quad \gamma = 8\, B_0^2\, a^2, \quad \kappa = \frac{100}{3}\, \lambda\, B_0^4\, a^4, \tag{2}$$

where g is the (chromo-)magnetic coupling, B_0 the location of the minimum and λ the strength of the corresponding continuum Higgs potential, and a is the lattice spacing. These parameters can be chosen to reproduce very well the infrared behaviour of QCD. The classical flux tube solutions of the Higgs model on sufficiently fine lattices give very good agreement with corresponding continuum calculations.[7] It is now an interesting question how quantum and string fluctuations (of the fluxoid string) do effect the physical quantities, e. g. the string tension.

For the set of parameters which yields the desired classical solution, the system is not in the Higgs phase any more on a reasonably fine lattice, if fluctuations are turned on. For this reason we introduce a "quantum scale" n, by substituting $\exp(-S)$ in the path integral by $\exp(-n\,S)$, which allows for a smooth interpolation between the classical limit ($n = \infty$) and an ordinary simulation ($n = 1$). The changing of the string tension as a function of n is depicted in Fig. 1 for two cases: Considering only quantum fluctuations but keeping the physical fluxoid string fixed, and also considering string fluctuations, which leads to an increase of the energy per length (until the system undergoes a phase transition to the Coulomb phase)

Physical arguments suggest that the dual superconductor should be approximately at the borderline between type-I and type-II. This is fulfilled for the chosen classical solution. Investigating the influence of fluctuations on an effective Landau-Ginzburg parameter (defined by the ratio of the individual contributions to the string tension) showed us that the system is pushed towards type-I behaviour by fluctuations.

4. Conclusions

By dual simulations of flux tubes one can address a lot of interesting questions concerning the confinement mechanism. Here we focussed on the discussion whether the confinement vacuum is a dual type-I or type-II superconductor, and on the meaning of fluctuations in dual theories. Similarly to the behaviour of $U(1)$ in the vicinity of the phase transition, these fluctuations can produce an effective type-I superconductor also in the full Abelian Higgs model.

Of course, it is another interesting question, which set of (bare) parameters has to be used to describe the same physics as with the classical solution of the Abelian Higgs model. For this task, however, it will be necessary to improve the control over string fluctuations in our simulations.

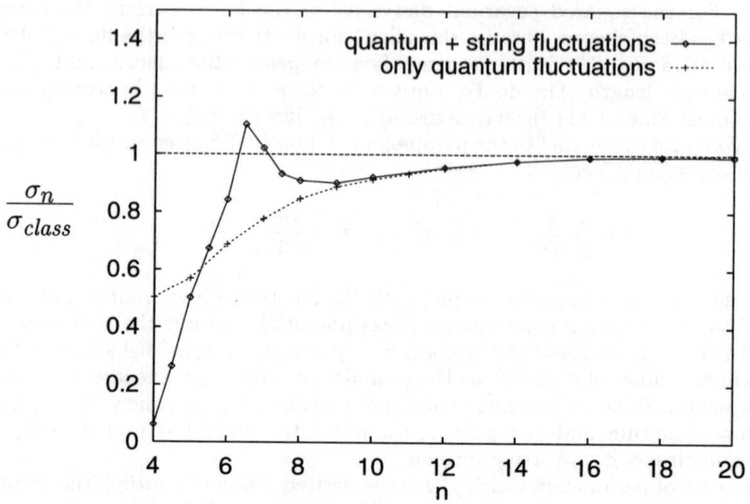

Fig. 1. The changing of the string tension σ with respect to the classical value σ_{class} under the influence of quantum fluctuations (dotted line), and of both quantum and string fluctuations (solid line). This calculation has been performed for a torelon on a 8^4 lattice.

Acknowledgments

This work was supported by the Fonds zur Förderung der wissenschaftlichen Forschung, Project No. P12495-TPH.

References

1. T. Banks, R. Myerson, J. Kogut, *Nucl. Phys.* B 129 (1977) 493.
2. J. Fröhlich, P. A. Marchetti, *Europhys. Lett.* 2 (1986) 933.
3. M. Zach, M. Faber, P. Skala, *Phys. Rev.* D 57 (1998) 123.
4. M. Zach, M. Faber, P. Skala, *Nucl. Phys.* B (1998), in press, hep-lat/9709017.
5. V. Singh, R. W. Haymaker, D. A. Browne, *Phys. Rev.* D 47 (1993) 1715.
6. For a review see e.g. M. Baker, J. S. Ball, F. Zachariasen, *Phys. Rev.* D 41 (1990) 2612.
7. M. Baker, N. Brambilla, H. G. Dosch, A. Vairo, *Phys. Rev.* D 58 (1998) 034010.

GLUON PROPAGATOR IN THE LANDAU GAUGE FIXED LATTICE QCD SIMULATION

Hideo Nakajima
Dept. of Infor. Sci., Utsunomiya Univ., Utsunomiya 321-8585 Japan
and

Sadataka Furui

School of Sci. and Engr., Teikyo Univ., Utsunomiya 320-8551, Japan

August 28, 1998

Abstract

We measured the gluon propagator in the Landau gauge fixed QCD Langevin simulation[2] and studied the infra-red behaviour of the gluon propagator. The $(4^3 \times 8)$ lattice siulation was done for quenched $\beta = 3, 4$ and 5 and unquenched $\beta = 4, \kappa = 0.1, 0.15$ and 0.2, using each 100 independent samples. The Landau gauge fixing was done by an extension of the Fourier acceleration method with the condition $|\mathrm{div}A| < 10^{-4}$, and the field A is related to the link variable by $U = \exp A$ instead of the usual U-linear definition. We confirmed gauge fixing with smearing preconditioning[3] works perfectly for the purpose of finding the global minimum of the squared norm of the gauge field when β is large (e.g. $\beta = 5$). Our simulation results suggests the possibility of a realization of the infrared behaviour of the Gribov-Zwanziger theory.

1 Introduction

In 1978 Gribov showed that the fixing of the divergence of the gauge field in non-Abelian gauge theory does not uniquely fix its gauge. The restriction of the gauge field such that its Faddeev-Popov determinant is positive (Gribov region) is necessary yet insufficient for excluding the ambiguity.

The restriction to the fundamantal modular region, i.e. minimizing the norm of the gauge field, cancels the infrared singularity of the perturbation theory and makes the gluon to have the complex mass.

$$\begin{aligned} D_{\mu\nu}(k) &= \frac{1}{n}\sum_{x=\mathbf{x},t} e^{-ikx} Tr\langle A_\mu(x)A_\nu(0)\rangle \\ &= G(k^2)(\delta_{\mu\nu} - \frac{k_\mu k_\nu}{k^2}) . \end{aligned} \tag{1}$$

$$G(k^2) = \frac{k^2}{k^4+\kappa^4} = \frac{1}{2}(\frac{1}{k^2+i\kappa^2} + \frac{1}{k^2-i\kappa^2}) \tag{2}$$

where $n = N^2-1$ is the dimension of the adjoint representation of the SU(N) group.

In order to measure the propagator on a lattice, it is necessary to avoid the Gribov copy. We confirmed that it is possible by the smeared gauge fixing[3], when $\beta = 5$ in the case of $4^3 \times 8$.

2 The gluon propagator

On the lattice we define an $SU(N)$ lattice action for $S = S_{gauge} + S_{fermi}$ in the pseudo-fermion scheme. We measure

$$\begin{aligned} D_T(\mathbf{k},t) &= \frac{1}{2n}\sum_{j=1}^{2}\sum_T Tr\langle \tilde{A}_j(\mathbf{k},T+t)\tilde{A}_j(-\mathbf{k},T)\rangle \\ &= \frac{1}{2n}\sum_{j=1}^{2}\sum_T\sum_{\mathbf{x}} e^{-i\mathbf{k}\mathbf{x}} Tr\langle A_j(\mathbf{x},T+t)A_j(\mathbf{0},T)\rangle \end{aligned} \tag{3}$$

where $\tilde{A}_j$ is the Fourier component of A_j, and j is summed over transverse polarization with respect to $\mathbf{k}$.

The gauge field on links is defined as $e^{A_{x,\mu}} = U_{x,\mu}$, where $A^\dagger_{x,\mu} = -A_{x,\mu}$ The gauge transformation is $e^{A^g_{x,\mu}} = g^\dagger_x e^{A_{x,\mu}} g_{x+\mu}$, $\phi^g_x = g^\dagger\phi_x$ where $g_x = e^{\epsilon_x}$ and ϵ is a traceless antihermitian

The connected part of the propagator is defined by the normalized Fourier transform

$$\tilde{A}^a_\mu(\mathbf{k},T) = \frac{1}{\sqrt{N_x^3}} \sum_x A^a_\mu(\mathbf{x},T) e^{-i\mathbf{k}\cdot\mathbf{x}} \tag{4}$$

as

$$\begin{aligned} G_{\mu\nu}(\mathbf{k},t)_c &= \sum_{T,a} \langle \tilde{A}^a_\mu(\mathbf{k},T+t)\tilde{A}^a_\nu(\mathbf{k},T)\rangle^* \\ &- \sum_{T,a} \langle \tilde{A}^a_\mu(\mathbf{k},T+t)\rangle\langle \tilde{A}^a_\nu(\mathbf{k},T)\rangle^* \end{aligned} \tag{5}$$

where $\langle \tilde{A}^a_\mu(\mathbf{k},T)\rangle$ specifies the expectation value of the gauge field in the Landau gauge. For this evaluation it is necessary to fix the global gauge transformation.

We fix the global gauge such that the average over T of the zero-mode of a sample: $\frac{1}{3N_T}\sum_{T,i} \tilde{A}^m_i(\mathbf{0},T)\lambda_m$, where λ is the SU(3) Gell-Mann matrix, is diagonalized and the smallest eigenvalue appears at the top and the largest eigenvalue at the bottom. We observed that the 2nd term of $G_{ii}(\mathbf{0},t)_c$ is almost independent of t and that the value is several % of the 1st term of $G_{ii}(\mathbf{0},0)_c$. The value is larger for the unquenched case as compared to the quenched case. The Fourier transform of $D_T(\mathbf{0},t)$ in the unquenched $4^3\times 8$ and in the quenched $8^3\times 16$ lattice simulation [5] suggest the suppression of the infrared part of the gluon propagator.

In the Stingl's factorizing-denominator rational approximants (FDRA)

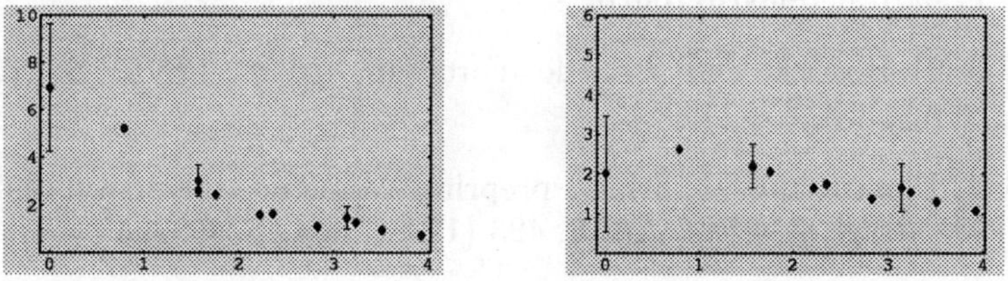

Figure 1: The gluon propagator $D_T(p)$. Ordinates are the euclidean 4-momentum p in the unit of a^{-1}. $\beta = 4, \kappa = 0$(left) and $\kappa = 0.2$(right). $4^3\times 8$ lattice.

method[4], the transverse gluon propagator is expressed in the 2nd order as

$$D_T(p^2)^{[r=3]} = \frac{\rho}{p^2 + u_+\Lambda^2} + \frac{\tau}{p^2 + v_+\Lambda^2} + c.c. \qquad (6)$$

We compared the lattice fourier transform of the above expression with $D_T(\mathbf{k}, t)$ and observed that in the confinement region there appears a pair of complex zero near $p^2 = 0$ as is suggested by the eq (2).

3 Discussion and outlook

In the Landau gauge, the transverse gluon propagator in the infrared region is finite at $p^2 = 0$ [6, 7]and the position of the zero of $D_T(p^2)$ approaches $p^2 = 0$ as the lattice size increases, while the propagator of the Faddeev-Popov ghost which is measured in the quenched $8^3 \times 16$ lattice shows the infrared singularity. These results support the possibility of a realization of the confinement mechanism of the Gribov-Zwanziger theory.

This work was supported by High Enery Accelerator Research Organization, as KEK Supercomputer Project (Project No.98-34).

References

[1] V.N. Gribov, Nucl. Phys. **B139**, 1 (1978); D. Zwanziger, Nucl. Phys. **B364**, 127 (1991).

[2] H. Nakajima and S. Furui, Nucl Phys. **B** (Proc. Suppl.) **63**, 976 (1998), hep-lat/9710028.

[3] J.E. Hetrick and Ph. de Forcrand, Nucl. Phys. **B**(Proc. Suppl.)**63**, 838 (1998).

[4] J. Fromm and M. Stingl, preprint, Münster University(1997); M. Stingl, Z. Phys. **A353**, 423 (1996), hep-th/9502157

[5] H. Nakajima and S. Furui, Lattice'98 Proceedings (1998)

[6] A. Cucchieri, Phys. Lett. **B422** 233 (1998),hep-lat/9709015.

[7] D.B. Leinweber, J.I Skullerud, A.G. Williams and C. Parrinello, Phys. Rev. **D58** (1998), hep-lat/9803015

AN ANALYSIS OF 4-QUARK ENERGIES IN SU(2) LATTICE MONTE CARLO

Sadataka Furui
School of Sci. and Engr., Teikyo Univ., Utsunomiya 320-8551, Japan
and
Bilal Masud
Centre for High Energy Physics, Punjab Univ., Lahore 54590, Pakistan

August 28, 1998

Abstract

Energies of four-quark systems with the tetrahedral geometry measured by the static quenched SU(2) lattice monte carlo method are analyzed by parametrizing the gluon overlap factor in the form $exp(-[b_s E\mathcal{A}+\sqrt{b_s}F\mathcal{P}])$, where $\mathcal{A}$ and $\mathcal{P}$ are the area and perimeter defined mainly by the positions of the four quarks, b_s is the string constant in the 2-quark potentials and E, F are constants.

1 Introduction

For the purpose of understanding meson-meson interactions from the QCD, we studied the ground and the 1st excited state energy of 4-quark system measured by the SU(2) lattice Monte-Carlo. The configuration of the four quarks that we consider are large square(LS), rectangular(R), tilted rectangular(TR), linear(L), quadrilateral(Q), non-planar(NP) and tetrahedral(TH).

There are three possible choices of the colour singlet pairs which are denoted by A(14,23), B(12,34) and C(13,24). (We interchange the definition ofA, B from[1].) Identifying these bases as P_1, P_2 and P_3, we obtain the eigen-energies by diagonalizing 3×3 matrix of

$$W_{ij}^T =< P_i|\hat{T}^T|P_j > \tag{1}$$

where $\hat{T} = exp(-a\hat{H})$ is the transfer matrix. Using a trial wave function $\psi = \sum_i a_i|P_i >$, the eigenvalue equation

$$W_{ij}^T a_j^T = \lambda^{(T)} W_{ij}^{T-1} a_j^T \tag{2}$$

is solved to get $exp(-aV_2) = \lambda^{(T)}$ for large T[1].

2 The model with the gluon overlap factor

We observed that among the three bases, one base can be treated as a linear combination of the other two, but within a base in the coarse lattice, when the shortest path between a quark and an antiquark is not along any link, there are several possible configurations of links, which are specified by the parity. When there are m(n) parity eigenstates in A(B), we solve;

$$det\left(\begin{bmatrix} V_A & \cdots & c_{A1m} & V_{AB}f_{11} & \cdots & V_{AB}f_{1n} \\ \vdots & \ddots & \vdots & \vdots & \ddots & \vdots \\ c_{Am1} & \cdots & V_A & V_{AB}f_{m1} & \cdots & V_{AB}f_{mn} \\ V_{AB}f^*{}_{11} & \cdots & V_{AB}f^*{}_{1m} & V_B & \cdots & c_{B1n} \\ \vdots & \ddots & \vdots & \vdots & \ddots & \vdots \\ V_{AB}f^*{}_{n1} & \cdots & V_{AB}f^*{}_{nm} & c_{Bn1} & \cdots & V_B \end{bmatrix}\right.$$

$$\left. -E\begin{bmatrix} 1 & \cdots & 0 & f_{11}/2 & \cdots & f_{1n}/2 \\ \vdots & \ddots & \vdots & \vdots & \ddots & \vdots \\ 0 & \cdots & 1 & f_{m1}/2 & \cdots & f_{mn}/2 \\ f_{11}^*/2 & \cdots & f_{1m}^*/2 & 1 & 0 & 0 \\ \vdots & \ddots & \vdots & \vdots & \ddots & \vdots \\ f_{n1}^*/2 & \cdots & f_{nm}^*/2 & 0 & 0 & 1 \end{bmatrix}\right) = 0$$

We introduce the gluon overlap factor in the form $f_{ij} = exp(-[b_s E\mathcal{A} + \sqrt{b_s}F_{ij}\mathcal{P}_{ij}])$, where $\mathcal{A}$ and $\mathcal{P}$ are the area and the perimeter. We assume that the length of the zig-zag perimeter has the fractal dimension 2, and so it is fixed from that of the coarse lattice, use the string constant b_s measured in the simulation of the 2-quark potentials and fit constants E and F_{ij}[2]. The length $\mathcal{P}_{ij}$ depends on the base but the minimal area $\mathcal{A}$ are chosen to be independent of the base and estimated by the analytical form derived in the regular surface approximation. (The analysis of NP of[2] is revised, where the area depended on the bases.) The perimeter dependent terms contain lattice artefacts and the physical quantities are obtained by subtracting the artefacts.

The parameter E for the area term and A_0 for the self energy in the Linear configuration are fixed in the previous model[2]. In the NP case we fitted three coefficients F corresponding to the three types of the perimeter lengths and a mixing parameter between the parity eigenstates[2]. The details of the revised fitting are in [4].

In the case of TH, when the length of the links are all equal($r = d$), the ground state energy is doubly degenerate. When they differ($r \neq d$) we evaluate minimal surface area for A-C($regular_1$) and for A-B($regular_2$) and redefine among B and C, the base B such that it is connected to A by the smaller area. We solve the secular equation of m=4 and n=4.

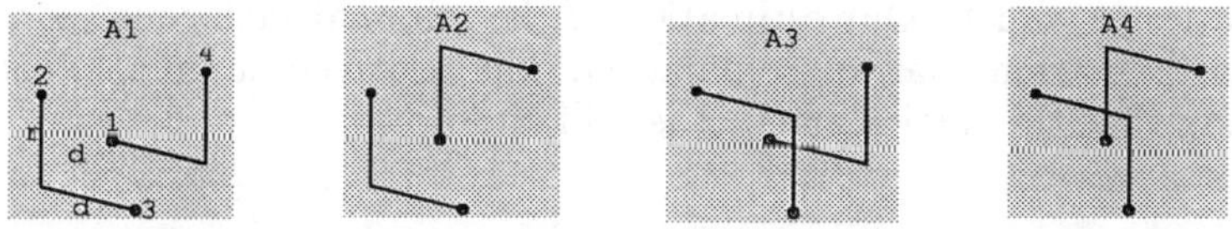

Figure 1: The bases A_1, A_2, A_3 and A_4 in the TH configuration.

We explain the degeneracy of the ground state energy for the $r = d$ cases and the ground and the first excited state eigen-energies of $r \neq d$ cases by introducing an additional coefficient F corresponding to the new type of perimeter length and a parameter that specifies the internal excitation of the two quark cluster.

3 Discussion and Conclusions

In an analysis of 4-quark energy of the tetrahedral geometry, Green and Pennanen[3] proposed a 6-basis model in which the ground ($A_{1g}A_{1g}$) and an excited state (E_uE_u) are considered for each of the configuration A, B and

Table 1: Minimal surface area in regular surface approximation and the sum of two triangles for the TH and the experimental and the fitted binding energy.

r,d	$regular_1$	$regular_2$	2 triangles	E_0	E_0(fit)	E_1	E_1(fit)
1 2	4.316	3.180	4.899	-0.043	-0.036	0.084	0.108
2 2	5.123	5.123	6.928	-0.021	-0.023	-0.021	-0.023
3 2	6.209	7.260	9.381	-0.008	-0.011	0.148	0.139
2 3	10.22	8.528	12.369	-0.040	-0.043	0.051	0.072
3 3	11.53	11.53	15.588	-0.028	-0.027	-0.028	-0.027
4 3	13.11	14.71	19.209	-0.007	-0.012	0.110	0.115
3 4	18.69	16.45	23.324	-0.041	-0.050	-0.032	-0.032
4 4	20.49	20.49	27.713	-0.030	-0.031	-0.030	-0.031
5 4	22.56	24.71	32.496	-0.010	-0.014	0.089	0.093

C. Our choice corresponds to $A_{1g}A_{1g}$, E_uE_u, $A_{1g}E_u$ and E_uA_{1g} for each of the configuration A and B. Our estimation of the minimal surface is more accurate than their triangular approximation. The relatively large coefficients of $\mathcal{P}$ suggests that the lattice artefact is still large.

References

[1] A.M.Green, C.Michael and J.E.Paton, Nucl.Phys. **A554**, 701 (1993); A.M.Green, C.Michael and M.Sainio, Z. Phys. **C67**, 291 (1995), hep-lat/9404004; A.M.Green, J. Lukkarinen, P. Pennanen, C.Michael and S.Furui, Nucl. Phys. **B**(Proc. Suppl.) **42**, 249 (1995).

[2] S.Furui, A.M.Green and B.Masud, Nucl Phys. **A582**, 682 (1995).

[3] A.M.Green and P. Pennanen, Preprint, HIP-1998-01/TH.

[4] S.Furui and B.Masud, to be published.

LATTICE CALCULATION OF THE GLUON PROPAGATOR AND THE STRANGENESS MAGNETIC MOMENT OF THE NUCLEON

Anthony G. Williams
CSSM and Department of Physics and Mathematical Physics
University of Adelaide, Australia 5005
awilliam@physics.adelaide.edu.au
http://www.physics.adelaide.edu.au/cssm

ABSTRACT

This contribution summarizes a presentation combining two topics in lattice QCD. Firstly, the gluon propagator in Landau gauge is calculated in quenched QCD on a large ($32^3 \times 64$) lattice at $\beta = 6.0$. New structure seen in the infrared region survives conservative cuts to the lattice data, and serves to exclude a number of models that have appeared in the literature. Secondly, I report on a recent lattice QCD calculation of the strangeness magnetic moment of the nucleon. The result is $G_M^s(0) = -0.36 \pm 0.20$. The strangeness Sachs electric mean-square radius $\langle r_s^2 \rangle_E$ was also found to be small and negative.

1. Introduction

Here I summarize the central results described in a single long parallel session talk, which was comprised of two distinct studies of lattice QCD. The first of these is concerned with the direct lattice calculation of the Landua gauge nonperturbative gluon propagator, an understanding of which is central to our understanding of the nature of confinement. The second topic is a lattice calculation of the strangeness magnetic moment of the nucleon. Different confining quark models predict a variety of results for this quantity and so lattice calculations provide a benchmark against which we can test our understanding of hadron structure.

2. Lattice Calculation of the Gluon Propagator

2.1. Motivation

The infrared behaviour of the gluon propagator is important for an understanding of confinement. Previous conjectures range from a strong divergence[1,2] to a propagator that vanishes in the infrared.[3,4] Lattice QCD should in principle be able to resolve this issue by first-principles, model-independent calculations. However, lattice studies have been inconclusive up to now,[5,6] since they have not been able to access sufficiently low momenta. The lower limit of the available momenta on the lattice is given by $q_{\rm min} = 2\pi/L$, where L is the length of the lattice. Here we will report results using a lattice with a length of 3.3 fm in the spatial directions and 6.7 fm in the time direction. This gives us access to momenta as small as 400 MeV.

2.2. *Summary of Formalism*

The gluon field A_μ can be extracted from the link variables $U_\mu(x)$ using $U_\mu(x) = e^{ig_0 a A_\mu(x+\hat{\mu}/2)} + \mathcal{O}(a^3)$. Inverting and Fourier transforming this, we obtain

$$\begin{aligned} A_\mu(\hat{q}) &\equiv \sum_x e^{-i\hat{q}\cdot(x+\hat{\mu}/2)} A_\mu(x+\hat{\mu}/2) \\ &= \frac{e^{-i\hat{q}_\mu a/2}}{2ig_0 a}\left[\left(U_\mu(\hat{q}) - U_\mu^\dagger(-\hat{q})\right) - \frac{1}{3}\mathrm{Tr}\left(U_\mu(\hat{q}) - U_\mu^\dagger(-\hat{q})\right)\right], \end{aligned} \tag{1}$$

where $U_\mu(\hat{q}) \equiv \sum_x e^{-i\hat{q}x} U_\mu(x)$ and $A_\mu(\hat{q}) \equiv t^a A_\mu^a(\hat{q})$. The available momentum values $\hat{q}$ are given by $\hat{q}_\mu = 2\pi n_\mu/(aN_\mu)$, $n_\mu = 0,\ldots,N_\mu - 1$, where N_μ is the number of points in the μ direction. The gluon propagator $D_{\mu\nu}^{ab}(\hat{q})$ is defined as $D_{\mu\nu}^{ab}(\hat{q}) = \langle A_\mu^a(\hat{q}) A_\nu^b(-\hat{q})\rangle / V$. The Landau gauge propagator in the continuum has the structure $D_{\mu\nu}^{ab}(q) = \delta^{ab}[\delta_{\mu\nu} - (q_\mu q_\nu/q^2)]D(q^2)$. At tree level, $D(q^2)$ will have the form $D^{(0)}(q^2) = 1/q^2$. On the lattice, this becomes

$$D^{(0)}(\hat{q}) = 1/\sum_\mu \left(\frac{2}{a}\sin\frac{\hat{q}_\mu a}{2}\right)^2 . \tag{2}$$

Since QCD is asymptotically free, we expect that up to logarithmic corrections, $q^2 D(q^2) \to 1$ in the ultraviolet. Hence we define the new momentum variable q by $q_\mu \equiv (2/a)\sin(\hat{q}_\mu a/2)$ and work with this throughout. The (bare) lattice gluon propagator is related to the renormalised continuum propagator $D_R(q;\mu)$ via $D^L(qa) = Z_3(\mu,a)D_R(q;\mu)$. The renormalisation constant $Z_3(\mu,a)$ can be found by imposing a momentum subtraction renormalisation condition $D_R(q)|_{q^2=\mu^2} = 1/\mu^2$. The asymptotic behaviour of the renormalised gluon propagator in the continuum is given to one-loop level by $D_R(q) \equiv D_{\rm bare}(q)/Z_3 = (1/q^2)\left[(1/2)\ln(q^2/\Lambda^2)\right]^{-d_D}$ with $d_D = (39 - \xi - 4N_f)/[4(33 - 2N_f)] = 13/44$, where both the gauge parameter ξ and the number of fermion flavours N_f are zero in this calculation.

2.3. *Simulation Parameters, Finite Size Effects and Anisotropies*

We have analysed three lattices, with different values for the volume and lattice spacing. The details are given in Table 1. In the following, we are particularly interested in the deviation of the gluon propagator from the tree level form. We will therefore factor out the tree level behaviour and plot $q^2 D(q^2)$ rather than $D(q^2)$ itself.

Fig. 1 shows the gluon propagator as a function of qa for the small and large lattices, with momenta in different directions plotted separately. For low momentum values on the small lattice, there are large discrepancies due to finite size effects between points representing momenta along the time axis and those representing

Name	β	$a_{\rm st}^{-1}$ (GeV)	Volume	$N_{\rm conf}$	$a\widehat{q}_{\rm Max}$	$\lvert\partial_\mu A_\mu\rvert$
Small	6.0	1.885	$16^3 \times 48$	125	$2\pi/4$	$< 10^{-6}$
Large	6.0	1.885	$32^3 \times 64$	75	$2\pi/4$	$< 10^{-6}$
Fine	6.2	2.63	$24^3 \times 48$	223	$2\pi/4$	$< 10^{-6}$

Table 1. Simulation parameters

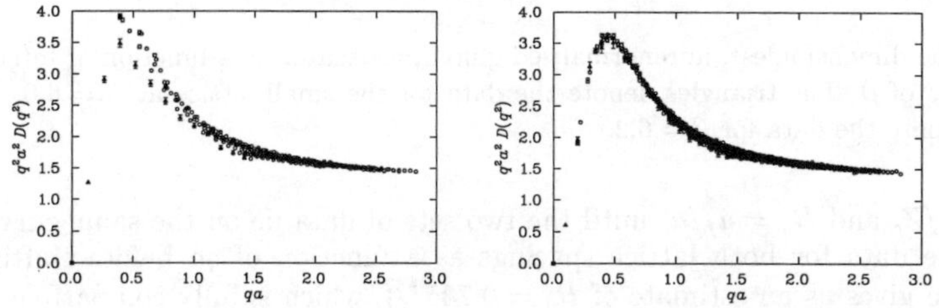

Fig. 1. Componentwise data, for the small lattice (left) and the large lattice (right). The filled triangles denote momenta directed along the time axis, while the filled squares denote momenta directed along one of the spatial axes.

momenta along the spatial axes. These discrepancies are absent from the data from the large lattice, indicating that finite size effects here are under control.

However, at higher momenta, there are anisotropies which remain for the large lattice data, and which are of approximately the same magnitude for the two lattices. In order to eliminate these anisotropies, which arise from large momenta (i.e., finite lattice spacing errors), we select momenta lying within a cylinder of radius $\Delta\hat{q}a = 2 \times 2\pi/32$ along the 4-dimensional diagonals in order to distribute large momenta across the four momentum components.[7]

2.4. *Scaling behaviour*

Since the renormalised propagator $D_R(q;\mu)$ is independent of the lattice spacing, we can derive a simple, q-independent expression for the ratio of the unrenormalised lattice gluon propagators at the same value of q:

$$\frac{D_c(qa_c)}{D_f(qa_f)} = \frac{Z_3(\mu, a_c)D_R(q;\mu)/a_c^2}{Z_3(\mu, a_f)D_R(q;\mu)/a_f^2} = \frac{Z_c}{Z_f}\frac{a_f^2}{a_c^2}, \tag{3}$$

where the subscript f denotes the finer lattice ($\beta = 6.2$ in this study) and the subscript c denotes the coarser lattice ($\beta = 6.0$). We can use this relation to study directly the scaling properties of the lattice gluon propagator by matching the data for the two values of β. This matching can be performed by adjusting the values for the ratios

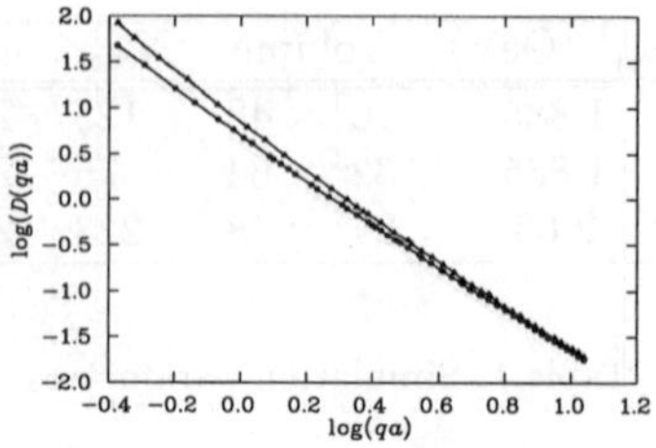

Fig. 2. The dimensionless, unrenormalised gluon propagator as a function of $\ln(qa)$ for the two values of β. The triangles denote the data for the small lattice at $\beta = 6.0$, while the circles denote the data for $\beta = 6.2$.

$R_Z = Z_f/Z_c$ and $R_a = a_f/a_c$ until the two sets of data lie on the same curve. Fig. 2 shows the data for both lattice spacings as a function of qa before shifting. This procedure gives us an estimate of $R_a = 0.745^{+32}_{-37}$, which is fully compatible with the string-tension value[8] of 0.716 ± 0.040. The corresponding estimate for the ratio of the renormalisation constants is $R_Z = 1.038^{+26}_{-21}$.

2.5. Model Fits

We have demonstrated scaling in our lattice data over the entire range of q^2 considered, and will now proceed with model fits. The following functional forms have been considered:

$$D(q^2) = \frac{Zq^2}{q^4+M^4}\left(\frac{1}{2}\ln\frac{q^2+M^2}{M^2}\right)^{-d_D} \qquad \text{(Gribov [3])} \tag{4}$$

$$D(q^2) = Z\frac{q^2}{q^4+2A^2q^2+M^4}\left(\frac{1}{2}\ln\frac{q^2+M^2}{M^2}\right)^{-d_D} \qquad \text{(Stingl [4])} \tag{5}$$

$$D(q^2) = \frac{Z}{(q^2)^{1+\alpha}+M^2} \qquad \text{(Marenzoni } et\ al \text{ [6])} \tag{6}$$

$$D(q^2) = Z\left[(q^2+M^2(q^2))\ln\frac{q^2+4M^2(q^2)}{\Lambda^2}\right]^{-1} \qquad \text{(Cornwall [9])} \tag{7}$$

$$\text{where} \quad M(q^2) = M\left\{\ln\frac{q^2+4M^2}{\Lambda^2}\Big/\ln\frac{4M^2}{\Lambda^2}\right\}^{-6/11}$$

$$D(q^2) = Z\left(\frac{A}{(q^2+M_{\rm IR}^2)^{1+\alpha}} + \frac{1}{q^2+M_{\rm UV}^2}\left(\frac{1}{2}\ln\frac{q^2+M_{\rm UV}^2}{M_{\rm UV}^2}\right)^{-d_D}\right) \tag{8}$$

$$D(q^2) = Z\left(\frac{A}{(q^2)^{1+\alpha}+(M_{\rm IR}^2)^{1+\alpha}} + \frac{1}{q^2+M_{\rm UV}^2}\left(\frac{1}{2}\ln\frac{q^2+M_{\rm UV}^2}{M_{\rm UV}^2}\right)^{-d_D}\right) \tag{9}$$

$$D(q^2) = Z\left(Ae^{-(q^2/M_{\rm IR}^2)^\alpha} + \frac{1}{q^2+M_{\rm UV}^2}\left(\frac{1}{2}\ln\frac{q^2+M_{\rm UV}^2}{M_{\rm UV}^2}\right)^{-d_D}\right) \tag{10}$$

We have also considered[10] special cases of the three forms (8)–(10), with $M_{\rm IR} = M_{\rm UV}$ or with specific values for the exponent α. Equations (4) and (5) are modified in

Model	χ^2/dof	Z	A	M_{IR}	M_{UV} or Λ	α
4	1972	2.19^{+31}_{-15}		0.23^{+14}_{-1}		
5	1998	2.2	0	0.23		
6	163	2.41^{+0}_{-12}		0.14^{+4}_{-14}		0.29^{+6}_{-2}
7	50.3	6.5^{+7}_{-9}		0.24^{+3}_{-16}	0.27^{+7}_{-7}	
8	2.96	1.54^{+10}_{-20}	1.24^{+21}_{-21}	0.46^{+2}_{-14}	0.96^{+47}_{-17}	1.31^{+16}_{-43}
8; $M_{\text{IR}}=M_{\text{UV}}$	3.73	1.71^{+9}_{-0}	0.84^{+0}_{-29}	0.48^{+2}_{-17}		1.52^{+12}_{-37}
9	1.57	1.78^{+45}_{-20}	0.49^{+17}_{-6}	0.43^{+5}_{-1}	0.20^{+37}_{-19}	0.95^{+7}_{-5}
9; $M_{\text{IR}}=M_{\text{UV}}$	4.00	1.62^{+3}_{-4}	0.58^{+5}_{-1}	0.40^{+9}_{-2}		0.92^{+17}_{-1}
10	47	2.09^{+30}_{-12}	29^{+166}_{-2}	0.22^{+0}_{-16}	0.14^{+11}_{-10}	0.49^{+0}_{-16}

Table 2. Parameter values for fits to models (4)–(10). The values quoted are for fits to the entire set of data. The errors denote the uncertainties in the last digit(s) of the parameter values which result from varying the fitting range. The fitting ranges considered when evaluating the uncertainties are those with a minimum of 40 points included and with the minimum value for qa no larger than 1.03 (point number 40).

order to exhibit the appropriate one-loop asymptotic behaviour. Models (8) and (9) are constructed as generalisations of (6) with the correct dimension and asymptotic behaviour. All models were fitted to the large lattice data using the cylindrical cut. The lowest momentum value was excluded, as the volume dependence of this point could not be assessed. In order to balance the sensitivity of the fit between the high- and low-momentum region, nearby data points within $\Delta(qa) < 0.05$ were averaged. χ^2 per degree of freedom and parameter values for fits to all these models are shown in Table 2. It is clear that model (9) accounts for the data better than any of the other models. The best fit to this model is illustrated in Fig. 3.

2.6. *Summary of Numerical Results*

We have evaluated the gluon propagator on an asymmetric lattice with a large physical volume. By studying the anisotropies in the data, and comparing the data with those from a smaller lattice, we have been able to conclude that finite size effects are under control on the large lattice. A clear turnover in the behaviour of $q^2D(q^2)$ has been observed at $q \sim 1\text{GeV}$, indicating that the gluon propagator diverges less rapidly than $1/q^2$ in the infrared, and may be infrared finite or vanishing. The data are consistent with a functional form $D(q^2) = D_{\text{IR}} + D_{\text{UV}}$, where

$$D_{\text{IR}} = \frac{1}{2}\frac{1}{q^4 + M^4}, \tag{11}$$

$M \sim 1$ GeV, and D_{UV} is the appropriate asymptotic form. A more detailed analysis[10] of the asymptotic behaviour reveals that the one-loop formula remains insufficient at $q^2 = 50\text{GeV}^2$. Issues for future study include the effect of Gribov copies and of dynamical fermions. We also hope to use improved actions to perform realistic simulations at larger lattice spacings. This would enable us to evaluate the gluon propagator on larger physical volumes, giving access to lower momentum values.

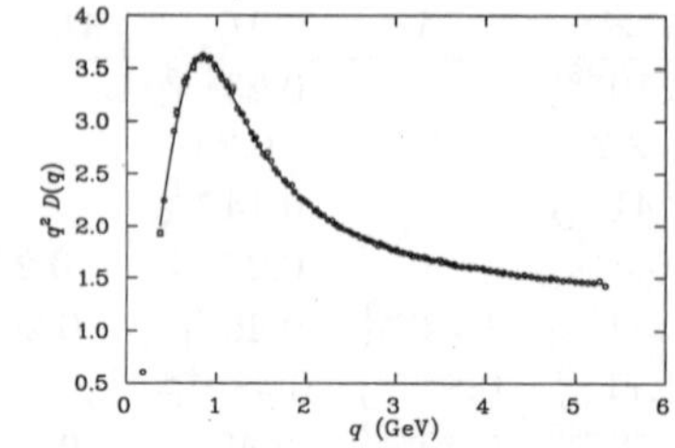

Fig. 3. The gluon propagator multiplied by q^2, with nearby points averaged. The line illustrates our best fit to the form defined in (9). The fit is performed over all points shown, excluding the one at the lowest momentum value, which may be sensitive to the finite volume of the lattice. The scale is taken from the string tension.[8]

3. Nucleon Strangeness Magnetic Moment

3.1. Motivation

We summarize some recent results[11] on nucleon electromagentic form factors, including the strangeness electric and magnetic form factors. The strangeness content of the nucleon has been a topic of considerable recent interest for a variety of reasons. The studies of nucleon spin structure functions in polarized deep inelastic scattering experiments at CERN and SLAC,[12] combined with neutron and hyperon β decays, have turned up a surprisingly large and negative polarization from the strange quark. In addition, there is a well-known long-standing discrepancy between the pion-nucleon sigma term extracted from the low energy pion-nucleon scattering[13] and that from the octect baryon masses.[14] This discrepancy can be reconciled if a significant $\bar{s}s$ content in the nucleon[14,15] is admitted. To address some of these issues, an experiment to measure the neutral weak magnetic form factor G_M^Z via elastic parity-violating electron scattering at backward angles was recently carried out by the SAMPLE collaboration.[16] The strangeness magnetic form factor is obtained by subtracting out the nucleon magnetic form factors G_M^p and G_M^n. The reported value is $G_M^s(Q^2 = 0.1\text{GeV}^2) = +0.23 \pm 0.37 \pm 0.15 \pm 0.19$ and does not yet provide a strong constraint on the sign.

Theoretical predictions of $G_M^s(0)$ vary widely. The values from various models and analyses range from -0.75 ± 0.30 in a QCD equalities analysis[17] to $+0.37$ in an $SU(3)$ chiral bag model.[18] While a few give positive values,[18,19] most model predictions are negative with a typical range of -0.25 to -0.45. Summaries of these predictions can be found in Refs.[17,20] A similar situation exists for the strangeness electric mean-square radius $\langle r_s^2 \rangle_E$. A number of the predictions are positive while a few are negative. Elastic $\vec{e}\ p$ and $\vec{e}\ {}^4He$ parity-violation experiments are currently planned at TJNAF[21] to measure the asymmetry A_{LR} at forward angles to extract $\langle r_s^2 \rangle_E$. Hopefully, they will settle the issue of its sign.

3.2. Numerical Results

The lattice formulation of the electromagnetic and other form factors has been given in detail in the past.[22,23] Here, we shall concentrate on the DI contribution, where the strangeness current contributes. In the Euclidean formulation, the Sachs EM form factors can be obtained by the combination of two- and three-point functions

$$G_{NN}^{\alpha\alpha}(t,\vec{p}) = \sum_{\vec{x}} e^{-i\vec{p}\cdot\vec{x}} \langle 0|\chi^\alpha(x)\bar{\chi}^\alpha(0)|0\rangle \tag{12}$$

$$G_{NV_\mu N}^{\alpha\beta}(t_f,\vec{p},t,\vec{q}) = \sum_{\vec{x}_f,\vec{x}} e^{-i\vec{p}\cdot\vec{x}_f+i\vec{q}\cdot\vec{x}} \langle 0|\chi^\alpha(x_f)V_\mu(x)\bar{\chi}^\beta(0)|0\rangle, \tag{13}$$

where χ^α is the nucleon interpolating field and $V_\mu(x)$ the vector current. With large Euclidean time separation, i.e. $t_f - t >> a$ and $t >> a$, where a is the lattice spacing,

$$\frac{\Gamma_i^{\beta\alpha} G_{NV_jN}^{\alpha\beta}(t_f,\vec{0},t,\vec{q})}{G_{NN}^{\alpha\alpha}(t_f,\vec{0})} \frac{G_{NN}^{\alpha\alpha}(t,\vec{0})}{G_{NN}^{\alpha\alpha}(t,\vec{q})} \longrightarrow \frac{\varepsilon_{ijk}q_k}{E_q+m} G_M(q^2), \tag{14}$$

$$\frac{\Gamma_E^{\beta\alpha} G_{NV_4N}^{\alpha\beta}(t_f,\vec{0},t,\vec{q})}{G_{NN}^{\alpha\alpha}(t_f,\vec{0})} \frac{G_{NN}^{\alpha\alpha}(t,\vec{0})}{G_{NN}^{\alpha\alpha}(t,\vec{q})} \longrightarrow G_E(q^2), \tag{15}$$

where $\Gamma_i = \sigma_i(1+\gamma_4)/2$ and $\Gamma_E = (1+\gamma_4)/2$.

We shall use the conserved current from the Wilson action which, being point-split, yields slight variations on the above forms and these are given in Ref.[23] Our 50 quenched gauge configurations were generated on a $16^3 \times 24$ lattice at $\beta = 6.0$. In the time direction, fixed boundary conditions were imposed on the quarks to provide larger time separations than available with periodic boundary conditions. We also averaged over the directions of equivalent lattice momenta in each configuration; this has the desirable effect of reducing error bars. Numerical details of this procedure are given in Refs.[23,24] The dimensionless nucleon masses $M_N a$ for $\kappa = 0.154$, 0.152, and 0.148 are 0.738(16), 0.882(12), and 1.15(1) respectively. The corresponding dimensionless pion masses $m_\pi a$ are 0.376(6), 0.486(5), and 0.679(4). Extrapolating the nucleon and pion masses to the chiral limit we determine $\kappa_c = 0.1567(1)$ and $m_N a = 0.547(14)$. Using the nucleon mass to set the scale to study nucleon properties,[22,24] the lattice spacing $a^{-1} = 1.72(4)$ GeV is determined. The three $\kappa's$ then correspond to quark masses of about 120, 200, and 360 MeV respectively.

The strangeness current $\bar{s}\gamma_\mu s$ contribution appears in the DI only. The full details of the extraction can be found in Ref.[11] and we satisfy ouselves here by simply quoting the results. We use a monopole form to extrapolate $G_M^s(q^2)$ with nonzero q^2 to $G_M^s(0)$, giving $G_M^s(0) = -0.36 \pm 0.20$. Correlations are taken into account and the error is from jackknifing the fitting procedure. A similar analysis is done for the strange Sachs electric form factor $G_E^s(q^2)$ and we find that $G_E^s(0)$ is consistent with zero as it should be. We find for the electric mean-square radius $\langle r_s^2\rangle_E = -6\frac{dG_E^s(q^2)}{dq^2}|_{q^2=0} = -0.061 \pm 0.003\,\text{fm}^2$.

Acknowledgments

Financial support from the Australian Research Council is gratefully acknowledged.

References

1. S. Mandelstam, *Phys. Rev.* **D 20**, 3223 (1979)
2. N. Brown and M.R. Pennington, *Phys. Rev.* **D 39**, 2723 (1989)
3. V.N. Gribov, *Nucl. Phys.* **B 139**, 19 (1978); D. Zwanziger, *Nucl. Phys.* **B 378**, 525 (1992)
4. M. Stingl, *Phys. Rev.* **D 34**, 3863 (1986); *Phys. Rev.* **D 36**, 651 (1987)
5. C. Bernard, C. Parrinello, A. Soni, *Phys. Rev.* **D49**, 1585 (1994)
6. P. Marenzoni, G. Martinelli, N. Stella, *Nucl. Phys.* **B 455**, 339 (1995); P. Marenzoni *et al*, *Phys. Lett.* **B 318**, 511 (1993)
7. D.B. Leinweber, J.I. Skullerud, A.G. Williams, and C. Parrinello*Phys. Rev.* **D 58**, 031501 (1998)
8. G.S. Bali and K. Schilling, *Phys. Rev.* **D 47**, 661 (1993)
9. J. Cornwall, *Phys. Rev.* **D 26**, 1453 (1982)
10. D.B. Leinweber, C. Parrinello, J.I. Skullerud, A.G. Williams, in preparation.
11. S.J. Dong, K.F. Liu, and A.G. Williams, hep-ph/9712483, to appear in Phys. Rev. D.
12. J. Ashman *et al.*, Nucl. Phys. **B328**, 1 (1989); B. Adeva *et al.*, Phys. Lett. B **329**, 399 (1994); K. Abe *et al.*, Phys. Rev. Lett. **74**, 346 (1995).
13. J. Gasser, H. Leutwyler, and M.E. Sainio, Phys. Lett. B **253**, 252, 260 (1991).
14. T.P. Cheng, Phys. Rev. **D 13**, 2161 (1976); **D 38**, 2869 (1988).
15. J. Gasser and H. Leutwyler, Phys. Rep. **87**, 77 (1982).
16. B. Mueller *et al.*, SAMPLE Collaboration, Phys. Rev. Lett. **78**, 3824 (1997).
17. D. B. Leinweber, Phys. Rev. **D 53**, 5115 (1996).
18. S. T. Hong, B. Y. Park, and D. P. Min, nucl-th/9706008.
19. P. Geiger and N. Isgur, Phys. Rev. **D 55**, 299 (1997).
20. C. V. Christov *et. al.*, Prog. Part. Nucl. Phys. **37**, 1 (1996).
21. Thomas Jefferson Lab. proposals # PR-91-010, M. Finn and P. Souder, spokespersons , # PR-91-017, D. Beck, spokesperson, and # PR-91-004, E. Beise, spokesperson.
22. S.J. Dong, J.-F. Lagaë, and K.-F. Liu, Phys. Rev. Lett. **75**, 2096 (1995); Phys. Rev. bf D 54, 5496 (1996).
23. T. Draper. R. M. Woloshyn, and K. F. Liu, Phys. Lett. **234**, 121 (1990); W. Wilcox, T. Draper, and K. F. Liu, Phys. Rev. **D 46**, 1109 (1992).
24. K. F. Liu, S. J. Dong, T. Draper, J. M. Wu, and W. Wilcox, Phys. Rev. **D 49**, 4755 (1994); K.F. Liu, S.J. Dong, T. Draper, and W. Wilcox, Phys. Rev. Lett. **74**, 2172 (1995).

Origin of Abelian Dominance in QCD in the Maximally Abelian Gauge

H. Suganuma[a], H. Ichie[b] and K. Amemiya[a]

a) *Research Center for Nuclear Physics, Osaka University, Osaka, 567-0047, Japan*
b) *Department of Physics, Tokyo Institute of Technology, Tokyo, 152-8551, Japan*

ABSTRACT

We study the origin of abelian dominance in the maximally abelian (MA) gauge in Euclidean QCD. In the MA gauge, the amplitude of the off-diagonal charged gluon $A_\mu^{\pm}(x)$ is strongly suppressed, and then the phase variable $\chi_\mu(x)$ of $A_\mu^{\pm}(x)$ tends to be a random variable due to the weak constraint from the QCD action. Abelian dominance for confinement is proved with the random-variable approximation for $\chi_\mu(x)$. Using the lattice QCD, we study the gluon propagator in the MA gauge, and find the generation of the effective charged-gluon mass $m_{\rm ch} \simeq 0.9$GeV, which provides the general origin of abelian dominance for the long-distance physics in the MA gauge. We consider the structure of the hadron flux-tube in the MA gauge, and conjecture the charged-gluon cloud around the abelian electric-flux core, which leads to the thickness difference about $m_{\rm ch}^{-1} \simeq 0.2$fm between SU(2) and abelian flux-tubes.

1. Gluon Properties in the Maximally Abelian Gauge

In solving problems of mechanics, a 'suitable choice' of the coordinate frame often leads to a transparent answer. Similarly, in gauge theories, there may be a 'suitable choice' of the internal coordinate frame, *i.e.* suitable gauge fixing, according to the physical purpose. For quark confinement in terms of the abelian dual-Higgs theory the maximally abelian (MA) gauge fixing seems the suitable gauge choice in QCD.

The MA gauge is defined to minimize $R_{\rm off}[A_\mu(\cdot)] \equiv \int d^4x \sum_\alpha {\rm tr}[\hat D_\alpha, \vec H]^\dagger [\hat D_\alpha, \vec H]$, with $\hat D_\mu \equiv \hat\partial_\mu + ieA_\mu$, in Euclidean QCD. In the MA gauge, the off-diagonal gluon component is forced to be as small as possible by the SU(N_c) gauge transformation.[1,2] Therefore, the gluon field $A_\mu \equiv A_\mu^a T^a$ mostly approaches the abelian gluon $\vec A_\mu \cdot \vec H$. In the MA gauge, the gauge symmetry ${\rm SU}(N_c)_{\rm local}$ is reduced into ${\rm U}(1)_{\rm local}^{N_c-1} \times {\rm Weyl}_{\rm global}^{N_c}$, where the *global Weyl symmetry* is the subgroup of SU(N_c) relating the permutation of the bases in the fundamental representation.[1,2] Since $\hat D^\mu$ obeys the adjoint gauge transformation, the local condition of the MA gauge fixing is $[\vec H, [\hat D_\mu, [\hat D_\mu, \vec H]]] = 0$.

In the lattice QCD described by the link variable $U_\mu(s) \equiv \exp\{iaeA_\mu(s)\}$, the MA gauge is defined by maximizing $R_{\rm diag}[U_\mu(\cdot)] \equiv \sum_{s,\alpha} {\rm tr}\{U_\alpha(s)\vec H U_\alpha^\dagger(s)\vec H\}$. For SU(2) QCD, the link variable is factorized as $U_\mu(s) = M_\mu(s) u_\mu(s)$ corresponding to the Cartan decomposition. Here, $u_\mu(s) \equiv \exp\{i\tau^3\theta_\mu^3(s)\} \in {\rm U}(1)_3$ denotes the abelian link variable, and the off-diagonal matrix $M_\mu(s) \in {\rm SU}(2)/{\rm U}(1)_3$ is parametrized as

$$M_\mu(s) \equiv e^{i\{\tau^1\theta_\mu^1(s) + \tau^2\theta_\mu^2(s)\}} = \begin{pmatrix} \cos\theta_\mu(s) & -\sin\theta_\mu(s) e^{-i\chi_\mu(s)} \\ \sin\theta_\mu(s) e^{i\chi_\mu(s)} & \cos\theta_\mu(s) \end{pmatrix},$$

where $\chi_\mu(s)$ corresponds to the phase variable of the off-diagonal charged gluon $A_\mu^\pm(x)$ in the continuum limit. The abelian projection is defined by the replacement as $U_\mu(s) \rightarrow u_\mu(s)$.

In the MA gauge, the *diagonal element* $\cos\theta_\mu(s)$ in $M_\mu(s)$ is maximized by a gauge transformation, *e.g.* $\langle\cos\theta_\mu(s)\rangle_{\rm MA} \simeq 0.93$ at $\beta = 2.4$. Accordingly, the *off-diagonal element* $e^{i\chi_\mu(s)}\sin\theta_\mu(s)$ in $M_\mu(s)$ has two relevant features in the MA gauge.[1,2]

(1) The off-diagonal amplitude $|\sin\theta_\mu(s)|$ is forced to be small in the MA gauge, which allows the approximate treatment on the off-diagonal element.

(2) The phase variable $\chi_\mu(s)$ of the charged gluon is not constrained by the MA gauge-fixing condition at all, and tends to be a random variable due to the weak constraint from the QCD action.

Of course, near the continuum limit, $\chi_\mu(s)$ may also approach to a continuous variable, however, the amplitude $|\sin\theta_\mu(s)|$ itself becomes almost zero in the MA gauge,[2] which would allow any approximation on $\chi_\mu(s)$. Hence, $\chi_\mu(s)$ can be regarded as a *random angle variable* in the MA gauge in a good approximation. (However, only near the monopole, there remains a large amplitude of $|\sin\theta_\mu(s)|$ even in the MA gauge,[1,2] and $\chi_\mu(s)$ is strongly constrained so as to make the QCD action small.[2])

2. Proof of Abelian Dominance for Confinement in the MA gauge

We analytically prove abelian dominance on the string tension in the MA gauge,[1,2] using the random-variable approximation for the phase variable $\chi_\mu(s)$ of the charged gluon. Here, we use $\langle e^{i\chi_\mu(s)}\rangle_{\rm MA} \simeq \int_0^{2\pi} d\chi_\mu(s)\exp\{i\chi_\mu(s)\} = 0$. In the Wilson loop $\langle W_C[U]\rangle \equiv \langle {\rm tr}\Pi_C U_\mu(s)\rangle = \langle {\rm tr}\Pi_C\{M_\mu(s)u_\mu(s)\}\rangle$, the off-diagonal matrix $M_\mu(s)$ is reduced as a c-number factor, $M_\mu(s) \rightarrow \cos\theta_\mu(s)\,\mathbf{1}$, and then the SU(2) link variable $U_\mu(s)$ is reduced to be a *diagonal matrix*, $U_\mu(s) \equiv M_\mu(s)u_\mu(s) \rightarrow \cos\theta_\mu(s)u_\mu(s)$, after the integration over $\chi_\mu(s)$.

We derive a formula for the *charged-gluon contribution to the Wilson loop* as[1,2]

$$W_C^{\rm off} \equiv \langle W_C[U]\rangle / \langle W_C[u]\rangle_{\rm MA} \simeq \exp\{L\langle \ln(\cos\theta_\mu(s))\rangle_{\rm MA}\}, \tag{1}$$

with the perimeter length L and the abelian Wilson loop $\langle W_C[u]\rangle_{\rm MA}$. This formula provides the relation between the *macroscopic* quantity $W_C^{\rm off}$ and the *microscopic* quantity $\langle\ln(\cos\theta_\mu(s))\rangle_{\rm MA}$. (Using the lattice QCD, we have checked this relation for large loops,[1,2] where such a statistical treatment is accurate.) Thus, $W_C^{\rm off}$ obeys the *perimeter law* in the MA gauge, and then the charged-gluon contribution to the string tension vanishes as $\sigma_{\rm SU(2)} - \sigma_{\rm Abel} \simeq -2\langle\ln\{\cos\theta_\mu(s)\}\rangle_{\rm MA} \lim_{R,T\to\infty}\frac{R+T}{RT} = 0$. In this way, ***abelian dominance for the string tension***, $\sigma_{\rm SU(2)} = \sigma_{\rm Abel}$, can be proved in the MA gauge only with the random-variable approximation for the phase variable $\chi_\mu(s)$ of the charged gluon.[1,2] (Since the electric current j_μ does not contribute to electric confinement, this abelian dominance leads to monopole dominance for confinement.)

Also, we can predict the deviation between $\sigma_{SU(2)}$ and σ_{Abel} as $\sigma_{SU(2)} > \sigma_{Abel}$ as the *finite size effect* on R and T in the Wilson loop.[1,2]

3. Effective Charged-Gluon Mass induced in MA Gauge

To clarify the general origin of infrared abelian dominance, we study the Euclidean gluon propagator $G^{ab}_{\mu\nu}(x-y) \equiv \langle A^a_\mu(x) A^b_\nu(y)\rangle$ in the MA gauge, using the SU(2) lattice QCD.[1,3] As for the residual $U(1)_3$ gauge symmetry, the $U(1)_3$ Landau gauge fixing is imposed to extract most continuous gauge configuration under the constraint of the MA gauge fixing. Due to the gauge fixing, $U_\mu(s)$ mostly approaches to **1** in the relevant gauge configuration in the lattice QCD, and it becomes possible to compare with the continuum theory. The continuum gluon field $A^a_\mu(x)$ is extracted from the link variable as $U_\mu(s) \equiv \exp\{iaeA^a_\mu(s)\frac{\tau^a}{2}\}$.

Here, the scalar combination $G^a_{\mu\mu}(r) \equiv \sum_{\mu=1}^4 \langle A_\mu{}^a(x) A_\mu{}^a(0)\rangle$ $(a=1,2,3)$ is useful to observe the interaction range of the gluon, because it depends only on the four-dimensional Euclidean radial coordinate $r \equiv (x_\mu x_\mu)^{\frac{1}{2}}$. In Fig.1, we show $G^3_{\mu\mu}(r)$ and $G^{+-}_{\mu\mu}(r) \equiv \sum_{\mu=1}^4 \langle A^+_\mu(x) A^-_\mu(0)\rangle = \frac{1}{2}\{G^1_{\mu\mu}(r) + G^2_{\mu\mu}(r)\}$ in the MA gauge using the SU(2) lattice QCD with $12^3 \times 24$ and $2.2 \le \beta \le 2.4$. We find *infrared abelian dominance for the gluon propagator in the MA gauge*: only the abelian gluon $A^3_\mu(x)$ propagates over the long distance and can influence the long-distance physics.[1,3]

Since the propagator of the massive vector boson with mass M takes a Yukawa-type asymptotic form as $G_{\mu\mu}(r) \sim \frac{M^{1/2}}{r^{3/2}}\exp(-Mr)$, the effective mass $m_{\rm ch}$ of the charged gluon $A^\pm_\mu(x)$ can be evaluated from the slope of the logarithmic plot of $r^{3/2}G^{+-}_{\mu\mu}(r) \sim \exp(-m_{\rm ch}r)$ as shown in Fig.1(b). The charged gluon $A^\pm_\mu(x)$ behaves as a massive field with $m_{\rm ch} \simeq 0.94$ GeV in the MA gauge for $r \gtrsim 0.2$ fm. Then, the *general origin of infrared abelian dominance* in the MA gauge can be physically interpreted by the *generation of the effective charged-gluon mass* $m_{\rm ch}$ induced by the MA gauge fixing.[1,3] Due to the effective mass $m_{\rm ch}$, the charged-gluon propagation is limited within the short range as $r \lesssim m_{\rm ch}^{-1}$ in the MA gauge, and hence charged gluons $A^\pm_\mu(x)$ cannot contribute to the long-distance physics, which leads to abelian dominance for NP-QCD.

4. Hadronic Flux-Tube Structure in the MA Gauge

Finally, we consider the structure of the hadronic flux-tube[5] in the MA gauge in terms of the effective mass of off-diagonal charged gluons $A^\pm_\mu(x)$. Here, $A^\pm_\mu(x)$ also contributes to the microscopic structure of hadrons, as well as the abelian gluon $A^3_\mu(x)$ even in the MA gauge, where $A^\pm_\mu(x)$ is negligible at the large scale.[1-4] Owing to the gluon self-interaction in QCD, there is a coupling between $A^3_\mu(x)$ and $A^\pm_\mu(x)$, and hence the charged-gluon cloud may be formed around the abelian electric-flux tube as shown in Fig.2. (This is similar to the pion cloud around the nucleon.) The charged-gluon cloud may provide the thickness difference about $m_{\rm ch}^{-1} \simeq 0.2$fm between SU(2) and abelian flux-tubes in the MA gauge, which is observed in the lattice QCD.[6]

References

1. H. Suganuma, H. Ichie, A. Tanaka and K. Amemiya, Prog. Theor. Phys. Suppl. **131** (1998) in press : hep-lat/9804027 and its references.
2. H. Ichie and H. Suganuma, hep-lat/9807025 ; hep-lat/9808054 ; this proceedings.
3. K. Amemiya and H. Suganuma, in *INNOCOM '97* (World Scientific, 1998).
4. K.-I. Kondo, Phys. Rev. **D** (1998) in press : hep-th/9805153.
5. H. Monden, H. Ichie, H. Suganuma and H. Toki, Phys. Rev. **C57** (1998) 2564.
6. G. S. Bali, this proceedings.

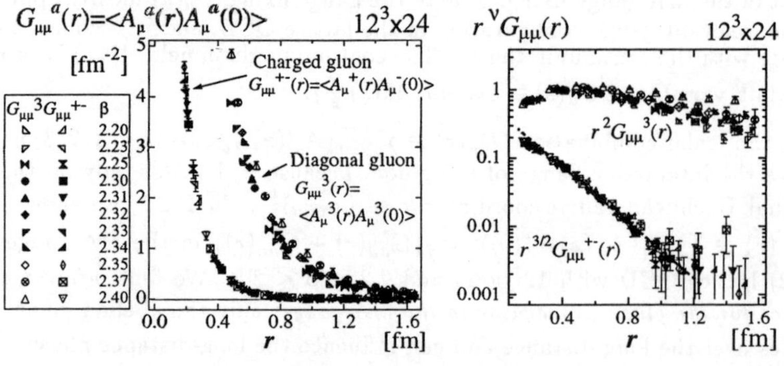

Figure 1: (a) The scalar correlation $G^a_{\mu\mu}(r)$ of the gluon propagator as the function of the 4-dimensional distance r in the MA gauge in the SU(2) lattice QCD with $12^3 \times 24$ and $2.2 \leq \beta \leq 2.4$. In the MA gauge, the off-diagonal (charged) gluon propagates within the short-range region as $r \lesssim 0.4$ fm, and cannot contribute to the long-range physics. (b) The logarithmic plot for the scalar correlation $r^\nu G^a_{\mu\mu}(r)$. The charged-gluon propagator behaves as the Yukawa-type function, $G^{+-}_{\mu\mu} \sim \frac{\exp(-m_{\rm ch} r)}{r^{3/2}}$. The effective mass of the charged gluon can be estimated as $m_{\rm ch} \simeq 0.9$GeV from the dotted line.

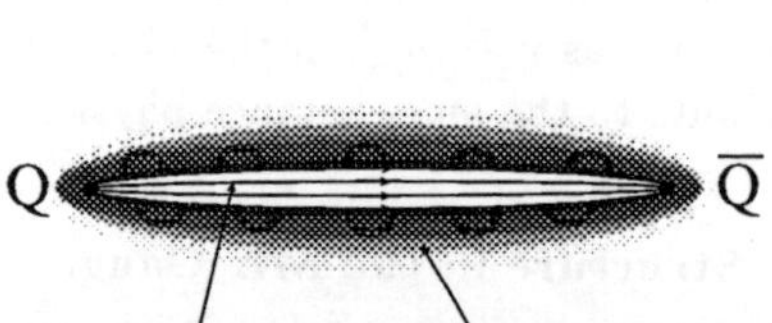

Figure 2: The structure of the hadronic flux-tube in the MA gauge. The charged-gluon cloud with the thickness of $m_{\rm ch}^{-1} \simeq 0.2$ fm would be formed around the abelian electric-flux core.

Comparison of $SU(3)$ Lattice Results on Heavy Quark Potentials with the Predictions of the Dual Superconductor Model

M. Baker

Department of Physics, University of Washington, Seattle, WA 98105, USA

Abstract

Bali, Schilling, and Wachter have recently calculated from lattice theory all of the heavy quark potentials – the central potential, all spin dependent potentials, and all velocity dependent potentials – through order (quark mass)$^{-2}$. In this talk we compare these potentials with the results obtained from the dual superconductor model of long distance QCD. This model has two parameters, the strong coupling constant α_s and the vacuum expectation value of the dual Higgs field ϕ_0. We determine the value of these parameters by fitting the lattice data for the central potential $V_0(R)$ to the expression given by the dual theory. This gives $\alpha_s \sim .2$, $\phi_0 \sim 215MeV$. The remaining 9 potentials are then uniquely predicted and are all compatible with the lattice data for distances greater than $.2fm$. This data shows that the QCD flux tube (string) can be described as a dual Abrikosov-Nielsen-Olesen vortex and provides quantitative support for the dual superconductor mechanism of confinement.

I. INTRODUCTION

Bali, Schilling, and Wachter [1] have recently determined the complete spin-dependent and momentum-dependent interquark potential from lattice simulations of QCD. We had previously computed the same potentials from a dual theory of long distance Yang-Mills theory [2] [3] [4]. In this talk, we will compare the lattice data with these predictions.

The dual theory is an effective theory of long-distance Yang-Mills theory described by a Lagrangian density $\mathcal{L}_{\text{eff}}$ in which the fundamental variables are an octet of dual potentials C_μ coupled minimally to three octets of scalar Higgs fields B_i carrying magnetic color charge [the gauge coupling constant of dual theory $g = \frac{2\pi}{e}$ where e is the Yang-Mills coupling constant.] The monopole fields B_i develop nonvanishing vacuum expectation values B_{0i} (monopole condensation) that give rise to massive C_μ and consequently to a dual Meissner effect. Dual potentials couple to electric color charge like ordinary potentials couple to monopoles. The potentials C_μ thus couple to a quark-antiquark pair via a Dirac string connecting the pair. The dual Meissner effect prevents the electric color flux from spreading out as the distance R between the quark-antiquark pair increases. As a result, a linear potential develops that confines the quarks in hadrons. The dual theory then provides a concrete realization of the Nambu-Mandelstam-'t Hooft dual superconductor picture of confinement [5].

Since the dual theory contains no Higgs fields in the fundamental representation of $SU(3)$, its gauge group is $\frac{SU(3)}{Z_3}$. We assume this symmetry then breaks down in stages:

$$\frac{SU(3)}{Z_3} \to U(1) \times U(1) \to U(1) \to \text{none}\,. \tag{1.1}$$

The lightest gauge boson arises from the spontaneous breakdown of the final $U(1)$ symmetry group. To very good approximation this $U(1)$ theory can be used (with a suitable color factor) to describe the Z_3 flux tube connecting the $q\bar{q}$ pair.

The Lagrangian $\mathcal{L}_{\text{eff}}$ can then be replaced by an Abelian Higgs Lagrangian:

$$\mathcal{L}_{\text{eff}}(G^S_{\mu\nu}) = \frac{4}{3}\left\{-\frac{(G_{\mu\nu}G^{\mu\nu})}{4} + \frac{|(\partial_\mu - igC_\mu)\phi|^2}{2} - \frac{\lambda}{4}\left(|\phi|^2 - \phi_0^2\right)^2\right\}, \tag{1.2}$$

where

$$G_{\mu\nu} \equiv \partial_\mu C_\nu - \partial_\nu C_\mu + G^S_{\mu\nu}\,, \tag{1.3}$$

and

$$G^S_{\mu\nu}(x) = -e\epsilon_{\mu\nu\alpha\beta}\int d\sigma \int d\tau \frac{\partial y^\alpha}{\partial \sigma}\frac{\partial y^\beta}{\partial \tau}\delta(x^\mu - y^\mu(\sigma,\tau)). \tag{1.4}$$

The integration in (1.4) is over the surface $y^\mu(\sigma,\tau)$ swept out by the Dirac string [6] attached to a $q\bar{q}$ pair moving along world lines Γ_1 and Γ_2. This surface $y^\mu(\sigma,\tau)$ is bounded by the loop Γ (See Fig. 1). The term $-\frac{1}{4}G_{\mu\nu}G^{\mu\nu}$ in (1.2) contains the coupling of the dual potentials to the $q\bar{q}$ pair via the Dirac string tensor $G^S_{\mu\nu}$. The notation $\mathcal{L}_{\text{eff}}(G^S_{\mu\nu})$ makes explicit this dependence. The values of the dual gluon mass M and the monopole mass M_ϕ obtained from (1.2) are

$$M = g\phi_0\,, \qquad\qquad M_\phi = \sqrt{2\lambda}\phi_0. \tag{1.5}$$

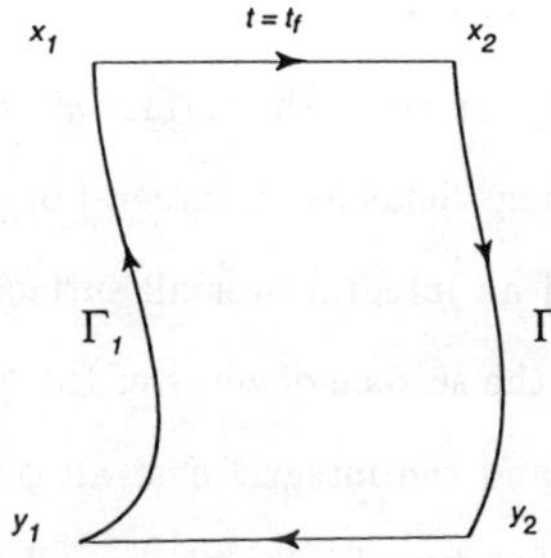

FIG. 1. Figure 1: The loop Γ.

For static quarks separated by a distance R there are classical solutions of the equation of motion having color flux confined to narrow tubes of radius $R_{FT} \sim \frac{1}{M}$ (Abrikosov-Nielsen-Olesen vortices) [7]. The monopole field $\phi(x)$ vanishes on the line connecting the $q\bar{q}$ pair and it approaches its vacuum value ϕ_0 at transverse distances greater than $\frac{1}{M_\phi}$. For small $q\bar{q}$ separations, $R < \frac{1}{M_\phi}$, the color electric field generated by the quarks expel the monopole condensate from the bulk of the region between them, and the field distribution becomes Coulomb-like.

The quantity $\frac{\lambda}{g^2}$ is the Landau-Ginzburg parameter of the dual theory. Its value can be estimated by relating the difference between the energy density at large distances from the flux tube and the energy density at its center to the gluon condensate. This give a value of $\frac{\lambda}{g^2} \approx \frac{1}{2}$, which corresponds to the border between a type I and a type II superconductor. There remain two free parameters in $\mathcal{L}_{\text{eff}}$ which we take to be $\alpha_s \equiv \frac{\pi}{g^2}$ and ϕ_0.

We denote $\mathcal{W}_{\text{eff}}(\Gamma)$ the Wilson loop of the dual theory:

$$W_{\text{eff}}(\Gamma) = \frac{\int \mathcal{D}C_\mu \mathcal{D}\phi \mathcal{D}\phi^* e^{i\int dx[\mathcal{L}_{\text{eff}}(G^S_{\mu\nu})+\mathcal{L}_{GF}]}}{\int \mathcal{D}C_\mu \mathcal{D}\phi \mathcal{D}\phi^* e^{i\int dx[\mathcal{L}_{\text{eff}}(G^S_{\mu\nu}=0)+\mathcal{L}_{GF}]}}, \tag{1.6}$$

where $\mathcal{L}_{GF}$ is a gauge-fixing term. The functional integral $\mathcal{W}_{\text{eff}}(\Gamma)$ determines in the effective dual theory the same physical quantities as the Wilson loop $W(\Gamma)$ in Yang-Mills theory, namely the effective action in the presence of a quark-antiquark pair moving along classical trajectories. The coupling of dual potentials to $G^S_{\mu\nu}$ in $\mathcal{W}_{\text{eff}}(\Gamma)$ plays the role of the factor $Pe^{-ie\oint_\Gamma dx^\mu A_\mu(x)}$ in the Wilson loop $W(\Gamma)$.

The Higgs fields $\phi(x)$ in (1.6) vanish on the surface $y^\mu(\sigma,\tau)$. We can then decompose the functional integral (1.6) over Higgs fields as an integral over all configurations which vanish on a fixed surface $y^\mu(\sigma,\tau)$, and an integral over all surfaces $y^\mu(\sigma,\tau)$. The surface $y^\mu(\sigma,\tau)$ in (1.4) can then be defined as the surface of zeros of the Higgs field. Hence the Higgs field configuration determines $G^S_{\mu\nu}$, and the integral over all ϕ in (1.6) produces an integration over all surfaces $y^\mu(\sigma,\tau)$. The functional integral over all field configurations which vanish on a given surface then defines the action for the effective string theory of Abrikosov-Nielsen-Olesen vortices [8]. We have evaluated $\mathcal{W}_{\text{eff}}(\Gamma)$ in the classical approximation, which comes from the contribution to (1.6) from the minimal energy configuration where the axis of the flux tube lies along the straight line connecting the $q\bar{q}$ pair.

To evaluate the long distance $q\bar{q}$ potential, we use $\mathcal{W}_{\text{eff}}(\Gamma)$ to approximate the Wilson loop $W(\Gamma)$ of Yang-Mills theory. Large distances means that the size R of the loop is large compared to the inverse of M and M_ϕ. However, since the classical dual field becomes

Coulomb-like for short distances, $\mathcal{W}_{\text{eff}}(\Gamma)$ will generate the lowest order perturbative interaction as $R \to 0$. The dual theory should then remain applicable down to distances where radiative corrections giving rise to a running coupling constant become important.

In the next section we compare the heavy quark potentials obtained from $\mathcal{W}_{\text{eff}}(\Gamma)$, calculated in the classical approximation to the dual theory, with the lattice data [1]. The contributions of the fluctuations of the flux tube which give rise to an effective string theory are not accounted for in these calculations.

II. THE HEAVY QUARK POTENTIAL

The heavy quark potential [1]

$$V = \bar{V}(R) + V_{SD}(\vec{R}, \vec{L}, \vec{S}_1, \vec{S}_2) + V_{MD}(\vec{R}, \vec{P}) \tag{2.1}$$

consists of a central part $\bar{V}$, a spin dependent part V_{SD}, and a momentum dependent part V_{MD}. The central part

$$\bar{V} = V_0(R) + \frac{1}{8}(\frac{1}{m_1^2} + \frac{1}{m_2^2})\nabla^2(V_0 + V_a) \tag{2.2}$$

includes the static potential $V_0(R)$ and corrections proportional to the sum of the inverse squares of the quark masses m_1 and m_2.

In lowest order perturbative QCD

$$V_0 = V_0^{PT} \equiv -\frac{4}{3}\frac{\alpha_s}{R}\,, \qquad\qquad V_a = 0\,. \tag{2.3}$$

The second term in (2.2) is then proportional to $\delta(R)$.

The non-perturbative part of $\nabla^2 V_a$ consists of an electric part $\nabla^2 V_a^E$ and a magnetic part $-\nabla^2 V_a^B$. The dual theory expression of the magnetic part is weakly divergent and requires a cutoff as does the lattice calculation. The quantity $\nabla^2(V_0 + V_a^E)$ is proportional to the divergence of the total color electric field [3], which in the dual theory is determined by the quarks alone. This is because the Higgs field ϕ carries only color magnetic charge and does not contribute to the divergence of the color electric field. Hence, in the dual theory,

$$\nabla^2(V_0 + V_a^E) = \nabla^2 V_0^{PT} = \frac{4}{3}\alpha_S \delta(R). \tag{2.4}$$

That is, $-\nabla^2 V_a^E$ is determined by the non-perturbative part of the central potential, $\nabla^2(V_0 - V_0^{PT})$.

The spin dependent potential V_{SD} contains a purely non-perturbative spin-orbit contribution of the form

$$\left(\frac{\vec{S}_1}{m_1^2} - \frac{\vec{S}_2}{m_2^2}\right) \cdot \vec{L}\, V_1'(R). \tag{2.5}$$

The remaining contributions to V_{SD} include a second spin-orbit interaction and two spin-spin interactions, determined by functions denoted by $V_2'(R)$, $V_3(R)$, and $V_4(R)$, respectively, all of which, in contrast to $V_1'(R)$, have perturbative contributions. The same is true of the four functions $V_b(R)$, $V_c(R)$, $V_d(R)$, and $V_e(R)$, which determine V_{MD} [1]. In Figs. 2 – 11 we compare the lattice data for these 10 functions of R determining the heavy quark potential with the predictions of the dual theory. The parameters α_s and ϕ_0 are chosen to give the best fit to the lattice data of the central potential shown in Fig. 2. These values are

$$\alpha_s \approx .2\,, \qquad \phi_0 \sim 215 MeV\,. \tag{2.6}$$

Thus $\phi_0 \sim \Lambda_{QCD}$, and the value of α_s is the value of the running of the coupling constant of Yang-Mills theory at a scale of about .1 fm. From (2.6) the dual gluon mass $M = g\phi_0 \sim 900 MeV$.

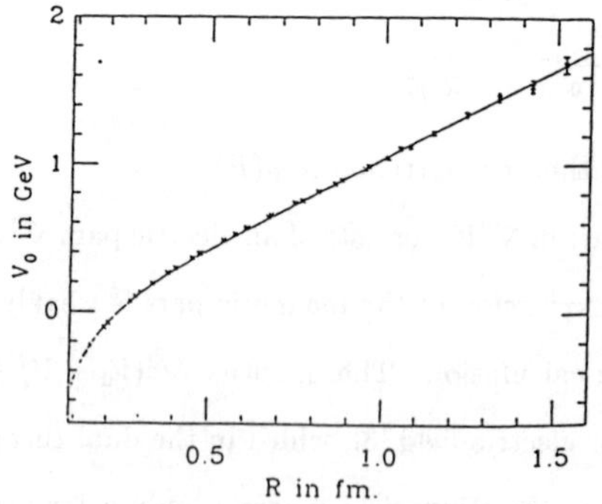

FIG. 2. Figure 2: Comparison of the dual theory potential (solid line) with the lattice central potential (points). (The dual theory parameters are $\phi_0 = 215 Mev, \alpha_s = .2$).

Figures 3 – 11 show the resulting parameter-free dual QCD predictions and the lattice data for the other potentials. All these agree remarkably well (within the lattice uncertainties). As is seen from Figures 5 – 11, all potentials having perturbative contributions are compatible with the lattice data down to distances of about $R \sim .1fm$. The predictions for V_1' and $\nabla^2 V_a^E$ (Figs. 3 and 4) deviate from the lattice data at distances $R \leq .2fm$. The derivations at short distance from the predictions of the dual theory first show up in these potentials because, unlike the other potentials, they have no perturbative contributions. (As explained above, the dual theory result for the heavy quark potentials approaches lowest order perturbation theory as $R \to 0$; this is the extent to which the theory is relevant at distances $< \frac{1}{M} \sim .2fm$.)

III. SUMMARY

The lattice data of Bali, Schilling, and Wachter on the heavy quark potentials shows that the QCD flux tube (string) can be described as a dual Abrikosov-Nielsen-Olesen vortex for quark-antiquark separations $R > .2fm$. Their data then provides quantitative support for the dual superconductor mechanism of confinement.

FIGURES 3 - 11

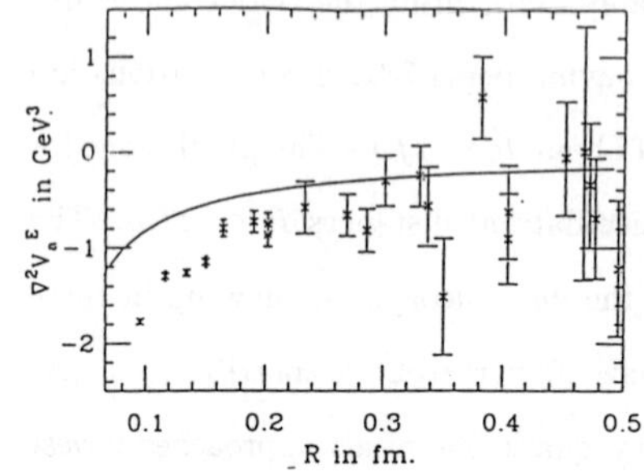

FIG. 3. Figure 3: A similar comparison (using the same parameters) for the quantity $\nabla^2 V_a^E$.

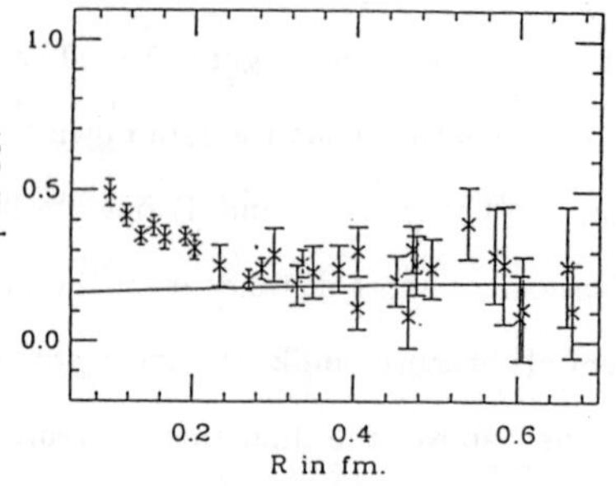

FIG. 4. Figure 4: Same as for Fig. 2 but for $-V_1'$. (Note the minus sign.)

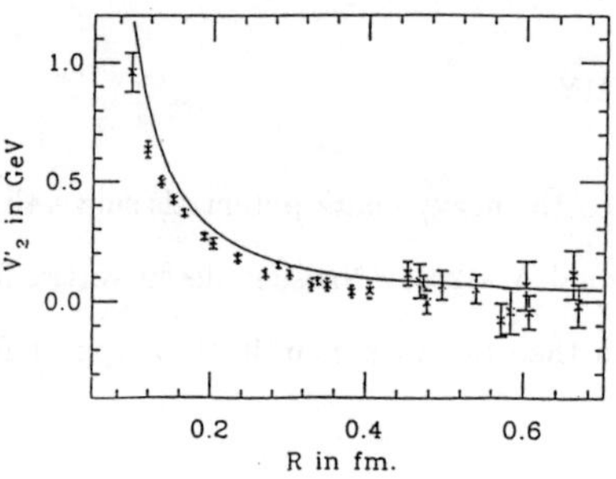

FIG. 5. Figure 5: Same as Fig. 2 but for V_2'.

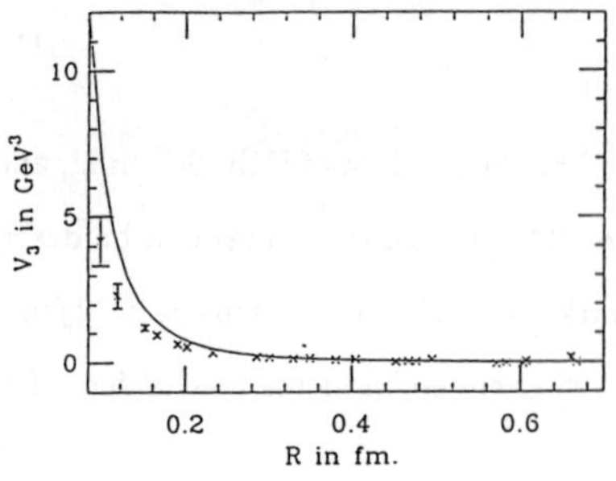

FIG. 6. Figure 6: Same as Fig. 2 but for V_3.

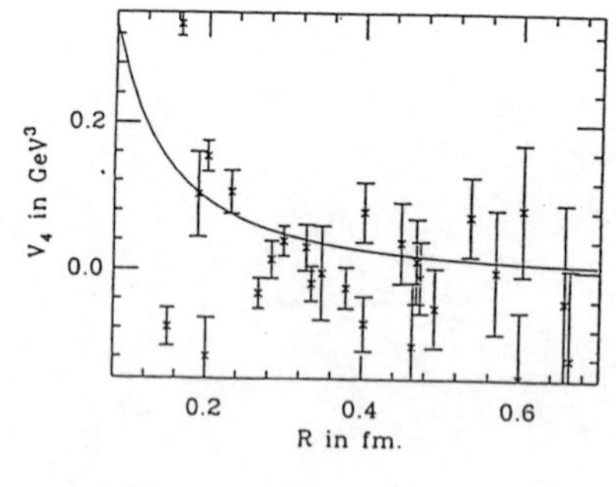

FIG. 7. Figure 7: Same as Fig. 2 but for V_4.

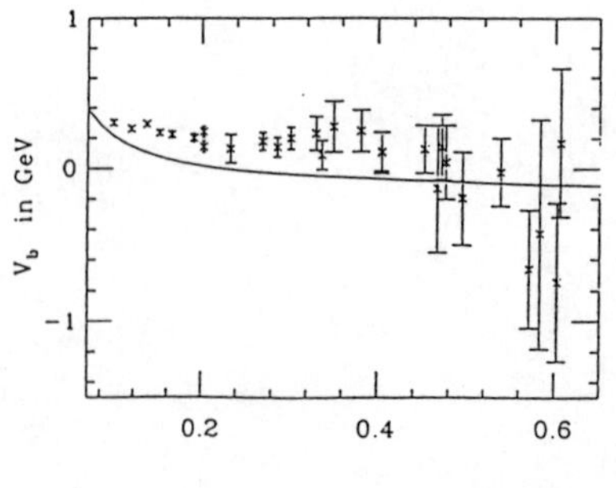

FIG. 8. Figure 8: Same as Fig. 7 but for V_b.

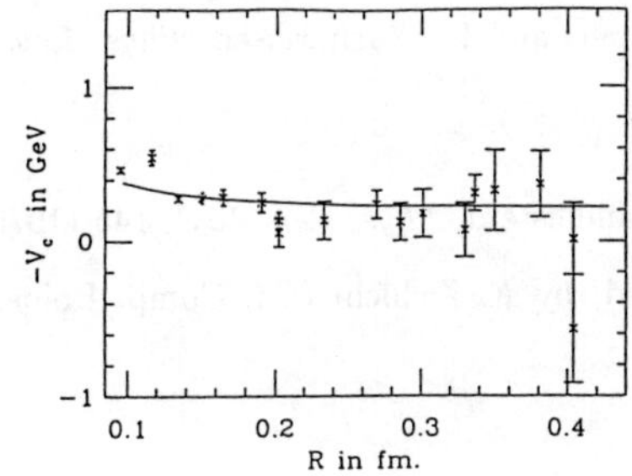

FIG. 9. Figure 9: Same as Fig. 7 but for V_c.

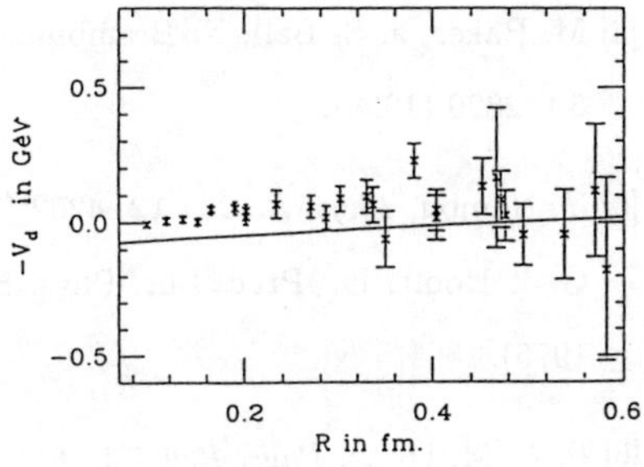

FIG. 10. Figure 10: Same as Fig. 7 but for V_d.

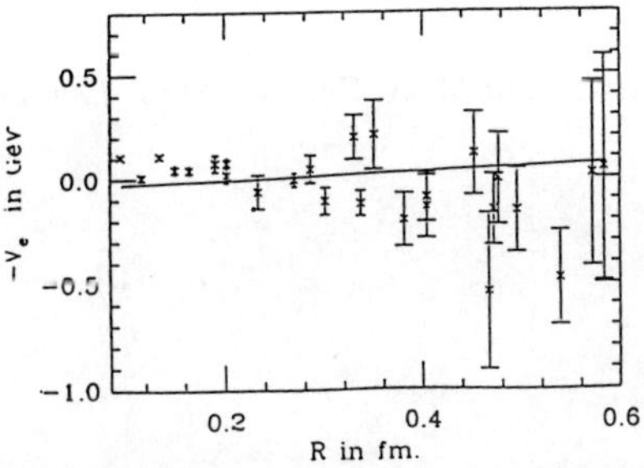

FIG. 11. Figure 11: Same as Fig. 7 but for V_e.

REFERENCES

[1] G. S. Bali, K. Schilling, and A. Wachter, *Phys. Rev.* D **56**, 2566 (1997).

[2] M. Baker, James S. Ball, and F. Zachariasen, *Phys. Rev.* D **51**, 1968 (1995).

[3] M. Baker, J. S. Ball, N. Brambilla, and A .Vairo, *Phys. Lett* B, **393**, 577 (1996).

[4] M. Baker, J. S. Ball, N. Brambilla, G. M. Prosperi and F. Zachariasen, *Phys. Rev.* D **54**, 2829 (1996).

[5] Y. Nambu, *Phys. Rev.* D **12**, 4262 (1974), S. Mandelstam, *Phys. Rep.* **23C**, 145 (1976). G. 't Hooft, in "Proc. Eur. Phys. Soc. 1975" ed. by A. Zichichi (Ed. Comp. Bologna 1976).

[6] P. A. M. Dirac, *Phys. Rev.* **74** , 817 (1948).

[7] A.A. Abrikosov *Sov. Phys. JETP* **32**, 1442 (1957); H .B. Nielsen and P. Olesen, *Nucl. Phys.* B **61**, 45 (1973).

[8] E. T. Akhmedov, M. N. Chernodub, M. I. Polikarpov and M. A. Zubkov, *Phys. Rev.* D **53**, 2097 (1996).

SECTION "C"

EFFECTIVE HAMILTONIANS AND QUARK MODELS

On Loop Corrections in the Linear Sigma Model

Nils A. Törnqvist
Physics Department, POB 9, FIN–00014, University of Helsinki, Finland

ABSTRACT

Quantum loops ares studied in the linear sigma model with scalar and pseudoscalar multiplets and a planar interaction term $\lambda\mathrm{Tr}[\Sigma\Sigma^\dagger\Sigma\Sigma^\dagger]$. It is shown how quantum loops break the OZI rule and the classical tree-level octet-singlet degeneracy, and how the Goldstone theorem can be maintained also in an iterative scheme including some multiloop diagrams.

The linear sigma model is the simplest renormalizable field theoretic model which incorporates the flavour and chiral symmetries of QCD, as well as many features of the light scalar and pseudoscalar experimental spectrum. Without explicit symmetry breaking terms the light pseudoscalars are Goldstone bosons of spontaneously broken chiral $U(3) \times U(3)$ as is easily verified at the tree level. The small pseudoscalar masses are then obtained by putting in, by hand, small light chiral quark masses, the origin of which is believed to come from interactions outside the domain of strong interactions.

In this note I study one-loop corrections to this model, and an iterative scheme to include some multiloop effects. In my talk I also discussed an apparent inconsistency with the Goldstone theorem, and asked the audience where my argument went wrong. Now I see how my original difficulty can be resolved, and therefore also my title has changed.

The Lagrangian[1–3] has the same flavour and chiral symmetries as massless QCD:

$$\mathcal{L} = \frac{1}{2}\mathrm{Tr}[\partial_\mu\Sigma\partial_\mu\Sigma^\dagger] + \frac{1}{2}m^2\mathrm{Tr}[\Sigma\Sigma^\dagger] - \lambda\mathrm{Tr}[\Sigma\Sigma^\dagger\Sigma\Sigma^\dagger] \ - \lambda'(\mathrm{Tr}[\Sigma\Sigma^\dagger])^2 \,. \tag{1}$$

Here, for $N_f = 3$, Σ is a 3×3 complex matrix, $\Sigma = \sum_{a=0}^{8}(s_a + ip_a)\lambda_a/\sqrt{2}$, where s_a are the scalar and p_a the pseudoscalar fields, and in which λ_a are the Gell-Mann matrices, normalized as $\mathrm{Tr}[\lambda_a\lambda_b] = 2\delta_{ab}$, and where for the singlet $\lambda_0 = (2/N_f)^{1/2}\mathbf{1}$ is included. Each classical meson in Eq. (1) has a definite $SU_f(3)$ symmetry content, which in the quark model means that it has a definite $q\bar{q}$ content. Thus the potential terms in Eq. (1) can be given a conventional quark line structure shown in Fig. 1. One can include into $\mathcal{L}$ conventional terms, which explicitely break the symmetries, $\mathcal{L}^{SB} = \epsilon_0 s_0 + \epsilon_8 s_8 + c[\det\Sigma + \det\Sigma^\dagger]$, which, by hand, give the pseudoscalars mass and break the flavour and $U_A(1)$ symmetries.

As discussed in Ref.[4,5] from fits to the scalar mesons one finds $\lambda \approx 16$ and $\lambda' \approx 0$, when the model was used as an effective theory with a symmetric 3-momentum cutoff 0.54 GeV/c. From now on I neglect the λ' term before renormalization.

Eq. (1) is clearly invariant under $\Sigma \to U_L \Sigma U_R^\dagger$ of $U(3) \times U(3)$. If $-m^2 < 0$ and $\lambda > 0$ the potential in Eq. (1) gives rise to an instability with vacuum condensate $< s_0 >= (2/N_f)^{1/2} f_\pi$. Then shifting as usual the scalar field, $\Sigma \to \Sigma + f_\pi \mathbf{1}$, one finds $f_\pi^2 = m^2/(4\lambda)$. Furthermore, the squared mass $(-m^2)$ of the classical 0^{++} nonet is replaced by $m_\sigma^2 = -m^2 + 12\lambda f_\pi^2 = 2m^2$, while the 0^{-+} classical nonet becomes massless, $-m^2 + 4\lambda f_\pi^2 = 0$.

Fig. 1. Graphical quark line representation of the λ and λ' terms of Eq. (1)

Also, after shifting the scalar singlet field, the λ term generates trilinear *spp* and *sss* couplings of the form $g\mathrm{Tr}[\Sigma\Sigma\Sigma^\dagger + h.c]/2$, where g is $4\lambda f_\pi$, and $g^2 = 2\lambda m_{a_0}^2$. With the flavour indices written out explicitly one has $g_{abc} = g\mathrm{Tr}[\lambda_a\lambda_b\lambda_c + \lambda_a\lambda_c\lambda_b]/(2\sqrt{8})$. These couplings obey the simply connected, Okubo-Zweig-Iizuka (OZI) allowed, quark line rules with flavour symmetry exact. Then the Lagrangian is reparametrized at the classical minimum:

$$\mathcal{L}' = \frac{1}{2}\mathrm{Tr}[\partial_\mu\Sigma\partial_\mu\Sigma^\dagger] - \frac{1}{2}m_\sigma^2\bar{s}^2 - \lambda\mathrm{Tr}[\Sigma\Sigma^\dagger\Sigma\Sigma^\dagger] + \frac{g}{2}\mathrm{Tr}[\Sigma\Sigma\Sigma^\dagger + h.c.] \,. \tag{2}$$

The m_{p_a} have vanished while the scalar nonet is massive and trilinear meson couplings have appeared. As is well known the linear sigma model is renormalizable,

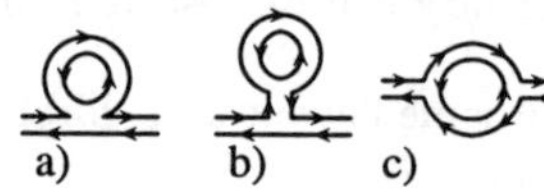

Fig. 2. The planar one loop tadpole diagrams of which (a) comes directly from the λ term in Eq.(1) and is for the pseudoscalar nonsinglet states, $8N_f\lambda[A(m_s^2) + A(m_p^2)]$, when the singlet-octet splitting is neglected in the loop masses. The tadpole diagram (b) with an intermediate scalar singlet is $-2N_f g^2/m_{s_0}^2[3A(m_s^2) + A(m_p^2)]$, while the planar s-channel loop diagram (c) is, for $s = 0$, $2N_f g^2 B(m_s^2, m_p^2)$ and more generally $N_f g^2 \mathrm{gaps}(s, m_s^2, m_p^2)/(8\pi)$.

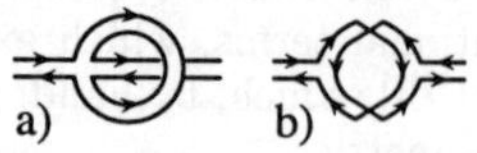

Fig. 3. The nonplanar, OZI rule violating one loop graphs, which break the tree-level scalar singlet-nonsinglet $(a_0 - \sigma)$ degeneracy. These are nonvanishing once the scalars are massive and nondegenerate with the pseudoscalars. The graph (a) gives when the singlet octet splitting is neglected for the pseudoscalars the expession $-4N_f\lambda[A(m_s^2) - A(m_p^2)]$, while (b) gives the same expression as given for Fig. 3c.

as first shown by Lee.[2] The renormalizability of more general sigma models including symmetry breaking terms, have been discussed by many authors.[7] Here I use Eq. (1) generally as an effective theory, where the cutoff has physical content in that the model does not describe contributions from near and above that cutoff, which comes from states, such as vector mesons, generally heavier than the pseudoscalar and scalar mesons. Also as my analysis includes only all one-loop diagrams the predictions are reliable only for low values of $s = p^2$ compared to the effective cutoff. These conditions are satisfied, since s is small at the 0^{-+} masses, and as we shall see, my predicted mass ratios depend very weakly on the cutoff (cf. Ref.[6]).

Consider now the quantum loops generated by the Lagrangian (1) after shifting the scalar field, Figs. 2a,b,c and 3a,b. As the theory is exactly flavour symmetric there exists always one flavour symmetric solution for the spectrum, also when the loops are included and summed over complete sets of flavour related thresholds.

From Fig.3a,b it is evident that these nonplanar diagrams split at the one loop level the singlet-nonsinglet degeneracy for the scalar multiplet. This implies that after renormalization a small tree level λ' term (See Crater[7]), must be included.

In these loop integrals one needs the function gaps*, which can be defined with a three momentum cutoff, $\Lambda = k_{cm}^{max}$, as

$$\text{gaps}(s, m_1^2, m_2^2) = \frac{1}{\pi} \int_{(m_1+m_2)^2}^{s_{max}(\Lambda)} ds' \lambda^{\frac{1}{2}}(s', m_1^2, m_2^2)/[s'(s-s')] \, . \tag{3}$$

Note that gaps is dimensionless and $\text{gaps}(s, m_1^2, m_2^2, \Lambda) = \text{gaps}(s/\Lambda^2, m_1^2/\Lambda^2, m_2^2/\Lambda^2, 1)$ and Λ is thus a scale parameter. With dimensional regularization (see e.g. Ref.[8]) Λ is related to the regularization scale $\mu = 2\Lambda/e$. It is only logaritmically dependent on Λ and is simply related to the two familiar functions $A(m^2)$ and $B(m_1^2, m_2^2)$, used in Ref.[4] and mentioned in the figure captions of Figs. 2 and 3: $A(m^2) - A(0) = m^2 \text{gaps}(0, 0, m^2)/(16\pi)$, and $B(m_1^2, m_2^2) = \text{gaps}(0, m_1^2, m_2^2)/(16\pi)$.

Consider now $N_f = 2$ and the one loop corrections to m_π, m_η, m_{a_0}, and m_σ. One finds from Figs. 2-3, that these mass corrections (which we denote by the masses themselves) to to the 0^{-+} masses and $m_{a_0}^2 - m_\sigma^2$ can be expressed entirely in terms of the gaps function:

$$\begin{aligned}
m_\pi^2 &= (8\pi)^{-1}[g_0^2 \text{gaps}(m_\pi^2, m_\sigma^2, m_\pi^2) + 2\lambda(m_\pi^2 - m_\sigma^2)\text{gaps}(0, m_\sigma^2, m_\pi^2) + \\
&\quad + g^2 \text{gaps}(m_\pi^2, m_{a_0}^2, m_\eta^2) + 2\lambda(m_\eta^2 - m_{a_0}^2)\text{gaps}(0, m_{a_0}^2, m_\eta^2)] \qquad (4) \\
m_\eta^2 &= (8\pi)^{-1}[g_0^2 \text{gaps}(m_\eta^2, m_\sigma^2, m_\eta^2) + 2\lambda(m_\eta^2 - m_\sigma^2)\text{gaps}(0, m_\sigma^2, m_\eta^2) + \\
&\quad 3g^2 \text{gaps}(m_\eta^2, m_{a_0}^2, m_\pi^2) + 6\lambda(m_\pi^2 - m_{a_0}^2)\text{gaps}(0, m_{a_0}^2, m_\pi^2)] \qquad (5) \\
m_{a_0}^2 - m_\sigma^2 &= (8\pi)^{-1}[g^2(\text{gaps}(m_{a_0}^2, m_\eta^2, m_\pi^2) + 9\text{gaps}(m_{a_0}^2, m_\sigma^2, m_{a_0}^2)) \\
&\quad - g_0^2/2(3\text{gaps}(m_\sigma^2, m_\pi^2, m_\pi^2) + \text{gaps}(m_\sigma^2, m_\eta^2, m_\eta^2)
\end{aligned}$$

*I use the name gaps for this function which frequently appears in gap and one-loop equations. It is also an acronym for "generalized analytic phase space" as the imaginary part of this function is simply proportional to two body phase space $\text{Im(gaps)} = -\lambda^{1/2}(s, m_1^2, m_2^2)/s = -2k_{cm}/\sqrt{s} \to -1$, when $s \to \infty$. This common function appears in the literature with a plethoria of different names, normalizations, subtractions and regularizations. Things would greatly simplify if one could agree on one convention.

$$+27\text{gaps}(m_\sigma^2, m_{a_0}^2, m_{a_0}^2) + 9\text{gaps}(m_\sigma^2, m_\sigma^2, m_\sigma^2))$$
$$-2g^2\text{gaps}(0, 0, m_{a_0}^2) + 2g_0^2 m_\pi^2/m_\sigma^2 \text{gaps}(0, 0, m_\pi^2)] \ . \qquad (6)$$

Here I have introduced a different trilinear coupling g_0 when the σ is involved, since for these the OZI rule of Eq.(2) can be violated at higher orders.

The terms proportional to λ come from the tadpole graphs Figs. 2a,b ans 3a, while the remaining terms come from the s-channel loops. The imaginary parts in Eq. (6) give finite widths to the σ and a_0 (but note that with my conventions $g^2_{\sigma\pi^0\pi^0} = g_0^2/2$).

Now using the relation $g_0^2 = 2\lambda m_\sigma^2$ for the σ couplings, and $g^2 = 2\lambda m_{a_0}^2$ for the a_0 couplings, and putting the pseudoscalar masses equal to zero in the r.h.s. of the two first equations one easily sees that the sum of the gaps functions vanishes. I.e. one has self-consistency with vanishing pion and eta masses also after one-loop effects independently of the splitting between a_0 and σ, which arises from the third equation. One can even iterate these equations, whereby one includes an infinite set of higher order graphs and still keeps the pion and eta massless. On the other hand, the OZI rule for the scalar meson masses, i.e. the degeneracy of the σ and the a_0, has been violated by the unitarization.

My difficulty with the Goldstone theorem as presented in the talk arose because I assumed the OZI rule relations for g in Eq.(2) to hold also in the iterative process. However, because of the breaking of the OZI rule at higher orders, the g_0 for the σ obtains nonplanar contributions, not present for the a_0 couplings. These conspire to keep the pseudoscalars massless. After the talk most people suggested that either my regularization scheme or the neglect of two-loop effects was the culprit. I did not beleive these because one should be able to renormalize order by order.

Finally in the talk I raised the question of whether the chiral symmetric solution is always unique? In some theories it could be unstable to infinitesimal perturbations and another stable solution could exist, This possibility I still believe is worth further study.

References

1. J. Schwinger, Ann. Phys. **2** (1957) 407, M. Gell–Mann and M. Levy, Nuovo Cim. XVI (1960) 705.
2. B. W. Lee, Nucl. Phys. B9 (1969) 649.
3. S. Gasiorowicz and D.A. Geffen, Rev. Mod. Phys. **41** (1969) 531.
4. N. A. Törnqvist (hep-ph/9712479) Physics Letters B**426** (1998) 105.
5. N. A. Törnqvist, Z. Physik **C68** 647 (1995); N. A. Törnqvist and M. Roos, Phys. Rev. Lett. **76** (1996) 1575.
6. S. Coleman and E. Weinberg, Phys. Rev, **D7** (1973) 1888.
7. L.-H. Chan and R. W. Haymaker, Phys. Rev. **D7** (1973) 402 and 415; H.C. Crater, Phys. ReV. **D 1** (1970) 3313.; A.J. Paterson, Nucl. Phys. B190 (1981) 188.
8. J.A. Oller and E. Oset, “The $0^\pm$ Nonet: Its Physical States and its Origin”, Valencia preprint, May 1998.

THE $O(4)_{D+1}$ NONLINEAR SIGMA MODEL: The Vacuum State and the Mass Gap

N.E. Ligterink, N.R Walet, and R.F. Bishop
Department of Physics, UMIST, PO Box 88, Manchester M60 1QD, UK

ABSTRACT

We study the ground state and low-lying excited states in the nonlinear sigma model with the coupled cluster method of quantum many-body theory.

1. Introduction

Although the Hamiltonian formulation of quantum field theory is more involved than its Euclidean equivalent, it has some fundamental advantages. These advantages start already at the level of the vacuum. Indeed, the *vacuum* is only clearly defined in the Hamiltonian formulation; it is the wave functional of the lowest-energy state. Given the vacuum, the particles are the low-lying excitations on this state, and therefore the properties of particles and their interactions are determined by the vacuum.[1] Even the absence of degrees of freedom, such as quarks, from the physical spectrum should almost automatically follow from the analysis of the vacuum.

Perturbations in single-quark degrees of freedom around the vacuum state should lead to high-energy states, probably a superposition of low-energy collective states. These perturbations can be considered the theoretical counterparts of deep-inelastic scattering, where an initial hit of a constituent leads to jets of particles in a "break-up" reaction. If the energy is high enough, the collective modes can move off freely.

Although the observables should follow easily when the vacuum state is determined, the practice is different. The vacuum problem is one of the hardest to solve. The number of non-perturbative methods for treating the vacuum state in the Hamiltonian formulation is restricted. Moreover, in QCD there are additional questions arising from the gauge invariance and the required gauge fixing.

The vacuum is invariant under space symmetries: rotations and translations. Therefore, the wave function can be restricted from the start. Usually it is restricted to mean-field states, where fields on each point act independently. However, in the true physical system correlations play an important role, not least to imprint the length scale of the collective modes which live on the vacuum.

If one wants to recover the global properties of a non-perturbative vacuum in an ab-initio calculation, one needs to retain as many as possible of the correlations allowed in the vacuum. For this purpose we investigate the possibility to apply the coupled-cluster method (CCM), a many-body technique, to nontrivial field theories. The CCM has a number of properties, which, in principle, makes it very suitable for the study of the vacuum state and low-lying excitations. As a many-body technique it is designed to deal with a large number of degrees of freedom, which can be chosen to satisfy the symmetries of the vacuum. Its quasi-analytical nature allows for a linear reponse approximation (RPA) upon the vacuum state to recover the excitation spectrum.

For an initial investigation we study the nonlinear sigma model. It is known to exhibit symmetry breaking, beyond which the whole nature of the ground state changes. The model is used as an effective field theory for QCD from different perspectives. It is best known for the dynamics of pions, dominated by the chiral symmetry breaking expressed in the low-energy theorem.

2. The Nonlinear Sigma Model

The nonlinear sigma model consists of a single field ϕ_α with four components ϕ_1, ϕ_2, ϕ_3, and ϕ_4 constrained by $\phi_\alpha^2 = 1$. The field is considered free and massless. The self-interaction among the components is the result of the constraint, which ties them together. The Lagrangian density

$$\mathcal{L} = \partial_\mu \phi_\alpha \partial^\mu \phi_\alpha \ , \tag{1}$$

leads to the corresponding $O(4)$ spin-lattice Hamiltonian

$$H = \frac{1}{2} \sum_i \mathbf{I}_i^2 + \lambda \sum_{\langle i,j \rangle} (1 - \phi_{\alpha,i} \phi_{\alpha,j}) \ , \tag{2}$$

where λ is proportional to $a^{2(D-1)}$ for a D-dimensional cubic lattice with spacing a,[2] and $\mathbf{I}$ is the angular momentum operator in four dimensions. The interaction term is summed over all nearest neighbour pairs of lattice sites $\langle i, j \rangle$.

If the model is thus regularised by considering only fields at lattice points it has two phases, although the case $D = 1$ is pathological since the coupling constant does not then depend on lattice spacing. As the coupling constant is small, the fields at each lattice point are virtually independent and the lowest-energy state consists of free rotors in the S-state at each site. If the coupling constant increases, the fields, which are vectors on a four-dimensional unit sphere, will align, and, in terms of the free-rotor spectrum, higher states will be filled. This behaviour is general for many systems: the kinetic and potential terms tending to drive the system into different phases.

The mean-field results give a rough indication of the behaviour of the system. In this case one has to introduce a direction into the system and investigate whether the vacuum aligns with respect to this direction. Generally, the mean-field state is

$$\langle \{\phi_{\alpha,j}\} | \Psi \rangle = \prod_i f(\phi_i \cdot \hat{e}) \ , \tag{3}$$

where $\hat{e} = (0, 0, 0, 1)$ and f is an arbitrary function. However, the position of the phase transition and the global behaviour is already given with the simple linear trial function: $f(\phi \cdot \hat{e}) = \cos\alpha + \phi \cdot \hat{e} \ \sin\alpha$. Below the critical coupling $\lambda_c = 3/D$ the ground

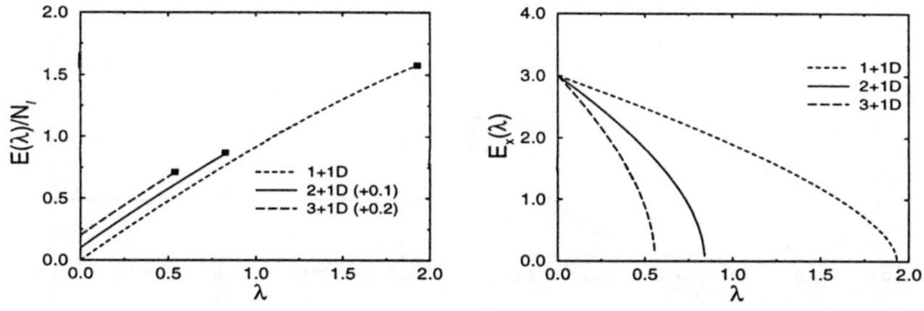

Fig. 1. The results using the functional NCCM: left, the ground-state energy per link in one, two, and three dimensions; and right, the corresponding lowest excitation energy. The solutions terminate as the excitation energy vanishes.

state is the constant state. However, above the critical coupling the ground state has a nontrivial solution,

$$f(x) = \sqrt{(\lambda D + 3)/(2\lambda D)} + \sqrt{(\lambda D - 3)/(2\lambda D)}\ x \ . \tag{4}$$

The exact mean-field result above the critical coupling is given by the first odd characteristic function ce_2 of the Mathieu equation. These results extend easily to the corresponding $O(N)$ model, yielding a critical coupling of $\lambda_c = 2(N-1)/D\sqrt{N}$. The mean-field results predicts essentially the same behaviour for any dimension, although it is known that in the one-dimensional system the symmetry is never broken.

3. The Coupled Cluster Method

The CCM is a powerful technique for quantum many-body calculations. Here we will review only the essential features, and we refer to the literature[3] for the details. There are three key features of the CCM: firstly, the exponentiated wave functional

$$|\Psi\rangle = e^{\hat{S}}|\Phi\rangle \ , \tag{5}$$

where $|\Phi\rangle$ is an uncorrelated model state, and $\hat{S}$ is a linear combination of all multi-configurational creation correlation operators, which commute among themselves and annihilate the model bra state. Due to the exponentiation the independent correlations are automatically summed in the correct way. Secondly, the bra state is parametrised in a way which builds in a similarity transform,

$$\langle\tilde{\Psi}| = \langle\Phi|(1+\hat{\tilde{S}})e^{-\hat{S}} = \langle\Phi|e^{\hat{\tilde{\Sigma}}}e^{-\hat{S}} \ , \tag{6}$$

where the first form is known as the *normal* CCM (NCCM), and the second as the *extended* CCM (ECCM).[4] The third key feature of the CCM is the choice of operators: $\tilde{S}$ and $\tilde{\Sigma}$ contain only destruction operators; they are conjugate to the creation operators

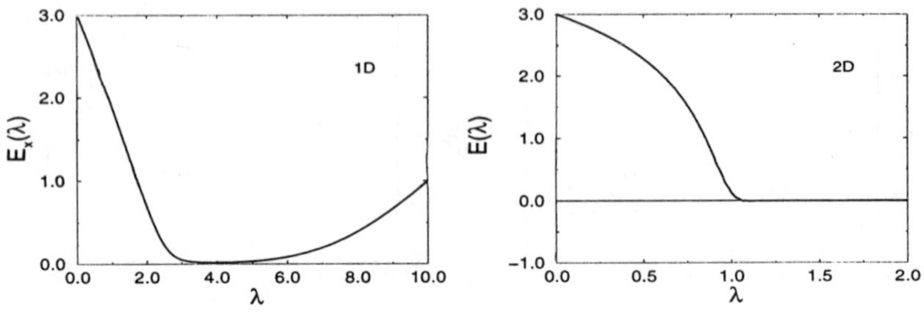

Fig. 2. The excitation energy, for operatorial ECCM, in one dimension (left) and two dimensions (right) (preliminary results).

in $\hat{S}$. The similarity transform then has the advantage that only correlations linked to the Hamiltonian are retained. The bi-variational equations in S and $\tilde{S}$ (or S and $\tilde{\Sigma}$) are therefore only of moderate complexity.

The NCCM SUB2 form, where only two-body correlation are retained, allows the use of *functions* as correlation operators

$$\hat{S} = \sum_{i,j} S_{i,j}(\phi_i \cdot \phi_j) \ , \tag{7}$$

which yield the results shown in Fig. 1.[5] However, the NCCM has the same failure as the mean-field result; it finds a phase transition in one dimension where there should not be one. The NCCM also fails to track across the phase transition, which is the ultimate goal, as we want to be able to describe the collective low-lying modes in terms of local degrees of freedom.

The ECCM seems to fulfill above demands.[5] Preliminary investigations show that we are indeed able to cross the phase transition and in one dimension the phase transition disappears, as we can see in Fig. 2, where the excitation energy remains massive in one dimension, but still vanishes, as expected, in two dimensions.

4. Conclusion and Outlook

The CCM allows an ab-initio study of the vacuum state and low-lying excitations in the nonlinear sigma model. However, to cross the phase transition requires the more elaborate *extended* CCM in an operatorial form. This work is in progress.

References

1. P. v. Baal, Nucl. Phys. B (Proc. Suppl.) **63A-C**, 126 (1998).
2. J.B. Kogut, Rev. Mod. Phys. **51**, 659 (1979).
3. R.F. Bishop, Theor. Chim. Acta **80**, 95 (1991).
4. J. Arponen, Ann. Phys. (N.Y.) **151**, 311 (1983).
5. N.E. Ligterink, N.R. Walet, and R.F. Bishop, Nucl. Phys. B (Proc. Suppl.) **63A-C**, 667 (1998); Ann. Phys. (N.Y.) **267**, 97 (1998) (and references therein).

The Light-Cone Wavefunction of the Pion

Thomas Heinzl
Institut für Theoretische Physik,
Universität Regensburg, D-93040 Regensburg, FRG

ABSTRACT

The light-cone wavefunction of the pion is calculated within the Nambu and Jona-Lasinio model.

1. Introduction

The question how to describe a hadron has a seemingly straightforward answer: determine its wavefunction. To do so suggests using a Hamiltonian Fock space approach in which one would write e.g. for a pion, $|\pi\rangle = \psi_2|q\bar{q}\rangle + \psi_3|q\bar{q}g\rangle + \psi_4|q\bar{q}q\bar{q}\rangle + \ldots$. The ψ_n denote the n-particle Fock-space wavefunctions, $\psi_n = \langle n|\pi\rangle$. Of course, calculating the ψ_n by solving the "QCD Schrödinger equation" represents a non-perturbative, relativistic, many-body bound-state problem. Each of these attributes constitutes a problem in its own right, and the culmination of them makes it so difficult to understand hadron structure. In this contribution, however, we want to show that there exists a particular approach providing ideal tools for dealing with most of these problems.

2. Light-Cone Wavefunctions

As Dirac has pointed out long ago,[1] the choice of time for a relativistic system is not unique. Instead of the ordinary cartesian time $\tau = x^0$ one might as well chose the light-cone (LC) time $\tau = x^+ = x^0 + x^3$ to describe the time evolution of the wavefunctions ψ_n. The generator of this time development is the LC Hamiltonian $P^- = P^0 - P^3$. One then has to solve the "LC-Einstein-Schrödinger equation"

$$(P^+P^- - P_\perp^2)\,|\pi\rangle = M_\pi^2\,|\pi\rangle \tag{1}$$

for the hadronic eigenstates (e.g. $|\pi\rangle$) and eigenvalues (e.g. the pion mass squared, M_π^2). In dealing with the many-body aspects one has the advantage that the LC vacuum is trivial and an eigenstate of the fully interacting Hamiltonian, $P^-|0\rangle = 0$. As a consequence, the transition matrix elements $\langle 0|P^-|n\rangle$ vanish, so that the wavefunctions are not polluted by disconnected vacuum contributions mixing with the ψ_n.[2] When dealing with the relativistic aspects of the problem, LC wavefunctions are again well suited, for the following reasons: within LC quantization, boosts become interaction independent ("kinematic"), so that wavefunctions are known in any frame when they are known at rest. Furthermore, due to a two-dimensional Galilei invariance, center-of-mass and relative coordinates separate. As a result, the ψ_n only depend on boost-invariant relative coordinates. A two-particle wavefunction, like e.g. the valence $q\bar{q}$-state of the pion with total longitudinal momentum P^+, will be of the form $\psi_2 = \psi_2(x, \mathbf{k}_\perp)$, where x is the longitudinal momentum fraction of one of the quarks, $x = p^+/P^+$, and $\mathbf{k}_\perp$ the relative transverse momentum.

As always, any simplification has its price, which for the case at hand is the fact that rotational symmetry and parity become dynamical and thus difficult to implement. The main difficulty of the LC approach, however, is the necessity to perform renormalization in a Hamiltonian context where one does not have explicit covariance. Accordingly, one has to deal with an abundance of counterterms which can even become non-local ($\sim 1/p^+$). In principle, one has to trade the cutoff-dependence of the wavefunctions for a scale-dependence, $\psi_n(\Lambda) \to \psi_n(Q)$, and one expects at least two distinct regimes, a perturbative one, in which $\psi_n(Q \gg \Lambda_{\rm QCD})$ describes the n-parton content of the hadron, and a non-perturbative regime, with $\psi_n(Q \simeq \Lambda_{\rm QCD})$ being the amplitude of finding n *constituent* quarks in a hadron. Ideally, in the latter case, ψ_n should be zero for $n > 2$ (mesons) or $n > 3$ (hadrons). Both these regimes and the associated LC wavefunctions should be related by a renormalization group equation, which, however, is largely unknown at the moment.[3]

At this point, of course, the question arises whether one can really calculate (hadronic) LC wavefunctions. The answer is positive, as will be shown in the next section.

3. Model Calculation

Our goal is to solve the bound-state equation (1) which can be rewritten as $(M_\pi^2 - M_0^2)|\pi\rangle = V|\pi\rangle$, where $M_0^2 = (k_\perp^2 + m^2)/x(1-x)$ is the free invariant mass squared (for a two-particle system), and $V \equiv P^-_{\rm int}P^+$ the interaction. Now, in general, it is technically involved to obtain the LC Hamiltonian and in particular V due to the presence of complicated constraints.[4] There is, however, an alternative way to proceed established by 't Hooft.[5] One can first solve the Schwinger-Dyson equation for the quark self-energy Σ and insert it in the Bethe-Salpeter equation (with kernel K). The latter is then solved for the Bethe-Salpeter amplitude which, upon projection onto $x^+ = 0$ (via k^--integration) yields the LC wavefunction $\psi_2(x, \mathbf{k}_\perp)$.

At the moment, this program has not been carried out for QCD, starting from first principles. It seems, however, feasible to tackle it within the framework of Schwinger-Dyson equations by modeling appropriate gluon propagators consistent with asymptotic freedom and confinement. Nevertheless, also in this approach, Σ and K are usually very complicated. As we are interested in the constituent region which is known to be dominated by the spontaneous breakdown of chiral symmetry[6] we have chosen a rather modest and simple approach: we use the well known low-energy effective Lagrangian of Nambu and Jona-Lasinio (NJL) with a four point fermion interaction of the form $\mathcal{L}_{\rm int} \sim g(\bar{\psi}\Gamma\psi)^2$, Γ being an appropriate, chirally symmetric Dirac structure.[7]

Let me recapitulate the most prominent features of the model: it is not renormalizable; therefore, the cutoff Λ is a parameter of the theory which is fixed by phenomenology. The model is not confining, but chiral symmetry is spontaneously broken above a critical coupling g_c. This phase transition is accompanied by dynamical mass generation, $m_0 \to m$ leading to a constituent mass m of the order of a few hundred MeV. In the chiral limit, $m_0 \to 0$, there is a massless pion in accordance with Goldstone's theorem.

The Schwinger-Dyson equation is solved in mean-field approximation yielding a self-energy proportional to the quark condensate, $\Sigma = m \sim \langle 0|\bar\psi\psi|0\rangle_\Lambda$. From the Bethe-Salpeter equation, which is solved in ladder approximation, one derives the (valence) LC wavefunction of the pion,

$$\psi_2(x, \mathbf{k}_\perp, \lambda, \lambda') = N\, D_0^{-1}\, \bar u_\lambda(x, \mathbf{k}_\perp)\, (M_\pi + \not P)\gamma_5\, v_{\lambda'}(1-x, -\mathbf{k}_\perp)\, \theta(\Lambda) \,. \qquad (2)$$

Here, N denotes a normalization constant, the "off-shell-ness" $D_0 \equiv M_\pi^2 - M_0^2$ is a measure of the binding, and $\theta(\Lambda)$ represents the invariant-mass cutoff, $M_0^2 \leq \Lambda^2$. Without this cutoff, the LC wavefunction would not be normalizable. As a crosscheck, it can be shown that the LC bound-state equation leads to a massles pion in the chiral limit, so that the Nambu-Goldstone mechanism is correctly reproduced.

With the LC wavefunction at hand one can start and calculate observables. As a technical prelude, however, we first have to determine its normalization. We do this by *enforcing* a constituent picture, that is, the probability that the pion consists of two constituent quarks is set to one, $||\psi_2|| = 1$. To keep things simple, we only consider the chiral limit, $M_\pi = 0$. As an additional approximation we will neglect terms of order $\epsilon^2 \equiv m^2/\Lambda^2$, i.e. we assume that the cutoff is large compared to the constituent mass. From the standard values, $\Lambda \simeq 1$ GeV, $m \simeq 300$ MeV, we expect that this assumption should induce an error of the order of 10%. The leading order will thus be independent of the constituent mass so that the cutoff Λ is our only parameter. Within the constituent picture, the normalization constant becomes $N = 2\sqrt{15}\pi/\Lambda$ which will be needed later on. We proceed by calculating the pion electromagnetic formfactor. It is defined by the matrix element of the electromagnetic current $J^\mu_{\rm em}$ between pion states. This leads to a Drell-Yan type formula which expresses the formfactor as the overlap of two wave functions.[2] Using our normalization the formfactor becomes (to leading order), $F(q_\perp^2) = 1 - 3\, q_\perp^2/\Lambda^2 + O(q_\perp^4)$ implying a pion charge radius $\langle r_\pi^2\rangle = 18/\Lambda^2$.

We want to fix the cutoff in terms of the pion decay constant, f_π, which is given by the 'wave function at the origin', $\psi_0 = f_\pi/2\sqrt{N_C}$.[2] This determines the cutoff, $\Lambda \simeq 12.7 f_\pi \simeq 1.2$ GeV, and thus the pion charge radius $\langle r_\pi^2\rangle = 18/\Lambda^2 = 0.50$ fm^2 $= (0.71$ fm$)^2$. The latter differs from the experimental value, 0.44 fm^2, by 13% and therefore is consistent with the expected accuracy.

The mean transverse momentum becomes $\langle \mathbf{k}_\perp^2\rangle = 3\Lambda^2/28 \simeq (0.43 \text{ GeV})^2$, which is a reasonable value. In particular, this value shows that the pion is a highly relativistic system as the momenta of its (valence) constituents exceed the pion mass.

The pion distribution amplitude is obtained by integrating the wave function over $k_\perp$. Due to the cutoff dependence of the wave function, the integration extends effectively only up to the cutoff scale $\Lambda^2 x(1-x) \simeq (600 \text{ MeV})^2$. The result is

$$\phi(x) = \frac{2\sqrt{3}}{f_\pi} \int \frac{d^2k_\perp}{16\pi^3}\, \psi_\Lambda(x, \mathbf{k}_\perp, \uparrow, \downarrow) = \frac{128}{3\pi}\, [x(1-x)]^{3/2} \,. \qquad (3)$$

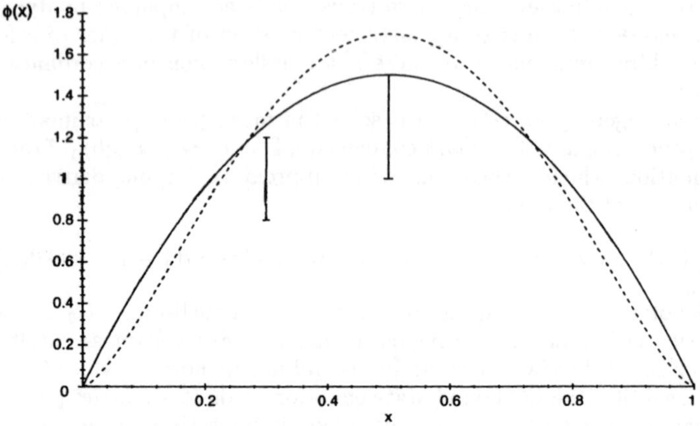

Fig. 1. The pion distribution amplitude. The full curve represents the asymptotic amplitude, the broken one the amplitude calculated within the NJL model. The vertical lines mark the constraints given by Belyaev and Johnson.[8]

This is qualitatively close to the asymptotic distribution amplitude, $\phi_{\text{as}}(x) = 6x(1-x)$ as is shown in Figure 1.

4. Conclusions

We have seen that it is actually possible to go beyond mere model building and calculate LC wavefunctions within a relativistic quantum field theory. Straightforward extensions will be to study the effects of finite cutoff, constituent-quark and pion mass. Instead of the NJL model one might use more realistic interactions in the Bethe-Salpeter kernel (to account e.g. for confinement) or the effective quark theory derived from the instanton model.[6]

Finally, I would like to express my gratitude to the organizer of the meeting, N. Isgur, and his team for all their efforts in providing such a stimulating atmosphere.

References

1. P. A. M. Dirac, Rev. Mod. Phys. **21**, 392 (1949).
2. S.J. Brodsky, G.P. Lepage, in: *Perturbative Quantum Chromodynamics*, A.H. Mueller, ed., World Scientific, Singapore, 1989, and references therein.
3. K. G. Wilson et al., Phys. Rev. D **49**, 6720 (1994).
4. V. A. Franke, Yu. V. Novozhilov, E. V. Prokhvatilov, Lett. Math. Phys. **5**, 437 (1981).
5. G. 't Hooft, Nucl. Phys. B **75**, 461 (1974).
6. D. Diakonov, in: *Selected Topics in Nonperturbative QCD*, A. Di Giacomo and D. Diakonov, eds., IOS Press, Amsterdam, 1996.
7. Y. Nambu, G. Jona-Lasinio, Phys. Rev. **122**, 345 (1961).
8. V.M. Belyaev, M.B. Johnson, hep-ph/9707329.

A Unified Effective Theory for Light Mesons and Heavy Mesons from Bargmann-Wigner Superfields

Lai-Him Chan

Department of Physics and Astronomy, Louisiana State University, Baton Rouge, Louisiana 70803-4001, USA

ABSTRACT

I present the formulation of a unified effective field theory for the light mesons and the heavy mesons which integrates many features of the existing low-energy hadron model into single universal effective theory. Its implications on the quark and meson masses and other meson phenomenology will be discussed

1. Introduction

Since the nonperturbative QCD corrections are difficult to obtained from the first principle calculation, our understanding of the low-energy hadron physics still relies heavily on phenomenological models, such as the quark model, vector dominance, chiral dynamics, Skyrmion model, the heavy quark effective theory,[1] for example. The important question is whether an effective relativistic quantum field theory can be constructed to encompass the essential elements of these models.

Recently I have proposed a unified effective theory of light mesons and heavy mesons[2] which integrates these important aspects into a single universal effective theory. In this talk I shall present a summary of this formulation. The underlying ideas of the theory is to identify the mesons fields as the Bargmann-Wigner superfields, localized direct product representations of quark and antiquark flavor spinors and to use the Bargmann-Wigner Lagrangian for the free fields.[3,4] The additivity law of the quark model is implemented by identifying the meson masses to be the sum of the corresponding quark masses. Chiral dynamics can be easily mapped from quark Lagrangian to the meson Lagrangian.

Low-lying meson are quark and antiquark s-wave ground states with spin-parity 0^- and 1^-. The light and heavy quarks are denoted by q and Q respectively. In the effective field theory, these meson fields are approximated by localized direct product quark and antiquark fields, $\mathcal{M}_{q\bar{q}} \sim q\bar{q}$ for the light mesons and $\{\mathcal{M}_{Q\bar{q}} \sim Q\bar{q}, \mathcal{M}_{q\bar{Q}} \sim q\bar{Q}\}$ for the heavy mesons. These spinor representations are related to the conventional tensor representations of the meson fields by:

$$\begin{aligned}
\mathcal{M}_{q\bar{q}} &= \mu^{\frac{1}{2}}\sigma - i\gamma_5\mu^{\frac{1}{2}}\phi - \gamma_5\gamma_\mu\frac{\phi^\mu}{\mu^{\frac{1}{2}}} + \gamma_\mu\mu^{\frac{1}{2}}\mathcal{V}^\mu + \sigma_{\mu\nu}\frac{\mathcal{V}^{\mu\nu}}{2\mu^{\frac{1}{2}}} \\
\mathcal{M}_{Q\bar{q}} &= \mu^{\frac{1}{2}}\mathcal{P}_0 - i\gamma_5\mu^{\frac{1}{2}}\mathcal{P} - \gamma_5\gamma^\mu\frac{\mathcal{P}_\mu}{\mu^{\frac{1}{2}}} + \gamma^\mu\mu^{\frac{1}{2}}\mathcal{P}^*_\mu + \sigma^{\mu\nu}\frac{\mathcal{P}^*_{\mu\nu}}{2\mu^{\frac{1}{2}}} \qquad (1)
\end{aligned}$$

where μ is the mass scaling factor.

The free constituent quark Lagrangian is

$$\mathcal{L}_0^q = -\bar{q}(\gamma\frac{1}{i}\partial + m_q)q + \bar{Q}(\gamma\frac{1}{i}\partial + m_Q)Q \tag{2}$$

where m_q and m_Q are the constituent light quark mass and the constituent heavy quark mass respectively. The mapping of the quark Lagrangian into the Bargmann-Wigner Lagrangian is accomplished by imposing the additivity law of the mass from the quark model, $\mathbf{m} = 2m_q$ and $\mathbf{m}_h = m_q + m_{Q_h}$, such that the free meson Lagrangian can be written,

$$\mathcal{L}_0^{\mathcal{M}} = \frac{1}{4}Tr\Big[\mathcal{M}_{q\bar{q}}(\gamma\frac{1}{2i}\partial + m_q)\mathcal{M}_{q\bar{q}} + \mathcal{M}_{q\bar{Q}}(\gamma\frac{1}{2i}\partial + m_{Q_h})\mathcal{M}_{Q\bar{q}} + \mathcal{M}_{Q\bar{q}}(\gamma\frac{1}{2i}\partial + m_q)\mathcal{M}_{q\bar{Q}}\Big] \tag{3}$$

and the free field equations become the Wigner-Bargmann equations,[3]

$$(\gamma\frac{1}{i}\partial + \mathbf{m})\,\mathcal{M}_{q\bar{q}} = 0\,, \quad (\gamma\tfrac{1}{i}\partial + \mathbf{m}_h)\,\mathcal{M}_{Q\bar{q}} = 0\,, \quad (\gamma\frac{1}{i}\partial + \mathbf{m}_h)\,\mathcal{M}_{q\bar{Q}} = 0$$
$$\mathcal{M}_{q\bar{q}}(-\gamma\frac{1}{i}\overleftarrow{\partial} + \mathbf{m}) = 0\,, \quad \mathcal{M}_{Q\bar{q}}(-\gamma\tfrac{1}{i}\overleftarrow{\partial} + \mathbf{m}_h) = 0\,, \quad \mathcal{M}_{q\bar{Q}}(-\gamma\frac{1}{i}\overleftarrow{\partial} + \mathbf{m}_h) = 0\,. \tag{4}$$

The symmetry group for the light quark sector is $SU(3)_L \otimes SU(3)_R$, which is spontaneously broken to $SU(3)$ and further to $SU(2)$. Chiral Lagrangian can be constructed by the standard trick of nonlinear represnetion using a transformation to separate the chiral phases,[5]

$$\sigma - i\gamma_5\phi = \xi_5^\dagger(S - i\gamma_5\frac{\eta_0}{\sqrt{3}})\xi_5^\dagger. \tag{5}$$

where η_0 is a pseudoscalar singlet and $\xi_5^\dagger = e^{-i\gamma_5\pi/f}$. Here $\pi = \sum_a \frac{1}{2}\pi_a\lambda^a$ is the octet of Goldstone bosons. f will be identified as the meson decay constant and may be flavor dependent in the presence of flavor symmetry breaking. I also define $\Sigma_5 = \xi_5^2 = e^{2i\gamma_5\pi/f}$.

I construct the chiral quark Lagrangian from a minimal extension of Eq. 2,

$$\mathcal{L}_q = -\bar{\mathbf{q}}\Big(\gamma\frac{1}{i}\mathcal{D} + m_G\Sigma_5 + \mu_q + \delta\tilde{g}_A\bar{\gamma}^\mu i(\Sigma_5^\dagger\partial_\mu\Sigma_5)\Big)\mathbf{q} - \bar{Q}\Big(\gamma\frac{1}{i}\mathcal{D} + \mu_Q\Big)Q \tag{6}$$

where μ_Q and μ_q are diagonal matrices of heavy quark and light quark current masses. From the spontaneous broken chiral symmetry, light quarks acquire additional contribution m_G to the mass term. The total quark mass $m_i = \mu_i + m_G$ becomes the constituent quark mass. We include a nonminimal interaction term in anticipation of a possible correction to the axial coupling constant.

The chirally neutral fields may be defined by removing the chiral phases:

$$q = \xi_5\mathbf{q} \tag{7}$$

$$M_{q\bar{q}} = \xi_5\mathcal{M}_{q\bar{q}}\xi_5 = \mu^{\frac{1}{2}}S - i\gamma_5\frac{\eta_0}{\sqrt{3}}\mu^{\frac{1}{2}} - \gamma_5\gamma_\mu\frac{\Phi^\mu}{\mu^{\frac{1}{2}}} + \gamma_\mu\mu^{\frac{1}{2}}\tilde{V}^\mu + \sigma_{\mu\nu}\frac{V^{\mu\nu}}{2\mu^{\frac{1}{2}}} \tag{8}$$

$$M_{Q\bar{q}} = \mathcal{M}_{Q\bar{q}}\xi_5 = \mu^{\frac{1}{2}}P_0 - i\gamma_5\mu^{\frac{1}{2}}P - \gamma_5\gamma^\mu\frac{P_\mu}{\mu^{\frac{1}{2}}} + \gamma^\mu\mu^{\frac{1}{2}}P_\mu^* + \sigma^{\mu\nu}\frac{P_{\mu\nu}^*}{2\mu^{\frac{1}{2}}}\,. \tag{9}$$

It is clear that the eight degrees of freedom from the octet of Goldstone mesons have been removed from $M_{q\bar{q}}$.

The quark Lagrangian Eq. 6 becomes:

$$\tilde{\mathcal{L}}_q = -\bar{q}\Big(\gamma^\mu(\frac{1}{i}\tilde{D}_\mu - \gamma_5 a_\mu) + (m_G + \xi_5^\dagger \mu_q \xi_5^\dagger - \delta\tilde{g}_A\gamma^\mu\gamma_5 a_\mu)\Big)q - \bar{Q}(\gamma\frac{1}{i}\partial + \mu_Q)Q \quad (10)$$

where I have defined

$$\begin{aligned} v_\mu &= \frac{i}{2}(\xi\partial_\mu\xi^\dagger + \xi^\dagger\partial_\mu\xi) = \frac{i}{2}[\frac{\pi}{f}, \partial_\mu\frac{\pi}{f}] + \mathcal{O}(\pi^4) \to U v_\mu U^\dagger + iU\partial_\mu U^\dagger \\ a_\mu &= \frac{i}{2}(\xi\partial_\mu\xi^\dagger - \xi^\dagger\partial_\mu\xi) = \partial_\mu\frac{\pi}{f} - \frac{1}{6}[\frac{\pi}{f}, [\frac{\pi}{f}, \partial_\mu\frac{\pi}{f}]] + \mathcal{O}(\pi^5) \to U a_\mu U^\dagger \quad (11) \end{aligned}$$

The dynamics of the quark Lagrangian can be mapped onto the meson Lagrangian either from Eq. 6 or Eq.10,

$$\begin{aligned} \mathcal{L}_M = \frac{1}{4}Tr\Big\{ & M_{q\bar{q}}\left(\frac{1}{2}\gamma^\mu(\frac{1}{i}\tilde{D}_\mu - \gamma_5 a_\mu) + (m_G + \xi_5^\dagger\mu_q\xi_5^\dagger - \delta g_A\gamma^\mu\gamma_5 a_\mu)\right) M_{q\bar{q}} \quad (12) \\ & + M_{q\bar{Q}}(\frac{1}{2}\gamma^\mu\frac{1}{i}\tilde{D}_\mu - \mu_Q)M_{Q\bar{q}} + M_{Q\bar{q}}\Big(\frac{1}{2}\gamma^\mu(\frac{1}{i}\tilde{D}_\mu - \gamma_5 a_\mu) + (m_G + \xi_5^\dagger\mu_q\xi_5^\dagger - \delta g_A\gamma^\mu\gamma_5 a_\mu)\Big)M_{q\bar{Q}}\Big\} \end{aligned}$$

Meson dynamics can be obtained directly from transcribing this Lagrangian into tensor representation. Through the auxiliary variables and the nonlinearity, even at the tree order the meson Lagrangian contains infinite number of interaction terms with higher derivatives and the explicit symmetry breaking from the current quark mass terms also induces $SU(3)$ symmetry breaking through wavefunction renormalization. Therefore the Lagrangian is far more complex than most of the phenomenological models. The details of the calculation are given in Ref.[2]

We have derived the well-known $SU(6)$ relation[6] $g_{\omega\rho\pi}\frac{m_\rho}{2} = g_{\rho\pi\pi} = \frac{g_A m_\rho}{2f_\pi}$ and the KSFR relation[7] with $g_A = \sqrt{2}$ which we obtain by imposing vector dominance conditions.[8] In the absence of vector fields, the leading terms (up to four-derivative) derivative expansion of the light meson Lagrangian reduces to the Skyrme Lagrangian[9] with the Skyrme parameter $e = g_{\rho\pi\pi} = 5.9$. With minor modification of the meson Lagrangian the light vector mesons become the nonabelian gauge fields of the local hidden $SU(3)$ symmetry. The light meson Lagrangian with the auxiliary fields $V_{\mu\nu} = 0$ reduces to the Lagrangian proposed by Bando.[10]

The heavy meson Lagrangian is more consistent and complete than that previously derived by piecing together the heavy quark effective theory and the quark model.[11] The Lagrangian in this work includes the chiral symmetry breaking effect due to current quark masses and has the correct off-mass-shell momentum dependent. We have shown that the wavefunction from this Lagrangian can be used directly to calculate the reduced matrix element of the heavy flavor changing weak current.[12]

If we assume that the vacuum is $SU(3)$ symmetric or nearly symmetric $\langle S_{ij}\rangle \approx \delta_{ij}\sqrt{2}f$, from the known meson masses we can calculate the current quark masses,[13]

$$\begin{aligned} m &= 379.0MeV, \quad & \mu_u &= 4.7MeV, \quad & \mu_d &= 7.2MeV, \\ \mu_s &= 131.0MeV, \quad & \mu_c &= 1379MeV, \quad & \mu_b &= 4685MeV\,. \end{aligned} \quad (13)$$

For meson masses we obtain the following formulae.

$$\begin{array}{ccccc} m_\omega & = & m_\rho & = & m_{K^{*+}} + m_{K^{*0}} - m_\phi \\ 781.94 \pm 0.12 MeV & = & 769.9 \pm 0.8 MeV & = & 768.3 MeV \end{array} \quad (14)$$

$$\begin{array}{rcl} (m_\rho^2 - m_{\pi^0}^2) - (m_{K^{*0}}^2 - m_{K^0}^2) & = & \frac{m_{K^0}^4}{4m_{K^{*0}}^2} - \frac{m_{\pi^0}^4}{4m_\rho^2}, \\ .01936 GeV^2 & = & .01896 GeV^2 \end{array} \quad (15)$$

and, $m_\eta = 466.2MeV$ and $m_{\eta'} = 949.3MeV$.

The effective bosonization, matching of the dynamics from the quark to the meson sector, is not complete without the corresponding mapping of the axial anomaly. It would be very interesting to investigate whether mesons in multispinor superfield formulation can generate same type of anomaly as the fermion.

References

1. E. Eichten and B. Hill, Phys. Lett. **B234**, 511 (1990) ; H. Georgi, Phys. Lett. **B240**, 447 (1990); A. F. Falk, B. Grinstein and M. Luke, Nucl. Phys. **B357**,185 (1991) .
2. Lai-Him Chan, Phys. Rev. **D 55**, 5362 (1997).
3. V. Bargmann and E. Wigner, Proc. Nat. Acad. Sci., Wash. **34**, 211 (1948).
4. B. Sakita and K. C. Wali, Phys. Rev. **139**, 1355 (1965); A. Salam, R. Delbourgo and J. Strathdee, Proc. Roy. Soc. **284**, 146 (1965); **285**, 312 (1965).
5. A. Manohar and H. Georgi, Nucl. Phys. B **234,** 189 (1984).
6. F. Gürsey and L. A. Radicati, Phys. Rev. Lett. **13**, 159 (1964); A. Pais, Phys. Rev. Lett. **13**, 162 (1964); F. Gürsey, A. Pais and L. A. Radicati, Phys. Rev. Lett. **13**, 165 (1964); B. Sakita, Phys. Rev. **136B**, 1756 (1964).
7. K. Kawarabayashi and M. Suzuki, Phys. Rev. Lett. **16**, 255 (1966); Riazuddin-Fayyazuddin, Phys. Rev. **147**, 1071 (1966).
8. J. J. Sakurai, Currents and Mesons (Univ. Chicago Press, Chicago, 1969).
9. T. H. R. Skyrme, Proc. Roy. Soc. **A260**, 127 (1961); **A262** (1961) 237; G. Adkins, C.R. Nappi and
10. M. Bando, T Kugo, S Uehara, K. Yamawaki, and T. Yanagida, Phys. Rev. Lett. **54**, 1215 (1985); M. Bando, T Kugo, and K. Yamawaki, Nucl. Phys. **B 259**, 493 (1985); M. Bando, T. Kugo, and K. Yamawaki, Phys. Rep. **164**, 217 (1988).

    E. Witten, Nucl. Phys. **228**, 552 (1983).
11. T. M. Yan *et al.*, Phys. Rev., 1148 **D46** (1992).
12. N. Isgur and M. B. Wise, Phys. Lett. **B232**, 113 ; ibid.**B237**, 527 (1990); Nucl. Phys. **B348**, 276 (1991); A. F. Falk, H. Georgi, B. Grinstein and M. B. Wise, Nucl. Phys. **B343**, 1 (1990); H. Georgi, Nucl. Phys. **B348**, 293 (1991); M. B. Voloshin and M. A. Shifman, Sov. J. Nucl. Phys. **45**, 292 (1987); H. D. Politzer and M. B. Wise, Phys. Lett. **B206**, 681 (1988); *ibid.* **B208**, 504(1988).
13. Particle Data Group, Phys. Rev. **D50**, 1173 (1994).

THE QUARK-ANTIQUARK WILSON LOOP FORMALISM IN THE NRQCD POWER COUNTING SCHEME

Antonio Vairo
Institut für Hochenergiephysik, Österreichische Akademie der Wissenschaften
Nikolsdorfergasse 18, A-1050 Vienna, Austria

ABSTRACT

The quark-antiquark interaction from the NRQCD Lagrangian is studied in the Wilson loop formalism.

1. The NRQCD Lagrangian

Large radii quarkonia (i.e. excited heavy mesons like $\bar{b}b$ and $\bar{c}c$) cannot be described in terms of perturbative QCD with the addition of leading nonperturbative effects encoded into local condensates.[1] This happens since in this situation the gluonic correlation length cannot be considered infinitely large with respect to the other scales of the system. A successful description needs therefore a systematic inclusion of non-local condensates. A solution is provided by the so-called Wilson loop formalism. The non-local quantities are field strength insertions in the Wilson loop made up by the quark trajectories.[2–4]

While a full relativistic QCD formulation in this formalism is still lacking,[5] such a formulation is possible for heavy quark bound states which are essentially described by non relativistic dynamics. These systems are characterized by a dynamical adimensional parameter, the quark velocity v, which is small and allows a classification of the energy scales of the problem in hard ($\sim m$), soft ($\sim mv$) and ultrasoft ($\sim mv^2$). Moreover, this provides a power counting scheme for the operators in the Lagrangian. The relation between v and the QCD parameters is unknown (for infinitely heavy quarks v coincides with α_s, for realistic quarks it is the result of perturbative and nonperturbative effects) but irrelevant once the power counting scheme works. Due to the heavy quark mass m, at a scale μ between m and mv the physics is still dominated by perturbative effects. Therefore, in order to describe heavy quark bound states, it is possible to substitute the QCD Lagrangian with an effective non relativistic Lagrangian via perturbative matching at that scale. The new Lagrangian is simpler, since the hard degrees of freedom have been integrated out explicitly, but equivalent to the QCD one at a given order in α_s and v. This effective Lagrangian is known as the NRQCD Lagrangian.[6] At order $O(v^4)$ the NRQCD Lagrangian describing a bound state between a quark of mass m_1 and an antiquark of mass m_2 is[7,8]

$$L = Q_1^\dagger \left(iD_0 + c_2^{(1)} \frac{\mathbf{D}^2}{2m_1} + c_4^{(1)} \frac{\mathbf{D}^4}{8m_1^3} + c_F^{(1)} g \frac{\sigma \cdot \mathbf{B}}{2m_1} + c_D^{(1)} g \frac{\mathbf{D}\cdot\mathbf{E} - \mathbf{E}\cdot\mathbf{D}}{8m_1^2} \right.$$

$$+ic_S^{(1)}g\frac{\sigma\cdot(\mathbf{D}\times\mathbf{E}-\mathbf{E}\times\mathbf{D})}{8m_1^2}\Bigg)Q_1 + \text{ antiquark terms } (1\leftrightarrow 2) + \frac{d_1}{m_1m_2}Q_1^\dagger Q_2 Q_2^\dagger Q_1$$
$$+\frac{d_2}{m_1m_2}Q_1^\dagger\sigma Q_2Q_2^\dagger\sigma Q_1 + \frac{d_3}{m_1m_2}Q_1^\dagger T^a Q_2Q_2^\dagger T^a Q_1 + \frac{d_4}{m_1m_2}Q_1^\dagger T^a\sigma Q_2Q_2^\dagger T^a\sigma Q_1, \quad (1)$$

where Q_j are the heavy quark fields. The coefficients $c_2^{(j)}$, $c_4^{(j)}$, ... are evaluated at the matching scale μ for a particle of mass m_j. They encode the ultraviolet regime of QCD order by order in α_s. The explicit expressions and a numerical discussion can be found in.[9] The power counting rules[6] for the operators of Eq. (1) are $Q \sim (mv)^{3/2}$, $\mathbf{D} \sim mv$, $gA_0 \sim mv^2$, $g\mathbf{A} \sim mv^3$, $gE \sim m^2v^3$ and $gB \sim m^2v^4$. Four quark operators which are apparently of order v^3 are actually suppressed by additional powers in α_s in the matching coefficients and the octet contributions by an additional power in v^2 on singlet states. Therefore in the following we will neglect these contributions with the exception of a term which mixes under RG transformation with the chromomagnetic operator contribution to the spin-spin potential.[10] We will call the corresponding matching coefficient d.

2. The Wilson Loop Formalism

The use of the Wilson loop formalism on the Lagrangian (1) was first suggested in.[10] Let us sketch the derivation of the heavy quark potential. The 4-point gauge invariant Green function G associated with the Lagrangian (1) is defined as

$$G(x_1,y_1,x_2,y_2) = \langle 0|Q_2^\dagger(x_2)\phi(x_2,x_1)Q_1(x_1)Q_1^\dagger(y_1)\phi(y_1,y_2)Q_2(y_2)|0\rangle,$$

where $\phi(x_2,x_1) \equiv \exp\left\{-ig\int_0^1 ds\,(x_2-x_1)^\mu A_\mu(x_1+s(x_2-x_1))\right\}$ is a Schwinger line added to ensure gauge invariance. After integrating out the heavy quark fields, G can be expressed as a quantum-mechanical path integral over the quark trajectories[5]:

$$G(x_1,y_1,x_2,y_2) = \int_{y_1}^{x_1}\mathcal{D}z_1\mathcal{D}p_1\int_{y_2}^{x_2}\mathcal{D}z_2\mathcal{D}p_2\exp\left\{i\int_{-T/2}^{T/2}dt\sum_{j=1}^{2}\mathbf{p}_j\cdot\dot{\mathbf{z}}_j - m_j - c_2^{(j)}\frac{p_j^2}{2m_j}\right.$$
$$\left.+c_4^{(j)}\frac{p_j^4}{8m_j^3}\right\}\frac{1}{N_c}\left\langle \mathrm{Tr\,P\,T_s}\exp\left\{-ig\oint_\Gamma dz^\mu A_\mu(z) + i\int_{-T/2}^{T/2}dz_{0j}\,c_F^{(j)}g\frac{\sigma\cdot\mathbf{B}}{2m_j}\right.\right.$$
$$\left.+c_D^{(j)}g\frac{\mathbf{D}\cdot\mathbf{E}-\mathbf{E}\cdot\mathbf{D}}{8m_j^2} + ic_S^{(j)}g\frac{\sigma\cdot(\mathbf{D}\times\mathbf{E}-\mathbf{E}\times\mathbf{D})}{8m_j^2}\right\}$$
$$\left.\times\exp\left\{\frac{i}{m_1m_2}\int_{-T/2}^{T/2}dt\,g^2 d\,T^{a(1)}\sigma^{(1)}T^{a(2)}\sigma^{(2)}\delta^3(\mathbf{z}_1-\mathbf{z}_2)\right\}\right\rangle$$
$$\equiv \int_{y_1}^{x_1}\mathcal{D}z_1\mathcal{D}p_1\int_{y_2}^{x_2}\mathcal{D}z_2\mathcal{D}p_2\exp\left\{i\int_{-T/2}^{T/2}dt\sum_{j=1}^{2}\mathbf{p}_j\cdot\dot{\mathbf{z}}_j - m_j - \frac{p_j^2}{2m_j} + \frac{p_j^4}{8m_j^3} - i\int_{-T/2}^{T/2}dt\,U\right\},$$

where the bracket means the Yang–Mills average over the gauge fields. Γ is the Wilson loop made up by the quark trajectories z_1 and z_2 and the endpoints Schwinger strings and $y_2^0 = y_1^0 \equiv -T/2$, $x_2^0 = x_1^0 \equiv T/2$. Reparameterization invariance[7] fixes $c_2 = c_4 = 1$. Assuming that the limit exists, we define the heavy quark-antiquark potential V as $\lim_{T\to\infty} \int_{-T/2}^{T/2} dt\, U$. For a discussion on the existence of a quark-antiquark potential we refer to.[11,12] Expanding in v (following the power counting given in the previous section) or in the inverse of the mass we get

$$
\begin{aligned}
V(r) &= \lim_{T\to\infty} \frac{i \log\langle W(\Gamma)\rangle}{T} \qquad (2)\\
&+ \left(\frac{\mathbf{S}^{(1)}\cdot\mathbf{L}^{(1)}}{m_1^2} + \frac{\mathbf{S}^{(2)}\cdot\mathbf{L}^{(2)}}{m_2^2}\right)\frac{2c_F^+ V_1'(r) + c_S^+ V_0'(r)}{2r} + \frac{\mathbf{S}^{(1)}\cdot\mathbf{L}^{(2)} + \mathbf{S}^{(2)}\cdot\mathbf{L}^{(1)}}{m_1 m_2}\frac{c_F^+ V_2'(r)}{r}\\
&+ \left(\frac{\mathbf{S}^{(1)}\cdot\mathbf{L}^{(1)}}{m_1^2} - \frac{\mathbf{S}^{(2)}\cdot\mathbf{L}^{(2)}}{m_2^2}\right)\frac{2c_F^- V_1'(r) + c_S^- V_0'(r)}{2r} + \frac{\mathbf{S}^{(1)}\cdot\mathbf{L}^{(2)} - \mathbf{S}^{(2)}\cdot\mathbf{L}^{(1)}}{m_1 m_2}\frac{c_F^- V_2'(r)}{r}\\
&+ \frac{1}{8}\left(\frac{c_D^{(1)}}{m_1^2} + \frac{c_D^{(2)}}{m_2^2}\right)\left(\Delta V_0(r) + \Delta V_{\mathrm{a}}^E(r)\right) + \frac{1}{8}\left(\frac{c_F^{(1)}}{m_1^2} + \frac{c_F^{(2)}}{m_2^2}\right)\Delta V_{\mathrm{a}}^B(r)\\
&+ \frac{c_F^{(1)} c_F^{(2)}}{m_1 m_2}\left(\frac{\mathbf{S}^{(1)}\cdot\mathbf{r}\mathbf{S}^{(2)}\cdot\mathbf{r}}{r^2} - \frac{\mathbf{S}^{(1)}\cdot\mathbf{S}^{(2)}}{3}\right)V_3(r) + \frac{\mathbf{S}^{(1)}\cdot\mathbf{S}^{(2)}}{3 m_1 m_2}\left(c_F^{(1)} c_F^{(2)} V_4(r) - 48\pi\alpha_s C_F\, d\, \delta^3(r)\right).
\end{aligned}
$$

$W(\Gamma) \equiv \mathrm{P}\exp\left\{-ig \oint_\Gamma dz^\mu A_\mu(z)\right\}$ is the non-static Wilson loop. The expansion of it around the static Wilson loop $W(\Gamma_0)$ (Γ_0 is a $r\times T$ rectangle) gives the static potential $V_0 = \lim_{T\to\infty} i\log\langle W(\Gamma_0)\rangle/T$ plus velocity (non-spin) dependent terms.[3] $\mathbf{S}^{(j)}$ and $\mathbf{L}^{(j)}$ are the spin and orbital angular momentum operators of the particle j. The matching coefficients are defined as $2c_{F,S}^{\pm} \equiv c_{F,S}^{(1)} \pm c_{F,S}^{(2)}$. The "potentials" V_1, V_2, ... are scale dependent gauge field averages of electric and magnetic field strength insertions in the static Wilson loop:

$$
\begin{aligned}
\Delta V_{\mathrm{a}}^E(r) &= 2\lim_{T\to\infty}\int_0^T dt\, \langle\langle \mathbf{E}(\mathbf{0},0)\mathbf{E}(\mathbf{0},t)\rangle\rangle - \langle\langle \mathbf{E}(\mathbf{0},0)\rangle\rangle\,\langle\langle \mathbf{E}(\mathbf{0},t)\rangle\rangle,\\
\Delta V_{\mathrm{a}}^B(r) &= 2\lim_{T\to\infty}\int_0^T dt\, \langle\langle \mathbf{B}(\mathbf{0},0)\mathbf{B}(\mathbf{0},t)\rangle\rangle,\\
\frac{r_k}{r}V_1'(r) &= \epsilon_{ijk}\lim_{T\to\infty}\int_0^T dt\, t\langle\langle B_i(\mathbf{0},0)E_j(\mathbf{0},t)\rangle\rangle,\\
\frac{r_k}{r}V_2'(r) &= \frac{1}{2}\epsilon_{ijk}\lim_{T\to\infty}\int_0^T dt\, t\langle\langle B_i(\mathbf{0},0)E_j(\mathbf{r},t)\rangle\rangle,\\
\left(\frac{r_i r_j}{r^2} - \frac{\delta_{ij}}{3}\right)V_3(r) &= 2\lim_{T\to\infty}\int_0^T dt\left[\langle\langle B_i(\mathbf{0},0)B_j(\mathbf{r},t)\rangle\rangle - \frac{\delta_{ij}}{3}\langle\langle \mathbf{B}(\mathbf{0},0)\mathbf{B}(\mathbf{r},t)\rangle\rangle\right],\\
V_4(r) &= 2\lim_{T\to\infty}\int_0^T dt\, \langle\langle \mathbf{B}(\mathbf{0},0)\mathbf{B}(\mathbf{r},t)\rangle\rangle,
\end{aligned}
$$

where $\langle\langle\ \rangle\rangle \equiv \langle\ W(\Gamma_0)\rangle/\langle W(\Gamma_0)\rangle$.

3. Conclusions

Two comments in order to conclude. The explicit expression for the potential (2) in terms of field strength insertions in a static Wilson loop is suitable for direct lattice[13] and analytic[14] evaluations. This makes the obtained result of particular importance. Different vacuum models can be easily compared on the heavy quark potential predictions, once the Wilson loop average has been evaluated.

The $O(v^2)$ leading order NRQCD Lagrangian $L = Q_1^\dagger(iD_0 + \partial^2/2m_1)Q_1+$ antiquark part does not contribute to (2) only with a static potential (with the exception of the perturbative contribution if evaluated in Coulomb gauge). Since the corresponding Wilson loop P exp $\left\{-ig \oint_\Gamma dz^0 A_0(z)\right\}$ is a function of the non-static loop Γ, its expansion produces in general velocity dependent terms as well.[3,13,14] This is not surprising since the power counting scheme of NRQCD has to be considered as a leading order power counting scheme. An exact value in v cannot be assigned to each term of the effective Lagrangian at least before the soft and ultrasoft degrees of freedom have been disentangled.[12]

An extensive analysis of the topic discussed here can be found in,[9] with particular emphasis on the relevance of the matching procedure in order to have a consistent potential in the perturbative regime.

References

1. F. J. Yndurain, Nucl. Phys. Proc. Suppl. **B 64**, 433 (1998).
2. L. S. Brown and W. I. Weisberger, Phys. Rev. **D 20**, 3239 (1979); E. Eichten and F. Feinberg, Phys. Rev. **D 23** (1981) 2724; M. E. Peskin in Proceeding of the 11th SLAC Institute, SLAC Report No. 207, 151, edited by P. Mc Donough (1983); D. Gromes, Zeit. Phys. **C 22**, 265 (1984); N. Brambilla, P. Consoli, G. Prosperi, Phys. Rev. **D 50**, 5878 (1994).
3. A. Barchielli, E. Montaldi and G. N. Prosperi, Nucl. Phys. **B 296**, 625 (1988).
4. N. Brambilla in these Proceedings.
5. N. Brambilla and A. Vairo, Phys. Rev. **D 56**, 1445 (1997).
6. G. P. Lepage, L. Magnea, C. Nakhleh, U. Magnea and K. Hornbostel, Phys. Rev. **D 46**, 4052 (1992).
7. A. V. Manohar, Phys. Rev. **D 56**, 230 (1997).
8. A. Pineda and J. Soto, hep-ph/9802365.
9. N. Brambilla, J. Soto and A. Vairo in preparation; N. Brambilla and A. Vairo in Proceedings of *QCD'98*, edited by S. Narison, to appear in Nucl. Phys. Proc. Supp. **B**.
10. Y. Chen, Y. Kuang and R. J. Oakes, Phys. Rev. **D 52**, 264 (1995).
11. G. P. Lepage in these Proceedings.
12. A. Pineda and J. Soto, Nucl. Phys. Proc. Suppl. **B 64**, 428 (1998); J. Soto in these Proceedings.
13. G. S. Bali, K. Schilling and A. Wachter, Phys. Rev. **D 56**, 2566 (1997).
14. N. Brambilla and A. Vairo, Phys. Rev. **D 55**, 3974 (1997).

The Soft Régime and β Function of NRQCD

Harald W. Grießhammer
Nuclear Theory Group, Department of Physics, University of Washington, Box 351 560, Seattle, WA 98195-1560, USA

ABSTRACT

Progress towards a complete velocity power counting in non-relativistic effective field theories, especially NRQCD, is motivated and summarised.

In Non-Relativistic QCD[1,2], a heavy quark is nearly on shell and its energy is dominated by its mass M, so that the resulting non-relativistic system can be treated perturbatively both in the coupling constant g and in the quark velocity v. The typical energy and momentum scales are the non-relativistic kinetic energy Mv^2 and the momentum Mv. The effective Lagrangean does not exhibit v explicitly, so that a power counting scheme has to be established which determines uniquely which terms in the Lagrangean must be taken into account to render predictive power to a given precision in v.

Identification of the relevant energy and momentum régimes has proven difficult: The first attempt by Lepage and co-workers[3] fell shot as working only in the Coulomb gauge and as being incomplete. In a recent article, Beneke and Smirnov[4] pointed out that the much simpler velocity rescaling rules proposed by Luke and Manohar for Coulomb interactions[5], and by Grinstein and Rothstein for bremsstrahlung processes[6], as united by Luke and Savage[7], and by Labelle's power counting scheme in time ordered perturbation theory[8], do not reproduce the two gluon exchange between particles near threshold. Recently, velocity power counting rules using dimensional regularisation were established for a toy model[9] and NRQCD[10] (see also[11]) following Beneke and Smirnov's threshold expansion[4]. Here, I summarise two recent publications[9,10].

The NRQCD propagators are read off from the Lagrangean as

$$Q : \frac{\mathrm{i\,Num}}{T - \frac{\vec{p}^2}{2M} + \mathrm{i}\epsilon}\;, \quad A^\mu : \frac{\mathrm{i\,Num}}{k^2 + \mathrm{i}\epsilon}\;, \tag{1}$$

where $T = \frac{\vec{p}^2}{2M} + \ldots$ is the kinetic energy of the quark. "Num" are numerators containing the appropriate colour, Dirac and flavour indices. Cuts and poles in scattering amplitudes stem from bound states and on-shell propagation of particles in intermediate states and give rise to infrared divergences. With energies and momenta of either scale, there are only three régimes in which quarks or gluons in (1) are on shell:

$$\begin{array}{rll} \text{soft régime:} & A^\mu_\mathrm{s}: & k_0 \sim |\vec{k}| \sim Mv \;, \\ \text{potential régime:} & Q_\mathrm{p}: & T \sim Mv^2 \;,\; |\vec{p}| \sim Mv \;, \\ \text{ultrasoft régime:} & A^\mu_\mathrm{u}: & k_0 \sim |\vec{k}| \sim Mv^2 \end{array} \tag{2}$$

Ultrasoft gluons A^μ_u are emitted as bremsstrahlung or from excited states in the bound system, and hence physical. Overall energy conservation forbids all processes with outgoing soft gluons but without ingoing ones. Gluons which change the quark momenta but keep them close to their mass shell relate the (instantaneous) Coulomb interaction:

$$A^\mu_\mathrm{p} \;:\; k_0 \sim Mv^2 \;,\; |\vec{k}| \sim Mv \tag{3}$$

When a soft gluon A^μ_s couples to a potential quark Q_p, the outgoing quark is far off its mass shell and carries energy and momentum of order Mv.

$$Q_\mathrm{s} \;:\; T \sim |\vec{p}| \sim Mv \tag{4}$$

As the potential quark has a much smaller energy than either of the soft particles, it can – by the uncertainty relation – not resolve the precise time at which the soft quark emits the soft gluon. So, a "temporal" multipole expansion is associated with this vertex. In general, the coupling between particles of different régimes will not be point-like but contain multipole expansions for the particle belonging to the weaker kinematic régime.

One may expand[4] the integrand of a loop integral in NRQCD about the various saddle points, i.e. about the values of the loop-momentum q where particles become on shell. For example, expanding about a physical gluon at the soft scale, the quark propagator is ($T_\mathrm{p} \sim \frac{\vec{p}^2}{2M} \sim Mv^2 \ll q_0 \sim |\vec{q}| \sim Mv$)

$$\frac{\mathrm{i}}{q_{0,\mathrm{s}} + T_\mathrm{p} - \frac{(\vec{p}+\vec{q})^2}{2M}} \longrightarrow \frac{\mathrm{i}}{q_{0,\mathrm{s}}} + \frac{\mathrm{i}}{q_{0,\mathrm{s}}}\,\mathrm{i}\Big(T_\mathrm{p} - \frac{(\vec{p}+\vec{q})^2}{2M}\Big)\,\frac{\mathrm{i}}{q_{0,\mathrm{s}}} + \dots \;. \tag{5}$$

Higher order terms in the expansion are interpreted as insertions into the soft quark propagator (static to lowest order) and as multipole expansions which modify the vertex rules. As the energy of potential gluons is much smaller than their momentum, the A_p-propagator becomes instantaneous. With these five fields $Q_\mathrm{s}, Q_\mathrm{p}, A^\mu_\mathrm{s}, A^\mu_\mathrm{p}, A^\mu_\mathrm{u}$ representing the relevant infrared-relevant degrees of freedom, i.e. quarks and gluons in the three different non-relativistic régimes, soft, potential and ultrasoft, NRQCD becomes self-consistent.

The quark propagator is gauge independent. Gauge fixing introduces gauge dependent denominators multiplying the gauge independent denominators in the gluon propagators. When they are non-zero for all but the above combinations of scales Mv and Mv^2 (2), the decomposition into the three régimes (2) remains unchanged, as do the rescaling properties of the fields and interactions. The standard gauges (axial, Weyl, Lorentz, Coulomb) will therefore all show the same power counting and vertex rules.

Details of the gluon propagator and its insertions are different in different régimes and gauges, and some gauges will not exhibit certain vertices, insertions and representatives; e.g. the Coulomb gauge is unique in having A_0 contribute only in the potential régime as physical fields are transverse by virtue of Gauß' law.

Finally, the regularisation scheme must be chosen such that expansion around one saddle point in the loop integral does not obtain any contribution from other saddle points. Cutoff regularisation jeopardises power counting and symmetries, and introduces unphysical power divergences. In contradistinction, using dimensional regularisation *after* the saddle point expansion preserves power counting and gauge symmetry. Its homogeneity guarantees that contributions from different saddle points and régimes do not overlap. (A simple example is given in Ref.[9].)

In order to establish explicit v counting in the NRQCD Lagrangean, one rescales the space-time coordinates such that typical momenta in each régime are dimensionless:

$$\begin{aligned} \text{soft:} \quad & t = (Mv)^{-1}\,T_{\rm s}\ ,\quad \vec{x} = (Mv)^{-1}\,\vec{X}_{\rm s}\ ,\\ \text{potential:} \quad & t = (Mv^2)^{-1}\,T_{\rm u}\ ,\quad \vec{x} = (Mv)^{-1}\,\vec{X}_{\rm s}\ ,\\ \text{ultrasoft:} \quad & t = (Mv^2)^{-1}\,T_{\rm u}\ ,\quad \vec{x} = (Mv^2)^{-1}\,\vec{X}_{\rm u}\ . \end{aligned} \tag{6}$$

The scaling of the fields is derived by normalising the propagator terms to be of order v^0. In order to maintain velocity power counting, corrections of order v or higher in the thus rescaled Lagrangean must be treated as insertions. Particles in the various régimes couple: On-shell (potential) quarks radiate bremsstrahlung (ultrasoft) gluons. In general, one must allow all couplings between the various régimes which obey "scale conservation" for energies and momenta. This excludes for example the coupling of two potential quarks ($T \sim Mv^2$) to one soft gluon ($q_0 \sim Mv$), but not to two soft gluons.

Amongst the fields introduced, two (six) interactions are allowed for example within and between the various régimes for the scalar coupling in the Coulomb (Lorentz) gauge whose v counting is read up from the rescaled Lagrangean[9–11], see table 1. Albeit both describing interactions with physical gluons, soft and ultrasoft couplings occur at different orders in v, and obey different multipole expansion rules. Double counting is prevented because in addition to most of the propagators, all vertices involve different multipole expansions. Because the vertex rules for the soft régime count powers of v with respect to the soft régime, one retrieves there the velocity power counting of Heavy Quark Effective Theory[12,13]. HQET becomes a sub-set of NRQCD, complemented by interactions between soft (HQET) and potential or ultrasoft particles. In NRQCD with two potential quarks as initial and final states, the power counting in a soft sub-graph has to be transfered to the potential régime[9,10]. Because soft loop momenta scale like $[\mathrm{d}^4\!k_{\rm s}] \sim v^4$, while potential ones like $[\mathrm{d}^4\!k_{\rm p}] \sim v^5$, each largest sub-graph which contains soft particles is enhanced by an additional factor $\frac{1}{v}$.

Matching between NRQCD and QCD proved that the soft quark and gluon are indispensable to reproduce the correct structure of collinear (infrared) divergences of the two gluon exchange contribution to Coulomb scattering between non-relativistic particles near threshold in a toy model and confirmed the proposed counting rules[9].

Because NRQCD is a well-defined field theory of quarks and gluons, one can investigate its renormalisation group equations under the assumption that perturbation

Table 1. *Velocity power counting and vertices for the interaction Lagrangean $-gQ^\dagger A_0 Q$ in the Lorentz gauge. In the Coulomb gauge, only the first two diagrams exist. Zigzagged, dashed and wiggly lines represent $A^\mu_{\rm s}$, $A^\mu_{\rm p}$ and $A^\mu_{\rm u}$, double and single lines $Q_{\rm s}$ and $Q_{\rm p}$, respectively.*

Vertex	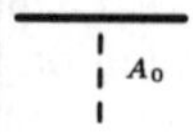	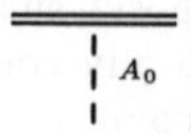	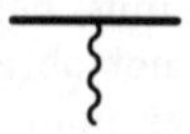	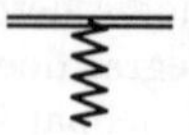		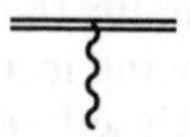
v power	$\frac{1}{\sqrt{v}}$	$\sqrt{v}$	v^0	v^0	v^0	v

theory is applicable at all scales. NRQCD must reproduce the QCD β function with $N_{\rm F}$ light quarks below the scale M. The result thus already anticipated,

$$\beta_{\rm NRQCD} = -\frac{g_{\rm R}^3}{(4\pi)^2}\left[\frac{11}{3}\,N - \frac{2}{3}\,N_{\rm F}\right] \tag{7}$$

to lowest order for the gauge group SU(N), the prime goal of this exercise[10] is not the result but to provide further insight: The Lorentz gauge is a legitimate gauge choice and simplifies renormalisation. The renormalised coupling strengths of all interactions steming from expanding the same term in the Lagrangean in the various régimes are the same, except that they have to be taken at different scales. Although the number of vertices is increased, the number of independent couplings is not. The soft régime is indispensable for NRQCD to describe especially the correct running coupling for the Coulomb gluons, as their sole contribution in the vacuum polarisation comes from the propagation of soft (on-shell) gluons in the loop. The computation also confirms the validity of the power counting proposed and is slightly simpler than its QCD counterparts. The non-Abelian nature of the gauge field enters to lowest order only via the non-zero contribution of the three gluon vertex to the vacuum polarisation. Non-Abelian vertex contributions are down by at least one power in v compared to Abelian vertex corrections and bare vertices. The quark self energy and vertex renormalisations differ from its QCD counterpart, but the Slavnov–Taylor identities hold.

From a technical point of view, the number of possible diagrams in power counted NRQCD is considerably larger than in the original theory because there are at least six vertices per interaction allowed by scale conservation, cf. table 1. The calculation is greatly simplified by diagrammatic rules[10] which follow from the homogeneïty property of dimensional regularisation. That integrals without scales set by external momenta or energies in the denominator vanish in dimensional regularisation,

$$\int \frac{{\rm d}^d q}{(2\pi)^d}\, q^\alpha = 0 \ , \tag{8}$$

allows one to recognise the majority of NRQCD (and threshold expansion[4]) graphs as zero to all orders in v just by drawing them, independently of details of the vertices involved. The concept is sensitive only to the multipole expansion of the vertices and independent of insertions and the chosen gauge. Only the denominators of the graph lowest order in v have to be looked at. It is usually only necessary to consider a subset of the whole diagram. One assigns the loop and external momenta to it, taking into account multipole expansions, and writes out the denominators (i.e. inverse propagators). If by shifting integration variables, one arrives at an energy or momentum integral without scale, the diagram vanishes, as do any diagrams which contain it as a sub-graph. As an example, the following rule may be quoted, which states that a potential gluon coupled to two soft particles, quark or gluon, vanishes because there is no scale in the q_0 part of the loop integration, even when potential or ultrasoft particles couple to the Coulomb gluon:

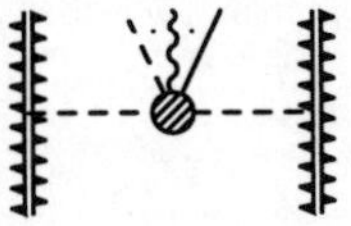

$$= 0 \tag{9}$$

In the case of the NRQCD β function, the number of vertices for each of the couplings of table 1 is reduced from three or more to one, and the number of vacuum polarisations also from three to one. This result holds for any gauge and any interaction involving loops with three particles, showing the general usefulness of these rules.

For a detailed account of the points outlined in this contribution, I refer to[9,10] and references therein. In conclusion, a complete velocity power counting of NRQCD has been suggested. The results are readily generalised to any effective field theory with two or more infrared scales.

References

1. W. E. Caswell and G. P. Lepage: *Phys. Lett.* B **167**, 437 (1986).
2. G. T. Bodwin, E. Braaten and G. P. Lepage: *Phys. Rev.* D **51**, 1125 (1995); *Phys. Rev.* D **55**, 5853 (1997).
3. G. P. Lepage et al.: *Phys. Rev.* D **46**, 4052 (1992).
4. M. Beneke and V. A. Smirnov: *Asymptotic Expansion of Feynman Integrals near Threshold*; CERN-TH/7-315, hep-ph/9711391, 1997 .
5. M. Luke and A. V. Manohar: *Phys. Rev.* D **55**, 4129 (1997).
6. B. Grinstein and I. Z. Rothstein: *Phys. Rev.* D **57**, 78 (1998).
7. M. Luke and M. J. Savage: *Phys. Rev.* D **57**, 413 (1998).
8. P. Labelle: *Effective Field Theories for QED Bound States: Extending Nonrelativistic QED to Study Retardation Effects*; MCGILL-96-33, hep-ph/9608491, 1996 .
9. H. W. Grießhammer, *Threshold Expansion and Dimensionally Regularised NRQCD*; NT@UW-98-3, hep-ph/9712467, 1997 (to be published in *Phys. Rev.* D).
10. H. W. Grießhammer, *Power Counting and β Function in NRQCD*; NT@UW-98-22, 1998 .
11. H. W. Grießhammer, *The Soft Régime in NRQCD*; NT@UW-98-12, hep-ph/9804251, 1998 (Proceedings of the Workshop "Nuclear Physics With Effective Field Theories" at Caltech, 26th – 27th February 1998, eds. R. Seki, U. van Kolck and M. J. Savage, World Scientific).
12. N. Isgur and M. B. Wise: *Phys. Lett.* B **232**, 113 (1989).
13. N. Isgur and M. B. Wise: *Phys. Lett.* B **237**, 527 (1990).

POTENTIAL NON-RELATIVISTIC QCD

Joan Soto
Departament d'Estructura i Constituents de la Matèria
and
Institut de Física d'Altes Energies
Universitat de Barcelona
Diagonal 647, 08028 Barcelona, Catalonia, Spain

ABSTRACT

Heavy quarkonia bound states have at least three dynamical scales. Namely, the mass of the heavy quarks (hard $\sim m$), the typical relative momentum (soft $\sim mv$) and the typical binding energy (ultrasoft $\sim mv^2$). The hierarchy $m >> mv >> mv^2$ allows for the introduction of two effective field theories after sequencially integrating out gluons of scales m and mv. Integrating out the scale m gives rise to the celebrated Non-Relativistic QCD (NRQCD), which contains gluons of energy and momentum $O(mv)$ or less. Integrating out the scale mv gives rise to an effective field theory which is local in time but non-local in space (i.e. with potential terms) and contains gluons of energy and momentum $O(mv^2)$ or less, which we have called potential NRQCD (pNRQCD). We shall present the pNRQCD lagrangian and some non-trivial checks of the ideas above in QED.

1. Introduction

Since all the results presented in my talk are already available in the literature, I shall restrict myself to direct the interested reader to the relevant material.

2. NRQCD

I would like to start with a couple of comments on NRQCD.[1]

First of all, I would like to mention an important point which has become clear during the last year, namely that the NRQCD and Heavy Quark Effective Theory (HQET)[2] lagrangians are the same.[3,4] Only the phenomenological use of them makes the difference. For instance, terms with two heavy two light quark fields are relevant in heavy-light systems but not in heavy-heavy systems and hence they are considered in HQET works[5] but not in NRQCD papers. Conversely, terms with four heavy quark fields are relevant in heavy-heavy systems but not in heavy-light systems and hence they are considered in NRQCD works but not in HQET papers. Terms with two heavy quarks and gluons are relevant for both heavy-light and heavy-heavy systems, and hence results obtained in NRQCD and HQET are interchangeble. We have calculated the one loop matching coefficients for the four heavy quark field terms in.[4]

Second, I would like to comment on Lorentz invariance in the NRQCD/HQET lagrangian. Any Effective Field Theory (EFT) must have the symmetries of the fundamental theory it is derived from. Hence NRQCD/HQET must have the symmetries of QCD, in particular Lorentz invariance. However, Lorentz invariance may be realised in a non-trivial way. This implies that there will be unsuspected relations between different terms of the lagrangian. It is well known in HQET that such relations do exist and are a consequence of the so called *reparametrization invariance*[6] (see[7] for recent work). This suggests that reparametrization invariance is nothing but the non-trivial implementation of Lorentz invariance in a non-relativistic EFT.[8]

3. pNRQCD

The pNRQCD lagrangian was presented in ref..[9] It consist of a singlet and an octet wave function fields, which depend on the relative coordinate, center of mass coordinate and time, potential terms and ultrasoft gluon fields, which only depen on the center of mass coordinate and time. In the limit $m \to \infty$, then $v \to \alpha_s$ and pNRQCD can be derived from NRQCD perturbatively. In this limit we have a situation similar to QED, which becomes a good laboratory to check these ideas. The pNRQED lagrangian was also introduced in ref..[9] We have used it to calculate the spectrum of Hydrogen-like atoms[10] and positronium[11] at $O(m\alpha^5)$ and have found complete agreement with the known results. This makes us confident the pNRQED is correct and strongly suggests that pNRQCD also is, at least in the weak coupling limit. Nonetheless, for most of the bottomonium and charmonium bound states mv is not large enough as to justify a perturbative matching between NRQCD and pNRQCD. If we assume that the dynamical fields in the pNRQCD lagrangian are still the same, the only difference would be that arbitrary (non-perturbative) potentials must be introduced, presumably along the lines of ref.[12] (see also[13]). These potentials may be estimated using vaccuum models[14] or fit from lattice data.[15] An interesting point about pNRQCD is that potential models emerge as a truncation of the pNRQCD lagrangian (by droping ultrasoft gluons and the octet wave function field). If this truncation can be rigorously justified, potential models could be regarded as a consequence of QCD. If not, the full pNRQCD lagrangian is to be considered, which would imply, in particular, that retardation effects are indeed important.[16] These issues are under current investigation.

References

1. W.E. Caswell and G.P. Lepage, *Phys. Lett.* **B167** (1986) 437. G.T. Bodwin, E. Braaten and G.P. Lepage, *Phys. Rev.* **D51** (1995) 1125, Erratum *ibid* **D55** (1997) 5853.
2. M. Neubert, *Phys. Rep.* **245** (1994) 259.
3. A.V. Manohar, *Phys. Rev.* **D56** (1997) 230.
4. A. Pineda and J. Soto, *Matching at one loop for the four-quark operators in NRQCD*, hep-ph/9802365, to appear in *Phys. Rev.* **D**.

5. B. Block, J. G. Körner, D. Pirjol, J. C. Rojas, *Nucl. Phys.* **B496** (1997) 358. C. Balzereit, *Spectator Effects in Heavy Quark Effective Theory at* $O(1/m_Q^3)$, hep-ph/9805503.
6. M. Luke and A. V. Manohar, *Phys. Lett.* **B286** (1992) 348.
7. M. Finkemeier, H. Georgi and M. McIrvin, *Phys. Rev.* **D55** (1997) 6933. R. Sundrum, *Phys. Rev.* **D57** (1998) 331. C. L. Y. Lee, *Reparametrization Invariance in Heavy Particle Effective Field Theories*, hep-ph/9709238.
8. Y.-Q. Chen and Y.-P. Kuang, *Z. Phys.* **C67** (1995) 627.
9. A. Pineda and J. Soto, *Nucl. Phys. B (Proc. Suppl.)* **64** (1998) 428.
10. A. Pineda and J. Soto, *Phys. Lett.* **B420** (1998) 391.
11. A. Pineda and J. Soto, *Potential NRQED: The Positronium Case*, hep-ph/9805424, to appear in *Phys. Rev.* **D**.
12. Y.-Q. Chen, Y.-P. Kuang and R.J. Oakes, *Phys. Rev.* **D52** (1995) 264.
13. A. Vairo, these proceedings.
14. N. Brambilla, these proceedings.
15. G.S. Bali, K. Schilling and A. Wachter, *Phys. Rev.* **D56** (1997) 2566.
16. G. P. Lepage, these proceedings.

THE DIRAC STRUCTURE OF CONFINEMENT

Eric Swanson
North Carolina State University, Raleigh, NC 27695-8202 USA

ABSTRACT

It has become traditional to assume that the Dirac structure of the phenomenological quark confinement potential is scalar ⊗ scalar. We use the heavy quark expansion of the Coulomb gauge QCD Hamiltonian and the Flux Tube model to demonstrate that this is true but in the *effective* sense only. The demonstration contains some surprises: confinement is actually vector ⊗ vector and it is nonperturbative mixing between ordinary and hybrid $Q\bar{Q}$ states which generates the scalar-like spin dependent potential at order $1/m_q^2$. Thus the existence of hybrids is crucial to establishing well-known spin splitting phenomenology. Finally, the resolution also indicates that the gluonic degrees of freedom in a hybrid must be of a collective nature.

1. Introduction

As evidenced by the discussions at this conference, the Dirac structure of the long range $q\bar{q}$ potential has engendered much confusion. Comparison with experiment certainly indicates that the effective heavy quark interaction is scalar (this will be detailed below). This has led some to suggest that the Lorentz covariance of QCD then implies that the $q\bar{q}$ interaction *must* be scalar. In the following we show that this is incorrect. Before moving on to the main discussion we illustrate the issue in slightly more detail.

For simplicity, consider the generating functional of QED, $Z = \int DAD\bar{\psi}D\psi \exp(iS) = \int D\bar{\psi}D\psi \exp(iS_{eff})$ where $S_{eff} = \int d^4x\bar{\psi}(i\partial\!\!\!/-m)\psi+e^2/2\int d^4xd^4yj^\mu(x)G_{\mu\nu}(x,y)j^\nu(y)$. It appears that the quark interaction is unequivocally vector! However this need not be the case because of nontrivial mixing with intermediate states. The point is that the kernel contains an integral over the relative time coordinate which spoils the simple interpretation of the interaction in S_{eff} as a potential. The same argument applies to QCD with a more complicated set of kernels.

Quenched lattice gauge theory and heavy quark phenomenology indicate that the static ($m_q >> \Lambda_{QCD}$) long range potential should be linear with a slope of $b \approx 0.18\,\mathrm{GeV}^2$. The order $1/m_q^2$ quark-antiquark long range spin-dependent (SD) structure is also of interest and is the subject of this talk. In a model approach, $Q\bar{Q}$ interactions are typically derived from a nonrelativistic reduction of a relativistic current-current interaction

$$V_{conf} = \int j_\Gamma(\mathbf{x}) K(\mathbf{x}-\mathbf{y}) j_\Gamma(\mathbf{y}). \tag{1}$$

In the heavy quark, nonrelativistic limit, this reduces to

$$V_{conf} \to \epsilon(r) + V_{SD} + \dots \tag{2}$$

where the ellipsis denotes the spin independent interaction at $\mathcal{O}(1/m^2)$ and terms of higher order in $1/m$. The spin dependent interaction may be parameterized as[1]

$$\begin{aligned} V_{SD}(r) = &\left(\frac{\sigma_q \cdot \mathbf{L}_q}{4m_q^2} - \frac{\sigma_{\bar{q}} \cdot \mathbf{L}_{\bar{q}}}{4m_{\bar{q}}^2}\right)\left(\frac{1}{r}\frac{d\epsilon}{dr} + \frac{2}{r}\frac{dV_1}{dr}\right) + \\ &\left(\frac{\sigma_{\bar{q}} \cdot \mathbf{L}_q}{2m_q m_{\bar{q}}} - \frac{\sigma_q \cdot \mathbf{L}_{\bar{q}}}{2m_q m_{\bar{q}}}\right)\left(\frac{1}{r}\frac{dV_2}{dr}\right) + \\ &\frac{1}{12m_q m_{\bar{q}}}\left(3\sigma_q \cdot \hat{\mathbf{r}}\, \sigma_{\bar{q}} \cdot \hat{\mathbf{r}} - \sigma_q \cdot \sigma_{\bar{q}}\right)V_3(r) + \\ &\frac{1}{12m_q m_{\bar{q}}}\sigma_q \cdot \sigma_{\bar{q}} V_4(r). \end{aligned} \tag{3}$$

Here $\epsilon = \epsilon(r)$ is the static potential, $r = |\mathbf{r}_q - \mathbf{r}_{\bar{q}}|$ is the $\bar{Q}Q$ separation and the $V_i = V_i(r)$ are potentials which depend on the nonperturbative structure of QCD. Assuming that the V_i simply follow from Eq. (1) yields the results shown in Table I.[2] Tensor and axial vector currents have spin-dependent static potentials and are therefore not useful for phenomenology in this context. Even the derivation of these simple results is beset with ambiguities and controversy. The reader is referred to the excellent review by Gromes in Ref..[2] Finally we note that covariance under Lorentz transformations leads to a constraint between the SD potentials: $\epsilon(r) = V_2(r) - V_1(r)$, known as Gromes' relation.[3]

Table 1. Nonrelativistic limits of Eq. (1).

Γ	ϵ_Γ	V_1	V_2	V_3	V_4
scalar	S	$-S$	0	0	0
vector	V	0	V	$V'/r - V''$	$2\nabla^2 V$
pseudoscalar	0	0	0	$P'' - P'/r$	$\nabla^2 P$

Twenty-two years ago Schnitzer[4] examined splittings in P-wave heavy quarkonia in an attempt to distinguish the possible spin dependence of the confinement potential. Specifically, he assumed a vector Coulomb short range potential and scalar or vector long range potentials. The ratio of P-wave splittings is then given by

$$\rho_{\substack{V\\S}} \equiv \frac{M(^3P_2) - M(^3P_1)}{M(^3P_1) - M(^3P_0)} = \frac{\langle \frac{16\alpha_s}{5r^3} \pm A\frac{b}{r}\rangle}{\langle \frac{4\alpha_s}{r^3} \pm B\frac{b}{2r}\rangle}, \tag{4}$$

where $A = 14$ (1) and $B = 4$ (1) for the vector (scalar) case. Experimentally, the splittings for the J/ψ and Υ systems are 0.48 and 0.66 respectively. These are shown as horizontal lines in Fig. 1. The upper and lower curves are ρ_V and ρ_S respectively (these have not been convoluted with the wavefunctions). Since the $b\bar{b}$ wavefunction is smaller than the $c\bar{c}$, the curve must drop with increasing r to fit the data. Clearly only scalar confinement meets this criterion. This conclusion has been supported by many other methods, including lattice gauge theory,[5] Wilson loop calculations,[6,7] and flux tube models.[8]

A consistent picture of a QCD-generated effective scalar confinement interaction appears to be emerging. It is therefore disconcerting that attempts to build models which are "closer" to QCD (typically relativistic models) seem to require vector confinement. For example, scalar Salpeter equations do not have stable solutions[9] and it appears to be impossible to construct a BCS-like vacuum of QCD when scalar confinement is assumed.[10,11] This is problematical if one wishes to dynamically generate constituent quark masses. Furthermore, attempts at modeling chiral pions will be hindered by the explicit lack of chiral symmetry in a scalar interaction. Finally, baryons anticonfine and colour singlet Tamm-Dancoff or RPA bound states are infrared divergent in a scalar potential.[12]

This issue was resolved in Ref.[13] by first performing a Foldy-Wouthuysen reduction of the full Coulomb gauge Hamiltonian of QCD. This immediately establishes that the Dirac structure of confinement for heavy quarks is of a timelike-vector nature. Furthermore, the spin dependent structure of the long range force is determined by nonperturbative mixing of the $q\bar{q}$ states with hybrids. Finally, if one assumes a flux tube-like configuration of the gluonic component of the hybrids, a "scalar" spin-dependent potential emerges in a natural way.

2. Conclusions

Spin splittings and lattice calculations indicate that confinement is scalar in nature. This conflicts with many relativistic models of QCD which require vector confinement. For example, a chirally symmetric interaction is needed if pseudo-Goldstone pions and spontaneous chiral symmetry breaking are to be generated dynamically. Furthermore it appears to be impossible to build a stable vacuum with a scalar kernel. We have examined this issue with the heavy quark limit of the Coulomb gauge QCD Hamiltonian. This approach is physically intuitive and is simpler to interpret and implement than methods based on the Wilson loop. We found that the static confinement potential must indeed be a Dirac timelike vector. Effective scalar interactions are generated at order $1/m^2$ by nonperturbative mixing with hybrid states.

We have argued that the long range spin-spin (V_3 and V_4) and the vector-like spin-orbit potentials (V_2) should all be zero since they involve field correlation functions evaluated between quark and antiquark. This statement follows by assuming that the gluonic degrees of freedom collapse into a flux tube-like configuration, as shown by lattice gauge theory. Alternatively, the scalar-like spin-orbit potential (V_1) is proportional to the matrix element of the electric and magnetic fields evaluated at the same point and hence is expected to be nonzero. Explicit calculations of the relevant matrix elements were carried out in the Flux Tube Model. The model was extended to include color degrees of freedom and to map the chromoelectric and chromomagnetic fields to flux tube phonon operators. The results obtained were in agreement with our general arguments and with Gromes' relation.

A consistent picture of the Dirac structure of confinement has emerged. The static central potential is timelike vector while the spin-dependent structure mimics the nonrelativistic reduction of an effective scalar interaction. This implies that it is incorrect to employ a scalar confinement kernel when doing calculations with light quarks. Note however that it would be acceptable to use scalar confinement when working explicitly in the chiral symmetry broken phase, *i.e.*, with constituent quarks in the nonrelativistic limit. The work presented here also implies that a constituent gluon picture of hybrids will yield incorrect results for certain observables. For example, V_1 and V_2 would be of comparable magnitude in a constituent gluon model. In general, these types of models must fail when nonlocal properties of the gluonic configuration are considered. Alternatively, it is possible that they perform quite well when evaluating global properties of gluonics such as the hybrid spectrum.

The author is grateful to Adam Szczepaniak for continuing discussions on these and related issues.

References

1. E. Eichten and F. Feinberg, Phys. Rev. **D23**, 2724 (1981).
2. W. Lucha, F. F. Schöberl, and D. Gromes, Phys. Rep. **200**, 127 (1991).
3. D. Gromes, Z. Phys. **C11**, 147 (1981); *erratum* **C14**, 94 (1982).
4. H.J. Schnitzer, Phys. Rev. Lett. **35**, 1540 (1975).
5. C. Michael, Phys. Rev. Lett. **56**, 1219 (1986).
6. E. Montaldi, A. Burchielli, and G.M. Prosperi, Nucl. Phys. **B296**, 625 (1988); M. Baker, J.S. Ball, N. Brambilla, G.M. Prosperi, and F. Zachariasen, Phys. Rev. **D54**, 2829 (1996); Yu.A. Simonov, Nucl. Phys. **B324**, 67 (1989).
7. N. Brambilla, P. Consoli, and G. M. Prosperi, Phys. Rev. **D50**, 5878 (1994).
8. W. Buchmüller, Phys. Lett. **112B**, 479 (1982). For extensions of this idea see M.G. Olsson, S. Veseli, and K. Williams, Phys. Rev. **D53**, 4006 (1996) and references therein.
9. J. Parramore and J. Piekarewicz, Nucl. Phys. **A585**, 705 (1995). See also J. Resag and C.R. Münz, Nucl. Phys. **A590**, 735 (1995).
10. A. Szczepaniak, E.S. Swanson, C.-R. Ji, and S.R. Cotanch, Phys. Rev. Lett. **76**, 2011 (1996).
11. A.P. Szczepaniak and E.S. Swanson, Phys. Rev. **D55**, 1578 (1997).
12. A.P. Szczepaniak and E.S. Swanson, unpublished.
13. A.P. Szczepaniak and E.S. Swanson, Phys. Rev. **D55**, 3987 (1997).
14. S. Perantonis and C. Michael, Nucl. Phys. **B347**, 854 (1990).
15. N. Isgur and J. Paton, Phys. Rev. **D31**, 2910 (1985).

HEAVY QUARK CONFINEMENT FROM DYSON-SCHWINGER EQUATIONS

Conrad Burden
Department of Theoretical Physics,
Research School of Physical Sciences and Engineering,
Australian National University, Canberra ACT 0200, AUSTRALIA

ABSTRACT

A formalism for studying the confinement of heavy quarks by considering the renormalised quark Dyson-Schwinger equation in the limit $m_Q \to \infty$ is described. We are particularly interested in studying the analytic structure of heavy quark propagators in the region of the complex momentum plane close to the bare fermion mass pole.

1. Introduction

Confinement and chiral symmetry breaking in quantum chromodynamics (QCD) must ultimately have a field theoretic cause. They are innately nonperturbative phenomena, demanding the use of appropriate techniques if they are to be fully understood. Lattice gauge theory and the Dyson-Schwinger (DS) equation technique[1] represent the two most promising lines of attack on QCD from a nonperturbative, field theoretic point of view.

Of these two, lattice gauge theory has the advantage that it is as close as one can find to an *ab initio* treatment of QCD. Approximations are clearly defined and in principle controllable. It has the disadvantage that a qualitative understanding of physical phenomena tends to be obscured by numerical brute force. Some progress is now being made, however, in interpreting lattice gauge fields in terms of monopoles and flux tubes.[2]

Dyson-Schwinger equations provide a complementary approach to QCD. The infinite tower of coupled integral equations relating Greens functions is an exact, nonperturbative description of QCD. Before any progress can be made, the tower must be truncated, generally at the level of quark and gluon propagator equations and bound state Bethe-Salpeter and Faddeev equations. Chiral symmetry breaking is signalled by the generation of a non-zero scalar part to the quark propagator, while a sufficient condition for quark confinement is the absence of real, timelike singularities in the propagator. These conditions ensue provided the dressed gluon propagator is sufficiently infrared enhanced.[3]

Numerical solution of the gluon propagator DS equation is a formidable task, and opinions differ about the resultant infrared behaviour of the gluon propagator.[4,5] It has been argued however that, if the three gluon and ghost-gluon vertices are modelled in a way free of singularities, a strongly infrared enhanced gluon propagator may result, leading to the required chiral symmetry breaking and quark confinement.[6] Within the DS picture, these two phenomena go hand in hand, both driven by the functional form of the quark propagator.

There is of course no necessary inconsistency between the DS picture and dual superconductor flux tube models. A strongly infrared enhanced gluon propagator can readily generate a Wilson area law.[7] A fortiori, one should not be perturbed that the flux tube picture is not immediately manifest in the interpretation of a timelike singularity free propagator as a signal of confinement. After all, that Wilson loops should satisfy an area law (which we interpret as the existence of a flux tube) is not an immediate, or even necessary consequence of the fact that Polyakov loops are zero valued (which we interpret as a signal of colour confinement). Obviously, both pictures of confinement can happily coexist.

In what follows we give a very brief description of a formalism designed to extend the DS picture to the realm of heavy quarks.[8–10] We shall see that the DS picture of quark confinement can be carried over to the extreme heavy quark limit in a natural and instructive way.

2. The Heavy Quark Limit

Our starting point is the quark DS equation

$$S(p)^{-1} = i\gamma\cdot pA(p^2)+B(p^2) \tag{1}$$

$$= Z_2[\gamma\cdot p+m_0(\Lambda)]+Z_1\int^{\Lambda}\frac{d^4q}{(2\pi)^4}g^2D_{\mu\nu}\gamma_\mu S(q)\Gamma_\nu(q,p), \tag{2}$$

with S, $D_{\mu\nu}$ and Γ_μ the renormalised dressed quark propagator, gluon propagator and proper quark-gluon vertex respectively. The renormalisation scale μ is set such that $S(p)|_{p^2=\mu^2}=1/[i\gamma\cdot p+m_R(\mu^2)]$. We use a Euclidean metric in which timelike vectors satisfy $p^2=-p^2_{\text{Minkowski}}<0$, and for which $\{\gamma_\mu,\gamma_\nu\}=2\delta_{\mu\nu}$.

The heavy quark limit is taken by assuming the expansions

$$A(p^2,\mu^2)=1+\frac{\Sigma_A(K,\kappa)}{m_R(\mu^2)}+\ldots,\qquad B(p^2,\mu^2)=m_R(\mu^2)+\Sigma_B(K,\kappa)+\ldots, \tag{3}$$

where we have defined the momentum variable K and renormalisation point κ by

$$K=\frac{p^2+m_R^2}{2im_R},\qquad \kappa=\frac{\mu^2+m_R^2}{2im_R}. \tag{4}$$

The change of independent variable $p^2\to K$ induced by the transformation Eq. (4) is discussed in detail in reference [9]. The main point to note here is that the real timelike p^2 axis in the vicinity of the 'would-be bare fermion mass pole', $p^2=-m^2$, becomes the imaginary K axis in the new coordinates. The aim of the exercise is to solve for the quark propagator in the vicinity of the bare mass pole $K=0$.

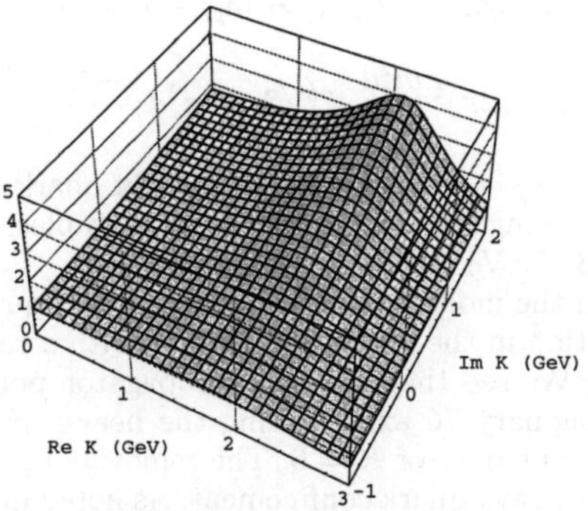

Fig. 1. The modulus $|\sigma_Q(K)|$ of the heavy quark propagator in the complex K plane obtained by solving the heavy quark DS equation in bare vertex approximation with a gaussian gluon propagator Δ_G.

For the purposes of expanding the DS equation in the inverse heavy quark mass we introduce the momentum substitutions generic to heavy quark effective theory,

$$p_\mu = im_R v_\mu + k_\mu, \tag{5}$$

where $v_\mu v_\mu = 1$. This gives $K = k \cdot v + k^2/(2im_R)$, and so in the vicinity of $k_\mu = 0$, or $K = 0$, the dressed quark propagator has the form,

$$\begin{aligned} S(p,\mu) &= \frac{1+\gamma\cdot v}{2}\frac{1}{iK + \Sigma_B(K,\kappa) - \Sigma_A(K,\kappa)} + O\left(\frac{1}{m_R}\right) \\ &\stackrel{\text{def}}{=} \frac{1+\gamma\cdot v}{2}\sigma_Q(K,\kappa) + O\left(\frac{1}{m_R}\right). \end{aligned} \tag{6}$$

The confinement criterion that $S(p)$ should be free from timelike singularities on the negative real p^2 axis translates in the heavy quark case to a requirement that σ_Q should be free from singularities on the imaginary K axis.

In reference [10] the heavy quark propagator is solved for using the bare vertex or "rainbow" approximation $\Gamma_\mu(q,p) = \gamma_\mu$, and Landau gauge smeared Frank and

Roberts[11] model gluon propagator $D_{\mu\nu}(k) = (\delta_{\mu\nu} - k_\mu k_\nu/k^2)\Delta(k^2)$, where

$$\Delta(k^2) = (2\pi)^4 \frac{m_t^2 d}{\alpha^2\pi^2} e^{-k^2/\alpha} + 4\pi^2 d \frac{1 - e^{-k^2/(4m_t^2)}}{k^2}. \tag{7}$$

Here $d = 12/(33 - 2N_f)$, $N_f = 3$ is the number of light quark flavours, $m_t = 0.69$ GeV is a parameter fitted to a range of calculated pion observables, and the Gaussian width α is chosen to be 0.5643 (GeV)^2.

In Fig. 1 is plotted the modulus $|\sigma_Q(K)|$ of the heavy quark propagator obtained by solving the DS equation in the vicinity of $K = 0$, with a renormalised heavy quark mass of $m_R = 5$ GeV. We see that the bare propagator pole at $K = 0$ has moved well away from the imaginary K axis, leaving the heavy quark propagator analytic over a large region in the vicinity of $K = 0$. The removal of propagator poles from the imaginary K axis signals heavy quark confinement. As noted in reference [9], a large shift of the pole portends well for the possibility of finding solutions to the Bethe-Salpeter equation for Q-$\bar{q}$ mesons.[8] Solution of the heavy meson Bethe-Salpeter equation, of which accurate determination of heavy quark propagators is a necessary first step, should prove helpful in impulse approximation studies of semileptonic decays.[12]

Acknowledgement

The author acknowledges helpful discussions with F. T. Hawes and C. D. Roberts, and the hospitality of the Special Research Centre for the Subatomic Structure of Matter in Adelaide for hosting the Workshop on Nonperturbative Methods in Quantum Field Theory where part of this work was completed.

References

1. C. D. Roberts and A. G. Williams, Prog. Part. and Nucl. Phys. **33**, 475 (1994).
2. G. Bali, *The Mechanism of Quark Confinement*, these proceedings.
3. C. D. Roberts, "Nonperturbative QCD with Modern Tools", nucl-th/9807026.
4. M. R. Pennington, "Calculating hadronic properties in strong QCD", hep-ph/9611242.
5. D. Atkinson and J. C. R. Bloch, Mod. Phys. Lett. A **13**, 1055 (1998), and references therein.
6. F. T. Hawes, P. Maris and C. D. Roberts, "Infrared behaviour of propagators and vertices", nucl-th/9807056.
7. G. B. West, Phys. Lett. **B115**, 468 (1982).
8. C. J. Burden and D.-S. Liu, Phys. Rev. D **55**, 367 (1997).
9. C. J. Burden, Phys. Rev. D **57**, 276 (1998).
10. C. J. Burden, "The effect of the ultraviolet part of the gluon propagator on the heavy quark propagator", hep-ph/9807438.
11. M. R. Frank and C. D. Roberts, Phys. Rev. C **53**, 390 (1996).
12. M. A. Ivanov, Yu. L. Kalinovsky, P. Maris and C. D. Roberts, Phys. Lett. B **416**, 29 (1998); Phys. Rev. C **57**, 1991 (1998).

Scalar meson exchange and the baryon spectra

P. Stassart and Fl. Stancu
Université de Liège, Institut de Physique B.5, Sart Tilman,
B-4000 Liège 1, Belgium
and
J.-M. Richard
Institut des Sciences Nucléaires, Université Joseph Fourier
IN2P3-CNRS, 53, Avenue des Martyrs
F-38026 Grenoble Cedex, France

Abstract

We explore the role of a scalar meson exchange interaction between quarks in a semirelativistic constituent quark model where the quarks are subject to a linear confinement. We search for a variational solution and show that the gap between the $N = 1$ band and the ground state $N = 0$ band increases with the strength of the scalar meson exchange interaction potential. This result has good implications on the description of the low-lying baryon masses, especially on the Roper vs. negative-parity resonances ordering.

It has been suggested [?, ?] that beyond the scale of spontaneous chiral symmetry breaking, nonstrange and strange baryons can be viewed as systems of three quarks interacting via exchange of Goldstone bosons (pseudoscalar mesons). It has also been shown that a properly parametrized interaction provides good baryon spectra with a correct order of positive and negative parity levels both in a nonrelativistic [?] or a semirelativistic [?] treatment. In both parametrizations the pseudoscalar meson exchange interaction has two distinct parts: a long-range Yukawa potential tail and a short-range part having opposite sign as compared to the Yukawa potential tail. It is the latter which plays a major role in describing the baryon spectra in the frame of Goldstone boson exchange (GBE) models.

Although the models [?, ?, ?] are thought to be a consequence of the spontaneous chiral symmetry breaking, the chiral partner of the pion, the sigma meson, is not considered explicitly. One can think of having mocked up its contribution in the parameters of the Hamiltonian [?, ?], e.g. in the regularization parameter of the short-range spin-spin term. The price which could have been paid is the large role played by the η' meson exchange which comes into the interaction with a strength $g_0^2/4\pi$ about equal [?] or larger [?] than the strength $g_8^2/4\pi$ associated with the pseudoscalar octet (π, K, η).

Here we study the explicit role of a sigma-exchange interaction by considering the following Hamiltonian

$$H_0 = \sum_i \left(m_i^2 + p_i^2\right)^{1/2} + \frac{1}{2}\sqrt{\sigma}\sum_{i<j}|\vec{r}_i - \vec{r}_j| - \frac{g_\sigma^2}{4\pi}\sum_{i<j}\frac{exp\left[-\mu_\sigma|\vec{r}_i - \vec{r}_j|\right]}{|\vec{r}_i - \vec{r}_j|} \quad (1)$$

where the second term is the confinement potential with a strength tension [?]

$$\sqrt{\sigma} = 1\ GeV\ fm^{-1} \quad (2)$$

μ_σ = 600 MeV is the sigma meson mass and $g_\sigma^2/4\pi$ is the coupling constant sigma-quark, taken as a variable parameter. ¿From the pion-nucleon coupling constant $g_{\pi NN}^2/4\pi \simeq 14$, one obtains a pion-quark coupling constant $g_{\pi qq}^2/4\pi = 0.67$ which has been used in Refs. 3 and 4. Assuming [?] a sigma-nucleon coupling $g_{\sigma NN}^2/4\pi \simeq 8$ one obtains, by scaling, $g_{\sigma qq}^2/4\pi \simeq 0.4$. In the following we study the role of the σ-meson exchange by considering values of $g_{\sigma qq}^2/4\pi$ up to 0.2. A qualitative argument that the σ-meson contributes attractively in (1) is that we view it as a two correlated pions, as in the nucleon-nucleon interaction. Moreover, one can consider more sophisticated arguments related to the differences appearing in the scalar and pseudoscalar propagators [?] in a Nambu-Jona-Lasinio model. From there one can argue that the Fourier transform would lead to a Yukawa + a contact term for a scalar and a Yukawa - a contact term, for a pseudoscalar exchange, as in Refs.

3 and 4. Due to the addition of the long and short range contributions in the scalar case, here we consider a simplified form of the σ-exchange potential, containing a single term as in (1).

The variational wave function has the form

$$\psi_n(\vec{r}_{12}, \vec{r}_{13}, \vec{r}_{23}) = F_{123}\left[\prod_{i<j} f(r_{ij})\, \phi_n(\vec{r}_{12}, \vec{r}_{13}, \vec{r}_{13})\right] . \tag{3}$$

where f and F_{123} are two- and three-body parts of the ground state wave function and Φ_n with $n \neq 0$ contain orbital excitations [?]. The function f is parametrized as

$$\begin{aligned} f(r) &= r^{\delta} exp\left\{-W(r)\gamma_1 r - [1 - W(r)]\gamma_{1.5} r^{1.5}\right\}, \\ W(r) &= \frac{1 + exp(-r_0/a)}{1 + exp[(r - r_0)/a]} \end{aligned} \tag{4}$$

The three-body part is chosen as

$$F_{123} = \left[1 - \beta\sqrt{\sigma}\left(\sum_i r_{i4} - \frac{1}{2}\sum_{i<j} r_{ij}\right)\right] . \tag{5}$$

The quantities γ_1, $\gamma_{1.5}$, a, r_0, δ and β are variational parameters. In the minimization procedure we found that the ground state expectation value of (1) varies smoothly with all these parameters but δ. It is quite natural because δ cares for the behaviour of the wave function near the origin, typical for the solution of a relativistic equation with a singular potential. The larger the coupling constant $g_\sigma^2/4\pi$ the more difficult reaching the minimum. The expectation value for the first two excited states are obtained with the optimal values of the ground state parameters. The result is exhibited in the table below.

$g_\sigma^2/4\pi$	ground st.	neg.parity	radial excit.
0.00	940	1287	1446
0.05	940	1309	1444
0.10	940	1333	1444
0.15	940	1380	1445
0.20	940	1431	1446

One can see that the mass difference between the first radially excited state and the negative parity state is positive up to the coupling constant value of 0.20 but tends to vanish. Note also that the difference between the radially excited state and the ground state remains practically constant as a function of the coupling constant, while the mass difference between the orbitally excited state and the ground state increases with $g_\sigma^2/4\pi$. The reason for this is that at $g_\sigma \neq 0$ all s states are lowered with respect to the p states. The wave function of the latter is small around the origin, so that it reduces some of the attraction in the expectation value.

Preliminary results indicate that the mass difference between the first radially excited state and the negative parity state becomes negative for $g_\sigma^2/4\pi > 0.20$. This is precisely the desired behaviour for reproducing the correct order of the experimental spectrum, as in Ref. 3. These calculations indicate that a model incorporating a potential, whose Laplacian is negative in a region around the origin, can yield the right ordering of the lowest positive and negative parity states. For the potential of Eq. (1) such a situation is achieved for suitable values of the coupling constant $g_\sigma^2/4\pi$. The Laplacian is related to the concavity of the two-body potential as discussed in Ref. 9.

The conclusion is that Goldstone boson exchange models should include explicitly the chiral partner of the pseudoscalar mesons. It might be misleading to mock up its effect in some nontrivial parameters of the model. We are presently studying the case of a potential with both scalar and pseudoscalar terms. However the question still remains about the role of the chromomagnetic interaction.

Finally, the scalar exchange term, which we have shown to be important for the level ordering, might also genearate a new spin-orbit contribution. It would be interesting to see whether or not that contribution could significantly improve our present knowledge of the spin-orbit problem in baryons [?].

References

[1] A. Manohar and H. Georgi, Nucl.Phys. **B234** (1984) 189

[2] L.Ya. Glozman and D.O. Riska, Phys. Rep. **268** (1996) 263

[3] L.Ya. Glozman, Z. Papp and W. Plessas, Phys. Lett. **B381** (1996) 311; L.Ya. Glozman, Z. Papp, W. Plessas, K. Varga and R.F. Wagenbrunn, Nucl.Phys. **A623** (1997) 90c

[4] L.Ya. Glozman, W. Plessas, K. Varga and R.F. Wagenbrunn, Phys.Rev. **D58**, in print, e-print hep-ph/9706507

[5] J. Carlson, J. Kogut and V.R. Pandharipande, Phys.Rev. **D28** (1983) 2808

[6] R. Machleidt, K. Holinde and Ch. Elster, Phys.Rep. **149** (1987) 1

[7] M. Jaminon, G. Ripka and P. Stassart, Nucl.Phys. **A504** (1989) 733

[8] R. Sartor and Fl. Stancu, Phys. Rev. **D31** (1985) 128

[9] A. Martin, J.-M. Richard and P. Taxil, Nucl. Phys. **B329** (1990) 327

[10] See, e.g. W. Lucha, F. Schöberl and D. Gromes, Phys.Rep. **203** (1991) 127 and references therein.

CONFINEMENT THROUGH A RELATIVISTIC GENERALIZATION OF THE LINEAR INTERACTION

Çetin Şavklı[1] and Franz Gross[1,2]

[1] *Department of Physics, College of William and Mary, Williamsburg, Virginia 23187, USA*

[2] *Jefferson Lab, Newport News, Virginia 23606, USA*

ABSTRACT

Generalizing a covariant framework previously developed, it is shown that confinement insures that meson $\rightarrow q + \bar{q}$ decay amplitudes vanish when both quarks are on-shell. Regularization of singularities in a covariant linear potential associated with nonzero energy transfers (i.e. $q^2 = 0, q^\mu \neq 0$) is improved.

1. Introduction

Even for a simple system such as a quark-antiquark (referred to collectively as "quarks") bound state, color confinement implies that the matrix element for meson decay, $\mu \rightarrow q + \bar{q}$, must vanish whenever this decay is kinematically possible. This trivial statement can be realized by two possible mechanisms. Either **i)** the quark propagators are free of timelike mass poles, as is usually assumed in Euclidean metric based studies,[1] or **ii)** the vertex function of the bound state **vanishes** when both quarks are on-shell. In an earlier work[2] it was assumed that the quark propagators have mass poles, but it was not shown that relativistic generalization of the linear confining potential guaranteed the correct vanishing of the vertex function. In this work it is shown that the vanishing of the vertex function is a general feature of any confining interaction, and that **insisting on the correct nonrelativistic limit leads naturally to the second option.**

2. The nonrelativistic linear potential

It was shown earlier[2] that the nonrelativistic linear interaction

$$\mathcal{V}(r) = -C + \sigma r, \tag{1}$$

can be expressed in momentum space by

$$\mathcal{V}(\vec{q}) = \lim_{\epsilon\rightarrow 0}\Big[V_A(\vec{q}) - \delta^3(\vec{q})\int d^3q' V_A(\vec{q}\,')\Big] - (2\pi)^3\delta^3(\vec{q})C, \tag{2}$$

where

$$V_A(\vec{q}) \equiv -8\pi\sigma\frac{1}{(\vec{q}^2+\epsilon^2)^2}\,. \tag{3}$$

In coordinate space

$$V_A(\vec{r}) = \int \frac{d^3q}{(2\pi)^2}\, e^{-iq\cdot r}\, V_A(\vec{q}) = -\sigma\frac{e^{-\epsilon r}}{\epsilon} \;\; \propto \;\; \lim_{\epsilon\to 0} \sigma(r - \frac{1}{\epsilon}), \tag{4}$$

and hence the delta function subtraction cancels the infinite $1/\epsilon$ term, insuring that the limit $\epsilon \to 0$ exists and that the potential confines. It also insures that the linear potential vanishes at the origin. In fact, the delta function subtraction could be used to construct confining potentials for any monotonically increasing $V_A(r)$ for which $V_A(r) - V_A(0) = \infty$ for some r. For example, after the delta function subtraction even the coulomb potential $V_A(r) = -1/r$ would result in a confining interaction.

Using the linear potential Eq. (2) the Schrödinger equation for two quarks of equal mass m becomes

$$\left[\frac{\vec{p}^2}{m} - E\right]\Psi(\vec{p}) = -\int \frac{d^3k}{(2\pi)^3} V_A(\vec{p}-\vec{k})\,\left[\Psi(\vec{k}) - \Psi(\vec{p})\right] + C\Psi(\vec{p})\,, \tag{5}$$

where E is the binding energy.

3. The relativistic generalization

A seemingly natural candidate for a relativistic generalization of the linear interaction (2) is to use the Bethe-Salpeter equation with a kernel given by

$$\mathcal{V}(\vec{q}) \stackrel{?}{\to} \lim_{\epsilon\to 0}\left[V_A(q) - \delta^4(q)\int d^4q' V_A(q')\right] - (2\pi)^3\delta^4(q)C\,, \tag{6}$$

where $\vec{q}^2 \to -q^2$ in Eq. (3). Unfortunately this form does not have the correct nonrelativistic limit. Therefore we rephrase the question: Can we find a covariant equation that reduces to the Schrödinger equation with linear interaction in the nonrelativistic limit? In order to motivate the relativistic equation, start by defining the Schrödinger *vertex* function $\Phi(\vec{p})$

$$\Phi(\vec{p}) \equiv \left[\frac{\vec{p}^2}{m} - E\right]\Psi(\vec{p})\,. \tag{7}$$

The Schrödinger equation with linear interaction is then

$$\Phi(\vec{p}) = -\int \frac{d^3k}{(2\pi)^3}\, mV_A(\vec{p}-\vec{k})\left[\frac{\Phi(\vec{k})}{\left(\vec{k}^2 - mE\right)} - \frac{\Phi(\vec{p})}{(\vec{p}^2 - mE)}\right] + mC\frac{\Phi(\vec{p})}{(\vec{p}^2 - mE)}\,. \tag{8}$$

Introducing the four-vectors P, k, and p, with $P = (2m + E, \vec{0})$ and $p^2 = k^2 = m^2$ (so that $E_k = \sqrt{m^2 + \vec{k}^2}$), the relativistic generalization of (8) which reduces to it in the nonrelativistic limit (when $m \to \infty$) is

$$\begin{aligned}\Phi(p) = &- \int \frac{d^3k}{(2\pi)^3}\frac{2m^2}{E_k} V_A(p-k)\left[\frac{\Phi(k)}{m^2 - (P-k)^2} - \frac{\Phi(p)}{m^2 - (P-p)^2}\right] \\ &+ 2mC\frac{\Phi(p)}{m^2 - (P-p)^2}\,.\end{aligned} \tag{9}$$

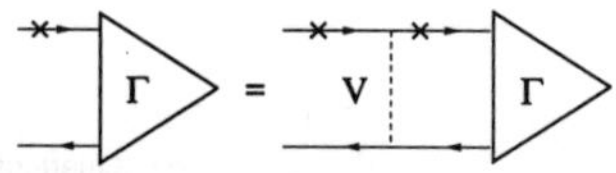

Fig. 1. The Gross equation. The $\times$ indicate that the particle is on-shell.

This is the Gross equation (see Fig. 1) and the discussion shows that this is the most natural relativistic equation obtained from a generalization of the nonrelativistic Schrödinger equation for confined particles.

In a previous application of the Gross equation[2] the kernel (3), with $\vec{q}^2 \to -q^2$, was used. This choice is undesirable for use with the two channel version of this equation, where the mass shell constraints which fix q_0 introduce singularities for non-zero momentum transfers (i.e. $q^2 = 0$, but $q^\mu \neq 0$). To correct this problem the kernel $V_A(q)$ is defined to be

$$V_A(q) \equiv -8\pi\sigma \frac{1}{q^4 + (P\cdot q)^4/P^4} \tag{10}$$

Advantages of this form are many: **i)** singularities are restricted to $q^\mu = 0$, **ii)** interaction strength *does not* depend on the bound state momentum P in the rest frame, **iii)** it has the correct nonrelativistic dependence on $\vec{q}^2$, and **iv)** the ultraviolet regularization used previously[2] is no longer needed.

4. Proof of confinement

It is sufficient to consider the case when the constant term $C = 0$. If the mass of the bound state $\mu > 2m$, then there exists a value of three momentum $|\vec{p}| = p_c$ when both quarks can be on-shell. In this case $(P - p)^2 = \mu^2 - 2\mu E_{p_c} + m^2 = m^2$, and the subtraction term in Eq. (9) appears to be singular. This singularity is not cancelled by the first term, and hence, if the equation is to have a solution, the vertex function must be zero at p_c. In particular, as $|\vec{p}| \to p_c \pm \epsilon$,

$$(P-p)^2 - m^2 \to \mp \frac{2\mu p_c\,\epsilon}{E_{p_c}}, \tag{11}$$

and the vertex function would have an infinite discontinuity at p_c unless it were zero there. We conclude that $\Phi(p) = 0$, when $2E_p = \mu$ (as illustrated in Fig. 2).

A numerical confirmation of this result is shown in Fig. 3. In the actual calculations to be presented in an upcoming paper, the bound state equation is averaged over positive and negative energy contributions and symmetrized by picking up the pole contributions of both constituents. The average over positive and negative energy contributions leads to a two channel equation with a two component vertex function.

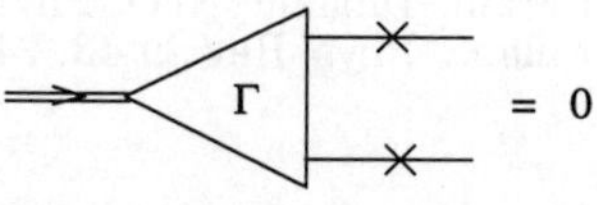

Fig. 2. When both constituents are on-shell, the vertex function vanishes.

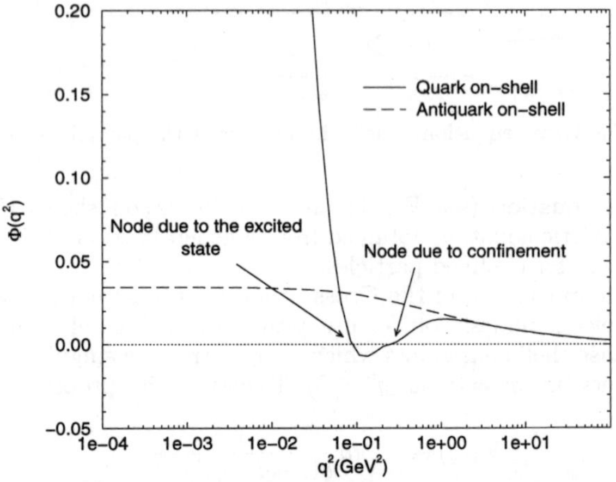

Fig. 3. The pion 1st excited state Gross vertex functions are shown. The first node is due to the excited state. The second node assures that the bound state does not decay.

In Fig. 3 the solid line is the (large) component in which the off-shell quark has positive energy and the dashed line is the (small) component in which the off-shell quark has negative energy. Since a physical decay must produce two positive energy quarks, only the large component must have the "confinement" node. Fig. 3 shows the vertex functions for an excited pion with mass $\mu = 1.2$ GeV composed of two quarks with masses $m = 0.34$ MeV. The large vertex function must have *two* nodes: one due to the excitation and another at $p^2 = 0.244$ GeV^2, exactly where both particles are on-shell.

5. Conclusions

The relativistic equation obtained from a generalization of the nonrelativistic Schrödinger equation for confined particles is of the Gross equation type rather than the Bethe-Salpeter equation type. The confinement mechanism arises not from the lack of quark mass poles, but through the vanishing of the vertex function when both quarks are on their positive energy mass-shell. Because of this confinement mechanism, the vertex functions (ground or excited state) have one additional node if the bound state is heavy enough.

This work was supported in part by the US Department of Energy under grant No. DE-FG02-97ER41032.

References

1. Çetin Şavklı and Frank Tabakin, Nucl. Phys. A **628**, 645 (1998).
2. F. Gross and J. Milana, Phys. Rev. D **43**, 2401 (1991); **45**, 969 (1992); **50**, 3332 (1994).

LOCAL TWO-BODY RELATIVISTIC SCALAR WAVE EQUATION*

John H. Connell†
Springfield Technical Community College, Armory Square, Springfield, Mass. 01105

ABSTRACT

A relativistic two-body bound-state wave equation, local in configuration space, is derived from the Bethe-Salpeter equation for a model system of two scalar particles of any masses bound by any static Lorentz-scalar interaction. It is proven algebraically that the first-order relativistic corrections of the local wave equation are exactly the same as those of its originating Bethe-Salpeter equation. For 13 states of a numerically-solvable harmonic-oscillator model, the Bethe-Salpeter energy levels are found exactly. Then the levels of its corresponding reduced Salpeter equation, and its corresponding local two-body wave equation, are compared to the exact BS levels. While the median absolute error of the reduced Salpeter equation's levels was 238 MeV, that of the local relativistic wave equation was only 9 MeV. We conclude that the new local two-body wave equation is more accurate than the widely-used reduced Salpeter ("spinless Salpeter") equation. It is also easier to solve in configuration space.

1. Introduction

In the two-thirds of a century since the discovery of the one-body Dirac and Klein-Gordon equations, no simple two-body counterpart to those equations has been discovered. There is no easy-to-solve relativistic analogue of the reduced-mass form of the non-relativistic Schrödinger equation. Instead, every reduction of the Bethe-Salpeter equation seems to end up with the non-local operator $\sqrt{-\nabla^2 + m^2}$. However we think it *should* be possible to derive simple two-body relativistic wave equations in configuration space from mainstream quantum field theory such as the Bethe-Salpeter equation. Here we do so for a scalar model.

2. Bethe-Salpeter Equation

The notation will be the following.[1] Let the constituent masses of the bound particles be m,M. Denote the CM energy by E, the bound-state wave number by b, as well as the particles' individual bound-state energies by t and T, as follows:

$$E = \sqrt{m^2 + b^2} + \sqrt{M^2 + b^2} \tag{1}$$

*Prepared for the Proceedings of the *Third International Conference on Quark Confinement and the Hadron Spectrum*, Newport News, Virginia, June 1998 (World Scientific, Singapore).
†connell@mail.stcc.mass.edu; http://physics.stcc.mass.edu

$$t \equiv \sqrt{m^2 + b^2} = \frac{E^2 + (m^2 - M^2)}{2E}, \tag{2}$$

$$T \equiv \sqrt{M^2 + b^2} = \frac{E^2 + (M^2 - m^2)}{2E} \tag{3}$$

The CM energy-momenta of the particles are written $(\mathbf{p}, t + p^0)$, $(-\mathbf{p}, T - p^0)$.

The Bethe-Salpeter equation for the bound-state vertex function Γ will be

$$\Gamma = I * S\, \Gamma \tag{4}$$

with

$$S(\mathbf{p}, p^0) \equiv -\frac{2m}{\mathbf{p}^2 + m^2 - i\epsilon - (t + p^0)^2} \frac{2M}{\mathbf{p}^2 + M^2 - i\epsilon - (T - p^0)^2} \tag{5}$$

The scalar interaction kernel is $4mMI(\mathbf{k}^2)$, or $4mMI(r)$ in configuration space. The star denotes an integration over 4-momentum with a factor $1/i$ present. The factors $2m$,$2M$ have been moved into the numerators of the scalar constituent particles' propagators in order to make all the scalar quantities analogous to those of spin-$\frac{1}{2}$ systems.

3. Salpeter Equation

Defining $\phi(\mathbf{p}) = \int dp^0 S(\mathbf{p}, p^0)\Gamma(\mathbf{p}, p^0)/2\pi i$, equation (4) when multiplied by S and integrated over p^0 gives the non-local Salpeter equation equivalent to (4):

$$\left[\sqrt{\mathbf{p}^2 + m^2} + \sqrt{\mathbf{p}^2 + M^2} + Z_p I(r)\right] \phi(\mathbf{r}) = E\phi(\mathbf{r}) \tag{6}$$

in which Z_p is a correction factor to the interaction $I(r)$:

$$Z_p = \frac{2mM\left[\sqrt{\mathbf{p}^2 + m^2} + \sqrt{\mathbf{p}^2 + M^2}\right]}{\sqrt{\mathbf{p}^2 + m^2}\sqrt{\mathbf{p}^2 + M^2}\left[\sqrt{\mathbf{p}^2 + m^2} + \sqrt{\mathbf{p}^2 + M^2} + E\right]} \tag{7}$$

The often-used[2,3] reduced Salpeter ("spinless Salpeter") equation is made by artificially replacing Z_p by 1 in (6).

4. Local Wave Equation

Because S peaks sharply at $p^0 = 0$ when $\mathbf{p}^2$ is not too large, it can be approximated by

$$S_0 \equiv -\frac{4mM}{2E} \frac{2\pi i \delta(p^0)}{(\mathbf{p}^2 - b^2 - i\epsilon)} \tag{8}$$

The error of the approximation is defined by $R \equiv S - S_0$. The Blankenbecler-Sugar[4] correction interaction U is defined by the equation $U = I + I * R\, U$. Then from the Bethe-Salpeter equation (4) it is easy to deduce the equation $\Gamma = U * S_0\, \Gamma$. Next, defining $\phi(\mathbf{p})$ by $S_0\, \Gamma = 2\pi i \delta(p^0)\phi(\mathbf{p})$, it follows that, with the notation $U = U(p; q)$, and omitting $i\epsilon$ for the bound-state problem,

$$(\mathbf{p}^2 - b^2)\phi(\mathbf{p}) = -\frac{4mM}{2E} \int \frac{d\mathbf{q}}{(2\pi)^3} U(\mathbf{p}, 0; \mathbf{q}, 0)\phi(\mathbf{q}) \tag{9}$$

Equation (9) is still exact in terms of the original Bethe-Salpeter equation (4). For a static kernel $I(\mathbf{k}^2)$, the equation $U = I + I * R\, U$ simplifies, because in its solution by iteration the integration over the relative-energy variable only acts on R. After that integration, the equation for U becomes

$$U(\mathbf{p},0;\mathbf{q},0) = I[(\mathbf{p}-\mathbf{q})^2] + \int \frac{d\,\mathbf{l}}{(2\pi)^3} I[(\mathbf{p}-\mathbf{l})^2] R(\mathbf{l}^2) U(\mathbf{l},0;\mathbf{q},0) \tag{10}$$

in which

$$R(\mathbf{l}^2) = \frac{mM}{E}\left\{\frac{1}{\sqrt{\mathbf{l}^2+m^2}\left(\sqrt{\mathbf{l}^2+m^2}+t\right)} + \frac{1}{\sqrt{\mathbf{l}^2+M^2}\left(\sqrt{\mathbf{l}^2+M^2}+T\right)}\right\} \tag{11}$$

where t and T were defined in (2) and (3).

Because R first appears in the *second* term of the iteration of (10), to first order it can be taken to be a constant,[1]

$$R_E \equiv R(\mathbf{l}^2)_{\mathbf{l}^2=0} = \frac{2M}{(m+E)^2 - M^2} + \frac{2m}{(M+E)^2 - m^2} \tag{12}$$

Then to the accuracy needed to give the first-order relativistic energy levels correctly, U can be written simply as

$$U_E(r) = I(r) + R_E I^2(r) \tag{13}$$

and the bound-state equation (9), expressed in configuration space, finally becomes the local wave equation

$$\left[\mathbf{p}^2 + 2\mu Z_E U_E(r)\right]\phi(\mathbf{r}) = b^2\phi(\mathbf{r}) \tag{14}$$

in which $Z_E \equiv (m+M)/E$ and $U_E(r)$ is given by (13). The energy E is expressed in terms of b^2 by (1).

5. Formal Equivalence Proof

The non-relativistic Schrödinger equation is as usual defined to be

$$\left[m + M + \frac{\mathbf{p}^2}{2\mu} + I(r)\right]\phi_0(\mathbf{r}) = E_0\phi_0(\mathbf{r}) \tag{15}$$

in which μ is the reduced mass. The non-relativistic bound-state wave number b_0 will be defined by $E_0 = m + M + b_0^2/2\mu$.

To find their first-order relativistic corrections to E_0, the Salpeter equation (6) and the wave equation (14) are each expanded in powers of $\mathbf{p}^2$ and b_0^2 in the usual way. Each gives the *same* first-order correction

$$\Delta E = -\frac{b_0^4}{8}\left(\frac{1}{m^3}+\frac{1}{M^3}\right) - \frac{b_0^2}{2mM}\langle\phi_0|\,I\,|\phi_0\rangle + \frac{1}{2\mu}\left(\frac{\mu^2}{m^2}+\frac{\mu^2}{M^2}\right)\langle\phi_0|\,I^2\,|\phi_0\rangle \tag{16}$$

which proves that our local wave equation has the same energy levels as the original Bethe-Salpeter equation when calculated to first relativistic order.

6. Numerical Test

At the Conference for which these are the Proceedings, F. Schöberl pointed out to the writer that if the interaction $I(r) = \frac{1}{2}kr^2 - B$ is used, the actual Bethe-Salpeter energy levels can be found, because the Salpeter equation (6) can be solved numerically as an ordinary differential equation in momentum space since r^2 becomes $-\nabla_{\mathbf{p}}^2$. In addition the reduced Salpeter equation can also be solved in momentum space. Our wave equation (14) can be solved in configuration space for almost any $I(r)$. So for this model interaction, with the values $k = 0.05$ GeV^3, $B = 800$ MeV (which gives an interaction close to a standard funnel potential at medium distances), and masses $m = 300$ MeV, $M = 500$ MeV, we have solved all three equations, and the resultant energy levels in MeV for 13 states are shown in the following table.

State	Energy Levels			Errors relative to BS	
	BS	Red. Salp.	Wave Eqn.	Red. Salp.	Wave Eqn.
1S	900	662	1006	-238	106
1P	1234	1052	1256	-182	22
1D	1465	1413	1476	-52	11
2S	1505	1376	1509	-129	4
2P	1695	1704	1696	9	1
2D	1852	2020	1858	168	6
3S	1882	1993	1871	111	-11
3P	2025	2286	2018	261	-7
3D	2150	2572	2158	422	8
4S	2173	2551	2159	378	-14
4P	2292	2820	2283	528	-9
4D	2398	3084	2399	686	1
5S	2417	3067	2403	650	-14

The median of the absolute values of the errors of the reduced Salpeter equation is 238 MeV, while that of the two-body relativistic wave equation is 9 MeV. The largest error in the wave equation levels is in the ground state, for which equation (10) must be solved for U more accurately.

Compared to the original Bethe-Salpeter equation, it appears that the two-body relativistic wave equation (14) is more accurate than the reduced Salpeter ("spinless Salpeter") equation. It is certainly easier to solve in configuration space.

I am most grateful for the hospitality of Prof. N. Isgur, the other local organisers, and the Jefferson Lab. staff at this Conference.

References

1. J.H. Connell, Phys. Rev. **D43**, 1393 (1991); SLAC-PUB-5633 (1991).
2. W. Lucha, F.F. Schöberl and D. Gromes, Phys. Reports **200**, 127-240 (1991).
3. L. P. Fulcher, Phys. Rev. **D50**, 447 (1994).
4. R. Blankenbecler and R. Sugar, Phys. Rev. **142**, 1051 (1966).

SECTION "D"

CONFINEMENT PHENOMENOLOGY AND EXPERIMENTS

INVESTIGATIONS ON THE TRANSITION FROM NON-PERTURBATIVE TO PERTURBATIVE QCD IN INELASTIC e^+P-SCATTERING AT HERA

Bernd Surrow
Experimental Physics Division, CERN
CH-1211 Geneva 23, Switzerland

ABSTRACT

Recent results on the measurement of the proton structure function F_2 from both HERA experiments H1 and ZEUS, allowed to investigate the behavior of F_2 in the transition region from photoproduction to deep-inelastic scattering. The x and Q^2 dependence of F_2 in the transition region is quantified using a simple phenomenological fit. A detailed phenomenological analysis as well as a QCD analysis have been carried out to investigate the limitations of a non-perturbative and a perturbative QCD description.

1. Introduction

Results on measurements of the proton structure function F_2 from the two HERA experiments H1 and ZEUS[1] revealed that the kinematic region of $Q^2 \geq 1.5\,\mathrm{GeV}^2$ exhibits a strong rise of F_2 with decreasing x or equivalently of the total virtual photon-proton cross-section $\sigma_{tot}^{\gamma^* p}$ with increasing W^2, where $\sigma_{tot}^{\gamma^* p} \approx (4\pi^2\alpha/Q^2) \cdot F_2$ at low x. pQCD allows to describe the rapid rise of F_2 with decreasing x down to $Q^2 = 1.5\,\mathrm{GeV}^2$. In contrast to $\sigma_{tot}^{\gamma^* p}$, the total cross-section for real photon-proton scattering $\sigma_{tot}^{\gamma p}$ shows only a modest rise with W^2. This behavior can be well described within the framework of non-perturbative QCD such as Regge theory.

Recent results at HERA from the experiments H1 and ZEUS[2] extended the kinematic region down to small values of $Q^2 \geq 0.1\,\mathrm{GeV}^2$. This allowed to investigate in detail the behavior of F_2 in the transition region from photoproduction to deep-inelastic scattering and thus from non-perturbative to perturbative QCD at low Q^2. A comparison to various phenomenological models has been carried out and can be found elsewhere.[2]

2. The x and Q^2 dependence of F_2 in the transition region

The x dependence of F_2 or equivalently the W^2 dependence of $\sigma_{tot}^{\gamma^* p}$ for fixed values of Q^2 for the above data sets has been discussed in detail.[2] To quantify the rise of F_2 at low x for fixed values of Q^2, a fit of $F_2 = C(Q^2) \cdot x^{-\lambda_{eff}}$ has been carried out to investigate the change in the rise of $\sigma_{tot}^{\gamma^* p}$ towards large values of W^2 to be

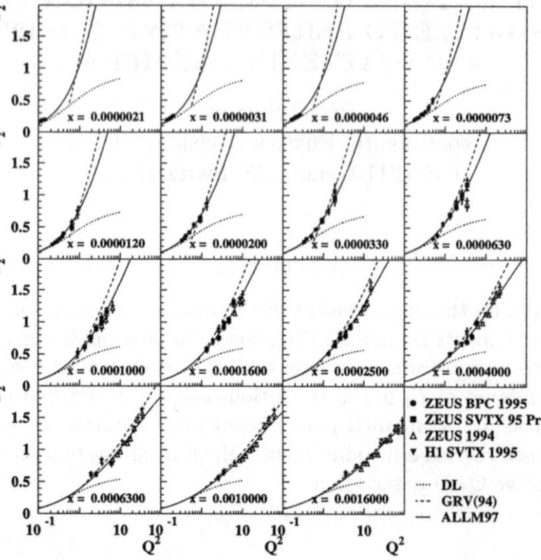

Fig. 1. F_2 as a function of Q^2 for different x values. Curves from the models of DL, GRV(94) and ALLM97 are overlaid.[4]

expected from the behavior of $\sigma_{tot}^{\gamma p}$ and $\sigma_{tot}^{\gamma^* p}$ at large values of Q^2. The slope of F_2, $\lambda_{eff} = d\ln F_2/d\ln(1/x)$, as a function of Q^2, is found to be constant for $Q^2 < 1\,\mathrm{GeV}^2$ indicating a 'soft' behavior in F_2. For $Q^2 \gtrsim 1\,\mathrm{GeV}^2$, λ_{eff} gradually increases with Q^2 indicating a 'hard' behavior in F_2 and $\sigma_{tot}^{\gamma^* p}$.

The dependence of F_2 for fixed values of x as a function of Q^2 is shown in Figure 1. F_2 exhibits, for fixed values of x with $x \geq 1.0 \cdot 10^{-4}$, an almost linear increase with the logarithm of Q^2. The violation of scaling is expected to increase at low x values and to decrease towards larger values of x. This behavior can be clearly inferred from Figure 1. For x bins of $1.2 \cdot 10^{-5} < x < 6.3 \cdot 10^{-5}$ with F_2 data points around $Q^2 = 1.0\,\mathrm{GeV}^2$, a change in the slope of F_2 as a function Q^2 is noticeable indicating the region where non-perturbative effects become dominant. In the limit $Q^2 \to 0$, F_2 is expected to behave like $F_2 \propto Q^2$. The four lowest x bins with data points of $Q^2 < 1\,\mathrm{GeV}^2$ due to the kinematical acceptance at HERA, clearly indicate a flattening in the slope of F_2 as a function of Q^2. The change in the slope of F_2 as a function of Q^2 for fixed values of x has been quantified using a fit of $F_2 = a + b \cdot \ln Q^2$.[5] The slope $b = dF_2/d\ln Q^2$ resembles the visual observation in Figure 1 of a slope at the lowest x bins which rises towards larger x values and eventually diminishes for large x values. Note that the rise in x is associated with a rise in Q^2 due to the kinematical acceptance at HERA.

3. Investigations on the behavior of F_2 in the transition region

A simple GVDM ansatz[6] has been used to describe the Q^2 dependence of the total cross-sections $\sigma_T^{\gamma^* p}$ and $\sigma_L^{\gamma^* p}$ using the ZEUS BPC F_2 results. The total cross-

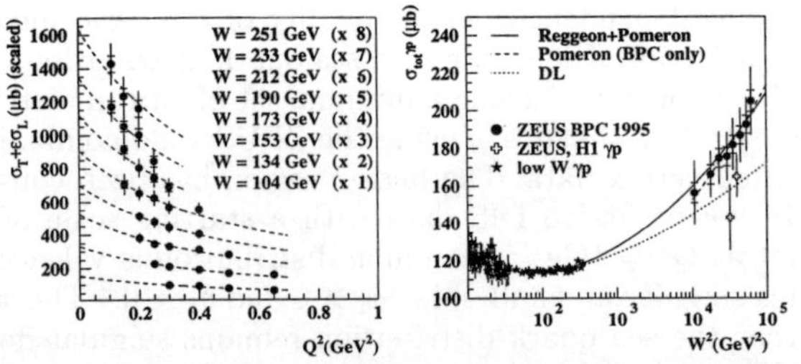

Fig. 2. Extrapolation to the total photoproduction cross-section using the BPC $\sigma_T^{\gamma^* p} + \epsilon\sigma_L^{\gamma^* p}$ results as a function of Q^2 (left). The GVDM-expression is overlayed (dashed curve). Extrapolated $\sigma_{tot}^{\gamma p}(W)$ values as a function of W^2 (right).[4]

sections $\sigma_T^{\gamma^* p}$ and $\sigma_L^{\gamma^* p}$ provide an equivalent description of the differential cross-section $d^2\sigma/dydQ^2 = \Gamma(\sigma_T^{\gamma^* p} + \epsilon\sigma_L^{\gamma^* p})$ with $\Gamma = \alpha(1+(1-y)^2)/(2\pi Q^2 y)$ and $\epsilon = 2(1-y)/(1+(1-y)^2)$, compared to $d^2\sigma/dydQ^2 = (2\pi\alpha^2 Y_+)/(yQ^4)\cdot[F_2 - (y^2/Y_+)F_L]$ with $Y_+ = 1+(1-y^2)$. At fixed W, the total cross-section $\sigma_T^{\gamma^* p}$ is related to the total photoproduction cross-section by: $\sigma_T^{\gamma^* p}(W^2, Q^2) = \left(\frac{m_0^2}{m_0^2+Q^2}\right)\sigma_{tot}^{\gamma p}(W)$. The contribution from the total longitudinal cross-section is expected to be small for $Q^2 < 1\,\mathrm{GeV}^2$ and has been therefore omitted for simplicity.[3] The result of the fit of the above expression $\sigma_T^{\gamma^* p}(W^2, Q^2)$ to the respective measured values yields $m_0^2 = 0.52 \pm 0.04$ (stat) $\pm\, 0.10$ (sys) GeV^2. The measured cross-section values are shown in Figure 2 (left) as a function of Q^2. The dashed line refers to the fitted GVDM expression of $\sigma_T^{\gamma^* p}(W^2, Q^2)$ which provides a good description of the ZEUS BPC F_2 results.

The energy dependence of the extrapolated total cross-section values $\sigma_{tot}^{\gamma p}(W)$ in eight bins of W is shown on Figure 2 (right). The extrapolated $\sigma_{tot}^{\gamma p}(W)$ values have been fitted to a simple Regge motivated ansatz as the sum of a Pomeron and a Reggeon contribution: $\sigma_{tot}^{\gamma p}(W) = A_R(W^2)^{\alpha_R - 1} + A_P(W^2)^{\alpha_P - 1}$. Fitting both terms to the low energy and BPC data yields $\alpha_P = 1.128 \pm 0.007$ (stat) $\pm\, 0.030$ (sys). The contribution from the Reggeon contribution is negligible in the kinematic region at HERA. Fitting the extrapolated BPC data using a Pomeron term only yields $\alpha_P = 1.145 \pm 0.020$ (stat) $\pm\, 0.040$ (sys). Both fitting results give a Pomeron intercept which is higher than the one given by Donnachie and Landshoff of $\alpha_P = 1.0808$.[7]

A NLO QCD fit has been performed based on the QCDNUM program[8] taking into account the ZEUS 1994 nominal vertex data and the ZEUS 1995 shifted vertex data in a restricted kinematical region of $3\cdot 10^{-5} < x < 0.7$ and $1 < Q^2 < 5000\,\mathrm{GeV}^2$. The high x region has been constrained using data from NMC and BCDMS. The input scale is chosen to be $Q_0^2 = 7\,\mathrm{GeV}^2$. The gluon distribution (xg), the sea quark distribution (xS) and the difference of up and down quarks in the proton ($x\Delta_{ud}$) are parameterized as follows: $xg(x, Q_0^2) = A_g x^{\delta_g}(1-x)^{\eta_g}(1+\gamma_g x)$, $xS(x, Q_0^2) = 2x(\bar{u}+\bar{d}+\bar{s}) = A_s x^{\delta_s}(1-x)^{\eta_s}(1+\epsilon_s\sqrt{x}+\gamma_s x)$ and $x\Delta_{ud}(x, Q_0^2) = x(u+\bar{u}) - x(d+\bar{d}) = A_\Delta x^{\delta_\Delta}(1-x)^{\eta_\Delta}$. The fit provides a good description of the F_2 data for $Q^2 \gtrsim 1\,\mathrm{GeV}^2$.

The qualitative features of a QCD analysis at low values of Q^2 have been investigated by studying the dependencies of the quality of the performed QCD fit and in particular the low x behavior of the gluon and sea quark distribution. These studies are based on the QCDNUM program[8] using a minimal set of parton distribution functions $xf_i(x, Q_0^2) = x^{-\lambda_i}(1-x)^{\eta_i}$ $i = u, d, s, g$ using the ZEUS 1994 nominal vertex data and the ZEUS 1995 shifted vertex data. The high x region has been constrained by using data from E665. In recent fits to DIS data with a starting scale of $Q_0^2 = 1\text{GeV}^2$ it has been found that at $Q^2 = 1\text{GeV}^2$ the gluon distribution is valence like but the sea quark distribution is singular at small x, i.e. $\delta_g > 0$ and $\delta_s < 0$.[9] The studies performed here have shown that the sea quark distribution remains singular down to a starting scale of $Q_0^2 \approx 1\,\text{GeV}^2$ which yields the observed rise in F_2 at low x, whereas the gluon distribution adopts a non-singular behavior for $Q_0^2 < 2.5\,\text{GeV}^2$.

4. Summary and conclusions

A simple ansatz within the GVDM model provides a good description of the Q^2 dependence of the measured total cross-sections $\sigma_T^{\gamma^* p} + \epsilon\sigma_L^{\gamma^* p}$ using the ZEUS BPC F_2 results. This allowed to extrapolate to the total photoproduction cross-section in eight bins of W. A Regge motivated fit to the extrapolated total photoproduction cross-section values yields a Pomeron intercept of $\alpha_P = 1.128 \pm 0.007$ (stat) $\pm\, 0.030$ (sys) which is larger than the one obtained by Donnachie and Landshoff of $\alpha_P = 1.0808$.[7] It is found that a Regge motivated fit being extended to larger values of Q^2 provides a good description of F_2 data for $Q^2 \lesssim 1\,\text{GeV}^2$. A NLO QCD fit allows to describe the F_2 data for $Q^2 \gtrsim 1\,\text{GeV}^2$. It is found that the low x behavior of the gluon and sea quark distribution exhibit a rather drastic change when approaching the low Q^2 region. A measurement of observables other than F_2 or a tighter theoretical constraint on the input parton distribution functions is required to gain a deeper insight in this field.

References

1. H1 Collaboration, S. Aid et al., Nucl. Phys. **B470** 3 (1996);
   ZEUS Collaboration, M. Derrick et al., Z. Phys. **C69** 607 (1996);
   ZEUS Collaboration, M. Derrick et al., Z. Phys. **C72** 399 (1996).
2. H1 Collaboration, C. Adloff et al., Nucl. Phys. **B497** 3 (1997);
   ZEUS Collaboration, J. Breitweg et al., Phys. Lett. **B407** 434 (1997);
   B. Surrow in *Proceedings of the Workshop on Deep Inelastic Scattering and QCD (DIS 97)*, edited by J. Repond (AIP, Chicago, 1997).
3. B. Surrow, DESY-THESIS-1998-004 (1998).
4. B. Surrow in *Proceedings of the Workshop on Deep Inelastic Scattering and QCD (DIS 98)*, edited by R. Roosen (World Scientific, Brussels, 1998).
5. A. Caldwell, DESY Theory Workshop, Hamburg, September 1997.
6. J.J. Sakurai and D. Schildknecht, Phys. Lett. **40** 121 (1972).
7. A. Donnachie and P.V. Landshoff, Phys. Lett. **296** 227 (1992).
8. M. Botje, in *Proceedings of the Workshop on Future Physics at HERA*, edited by G. Ingelmann et al. (Hamburg, Germany, 1995).
9. A.D. Martin et al., *DTP-96-44* (1996).

An Experimental Review of Exotic Mesons

D. P. Weygand
Jefferson Lab, Newport News, Virginia 23606, USA
and
Rensselear Polytechnic Institute, Troy, New York 12180, USA

ABSTRACT

Recently there have been exciting new results in the the search for exotic hadrons. In this note the current state of the search for $J^{PC} = 1^{-+}$ mesons, i.e., mesons with explicitly exotic quantum numbers, is reviewed.

1. Introduction

Much progress has been made in recent years in the theoretical description of hadrons which lie outside the scope of the constituent quark model. QCD predicts the existence of multi-quark $q\bar{q}q\bar{q}$ and hybrid $q\bar{q}g$ mesons as well as purely gluonic states. The most convincing experimental evidence for an exotic meson would be the determination of quantum numbers $J^{PC} = 0^{--}, 0^{+-}, 1^{-+}, 2^{+-}$,etc., as a $q\bar{q}$ pair cannot form a state with such quantum numbers.

Theoretical predictions for the mass of the lightest $J^{PC} = 1^{-+}$ hybrid meson are based on various models. The flux tube model[1] predicts 1^{-+} states at 1.8–2.0 GeV/c^2. Similar results are obtained in the calculations based upon lattice QCD in the quenched approximation.[2] Earlier bag model estimates suggest somewhat lower masses in the 1.3–1.8 GeV/c^2 range.[3] QCD sum rule predictions vary widely between 1.5 GeV/c^2 and 2.5 GeV/c^2.[4] The diquark cluster model[5] predicts the 1^{-+} state to be at 1.4 GeV/c^2. Finally, the constituent gluon model[6] concludes that light exotics should lie in the region 1.8–2.2 GeV/c^2. Most of these models predict the dominance of such decay modes of the 1^{-+} isovector hybrid meson as $b_1(1235)\pi$ or $f_1(1285)\pi$, with small (but non-negligible) $\rho\pi$ and $\eta\pi$ decay probability[7,8]. The decay of this state to $\eta\pi$, based purely on symmetry arguments may be prohibited.[0]

In this note we review the experimental evidence of several isovector $J^{PC} = 1^{-+}$ exotic candidates which have been reported recently. In particular, evidence for exotic signals have been reported in the flux-tube model inhibited channels $\eta\pi$, $\eta\pi$, $\rho\pi$, as well as the flux-tube preferred modes of $f_1\pi$ and $b_1\pi$.

2. $\eta\pi$ Decay mode

Several experiments have studied the $\eta\pi$ final state. The 1988 GAMS experiment[10] had claimed a narrow enhancement in the unnatural parity wave, but the analysis and conclusions were subsequently disputed by some of the members of the collaboration.[11] The 1993 KEK experiment[12] studied the reaction $\pi^- p \to \eta\pi^- p$ at 6.3

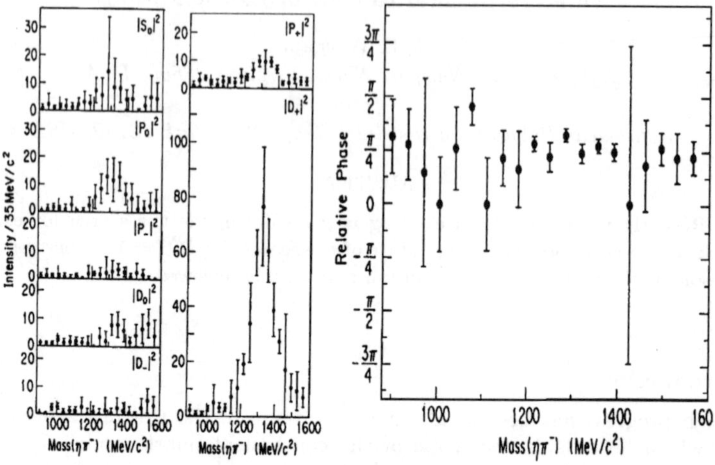

Fig. 1. KEK $\eta\pi^-$ results

GeV/c see a clear enhancement in the natural parity exchange P_+ wave* around 1.3 GeV/c^2, and the wave shows no phase motion with respect to the dominant D_+ wave, which is primarily the $a_2(1320)$. This lack of phase motion with respect to a known resonance could indicate that the P_+ wave is resonant also, with approximately the same mass and width.In addition, the P_+ peaks at approximately the same mass as the a_2. The partial wave results are shown in figure 1.

The 1993 VES experiment at Serpukov, (πN at 37 GeV/c on a beryllium target)[13] and E852 at the Brookhaven AGS (πp at 18 GeV/c)[14,15] have studied the same final state, with markedly different results. Both experiments see a significant P_+ exotic wave underneath the strong D_+ wave, but with considerable phase motion. The motion is nearly 100°, while 180° would be expected if there was only a single state. In both experiments the P_+ wave peaks at a slightly higher mass than the a_2, and the width is significantly wider. The VES data is shown in figure 2. The P-wave and phase for both E852 and VES is shown in figure 3. E852 interprets this wave and subsequent phase motion as an exotic state with a mass of $(1370\pm16^{+50}_{-30})$ MeV/c^2 and $(385\pm40^{+65}_{-105})$ MeV/c^2. The large systematic errors are a consequence of the ambiguities inherent in the two pseudo-scalar system[16,17].

*The partial waves in the two pseudo-scalar system are labeled by the orbital angular momentum of the decay. The subscript + indicates the natural-parity exchange wave with $|J_z| = 1$ in the Gottfried-Jackson frame, while the subscript − describes the unnatural-parity exchange wave with $|J_z| = 1$. The subscript 0 is the wave with $|J_z| = 0$, which is an unnatural-parity exchange wave. In the remainder of this paper, the waves are labeled by J^{PC} (isobar), the orbital angular momentum of the initial decay, $|J_z|$, and the naturality of the exchange particle.

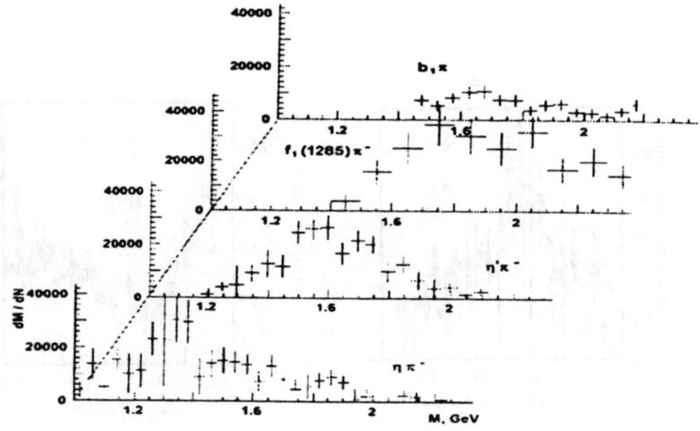

Fig. 2. VES $J^{PC} = 1^{-+}$ wave in $\eta\pi^-$, $\eta'\pi^-$, $f_1(1285)\pi^-$, and $b_1(1235)\pi^-$.

VES claims that the P-wave is consistent with a state with mass of $(1572 \pm 18_{stat} \pm 10_{syst}$ MeV$/c^2$ and a width of $(550 \pm 57_{stat} \pm 20_{syst})$ MeV$/c^2$.

The Crystal Barrel Collaboration at LEAR have recently reported results of a partial wave analysis of the annihilation $\bar{p}n \to \pi^-\pi^0\eta$ with the analysis assuming intermediate two-body states of $\pi^0\pi^-$ with a recoil η, or an $\eta\pi$ state off a recoil π. In their data, the resonant $\eta\pi$ P-wave contributes a significant fraction- over 11% plus interferences.[18,19] Resonance parameters extracted are consistent with E852, with a mass of $(1400 \pm 20_{stat} \pm 20_{syst}$ MeV$/c^2$ and a width of $(310 \pm 50_{stat} + 50/-30_{syst})$ MeV$/c^2$. The results of various χ^2 fits are shown in figure 4. The top line shows the contributions to the χ^2 for a fit without the 1^{-+} state (left shows positive contributions and right shows negative contributions). Note the obvious structure in the χ^2 contributions. The bottom line shows the fit with the 1^{-+} wave and the χ^2 contributions, which is structureless. The respective χ^2 per degree of freedom is 3.07 for the top fit, and 1.26 for the bottom fit.

3. $\pi^+\pi^-\pi^-$ Decay mode

While many experiments have looked at three pion final states in π^- induced interactions , again both VES and E852 have reported significant amounts of exotic $J^{PC} = 1^{-+}$ waves decaying to $\pi^+\pi^-\pi^-$, mainly in the channel $\rho\pi$. VES originally reported[22] this state as a resonance with a mass of 1.62 GeV$/c^2$ and width approximately 0.2 GeV$/c^2$, however, more recently the VES collaboration claims that this state is an artifact produced by approximations of their partial wave analysis model.[20] E852 however, has studied the same final state in the exclusive reaction $\pi^-p \to \pi^+\pi^-\pi^-p$[15,21].

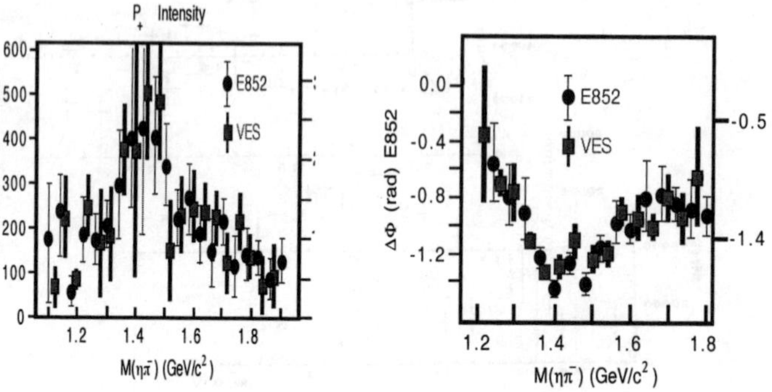

Fig. 3. E852 and VES $\eta\pi^-$ results, Intensity (left), and Phase (right)

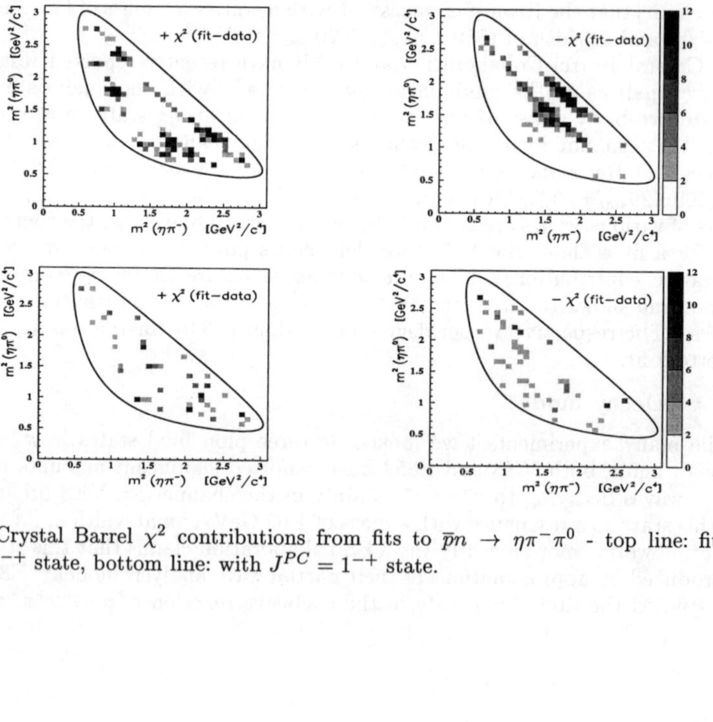

Fig. 4. Crystal Barrel χ^2 contributions from fits to $\overline{p}n \to \eta\pi^-\pi^0$ - top line: fit without $J^{PC} = 1^{-+}$ state, bottom line: with $J^{PC} = 1^{-+}$ state.

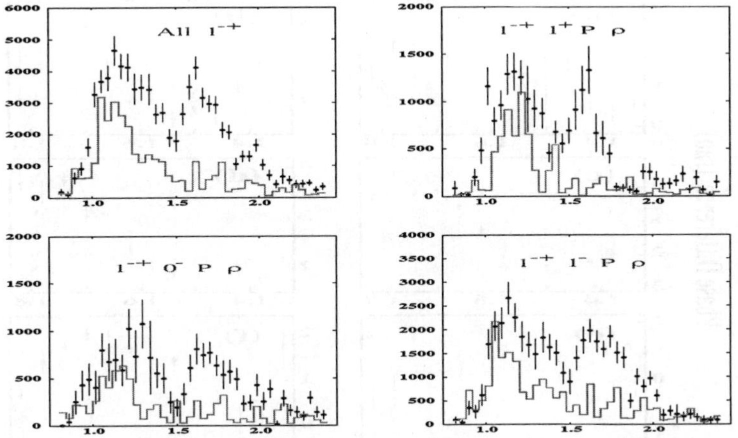

Fig. 5. E852 $\pi^+\pi^-\pi^-$ $J^{PC} = 1^{-+}$ partial wave. The points with error bars are the results of a PWA to the data, and the solid line is an estimate of leakage from other waves.

In general, the exclusive reaction should be described by a rank 2 density matrix, corresponding to spin-flip and spin-non-flip of the participating nucleon. However, the E852 collaboration claims an excellent description of the data with a rank 1 fit only. Twenty-one partial waves were sufficient to describe the data. All the major 3 π states are clearly seen in the partial wave analysis, and in addition there is a significant amount of $J^{PC} = 1^{-+}$ content shown in figure 5.

Two peaks are seen in the data, one around 1200 MeV/c^2 and the other around 1600 MeV/c^2. An estimate of the 'leakage' into this wave from other waves is also shown in the figure (solid line). Ambiguity between the $a_1(1160)$ and the 1^{-+} wave precludes any definite conclusions about the lower mass peak, but estimates of leakage into the high mass peak are small. The phase motion of the $J^{PC} = 1^{-+}$ state with respect to all other nearby states is clearly evident. (See figure 6). A mass-dependent fit was performed between the 1^{-+} wave decaying to $\rho\pi$ and the $\pi_2(1670)$, fitting both the magnitudes and phases of both waves (See figure 7), and as a result E852 claims a $J^{PC} = 1^{-+}$ exotic state with a mass of $(1593 \pm 8^{+29}_{-47})$ MeV/c^2 and $(168 \pm 20^{+150}_{-12})$ MeV/c^2.

4. $\eta'\pi$ Decay Mode

The $\eta'\pi^-$ system has also been studied at VES[13,22,20] (See figure 2) and BNL E852.[23] Both experiments agree in great detail, even though the VES data was taken with a Be target and at a higher energy.. Neither experiment sees significant unnatural parity production. The dominant wave in both experiments is P_+, and there is a significant amount of D_+. Again, the P-wave in $\eta'\pi$ is exotic $J^{PC} = 1^{+-}$. The P^+ wave peaks around 1600 MeV/c^2, and the $a_2(1320)$ is seen in natural parity exchange in the D_+ wave. A fit to the bump in the P_+ wave by VES[20] give a mass of $1572 \pm 18 \pm 20$ MeV/c^2, and a width of $550 \pm 57 \pm 40$ MeV/c^2. The E852 collaboration has fit the

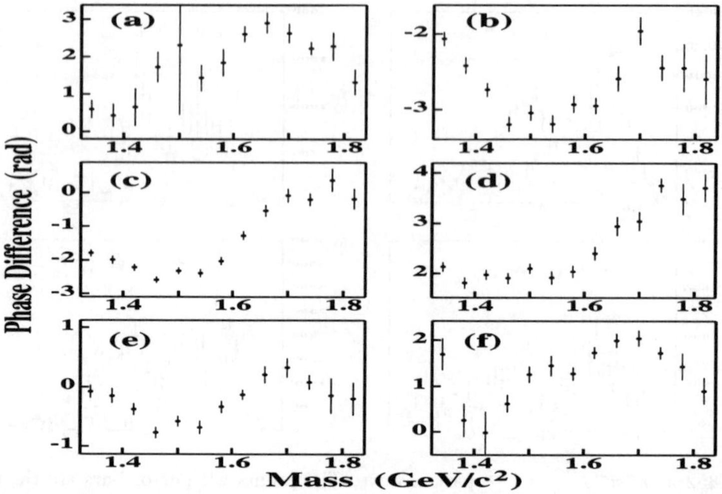

Fig. 6. E852 $\pi^+\pi^-\pi^-$ data, phase difference between the $1^{-+}[\rho(770)]P1^+$ wave and (a) the $0^{-+}[f_0(980)]S0^+$ wave, (b) the $2^{++}[\rho(770)]D1^+$ wave, (c) the $1^{++}[\rho(770)]S0^+$ wave, (d) the $1^{++}[\rho(770)]S1^+$ wave, (e) the $2^{-+}[\rho(770)]P0^+$ wave, (f) the $2^{-+}[f_2(1270)]D0^+$ wave.

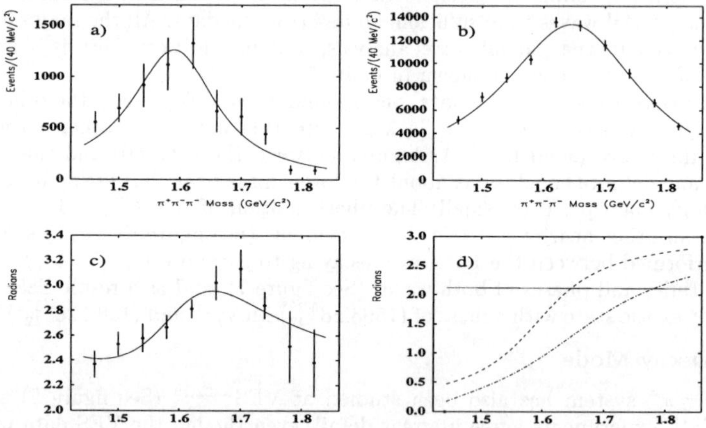

Fig. 7. A coupled mass-dependent fit of $1^{-+}1^+P\rho$ and $2^{-+}0^+Sf_2$ waves - E852 data. (a) $1^{-+}1^+P\rho$ wave intensity. (b) $2^{-+}0^+Sf_2$ wave intensity. (c) Phase difference between $1^{-+}1^+P\rho$ and $2^{-+}0^+Sf_2$ waves. (d) Phase motion of $1^{-+}1^+P\rho$ wave (dashed line), $2^{-+}0^+Sf_2$ wave (dot-dashed line) and production phase between them (solid line).

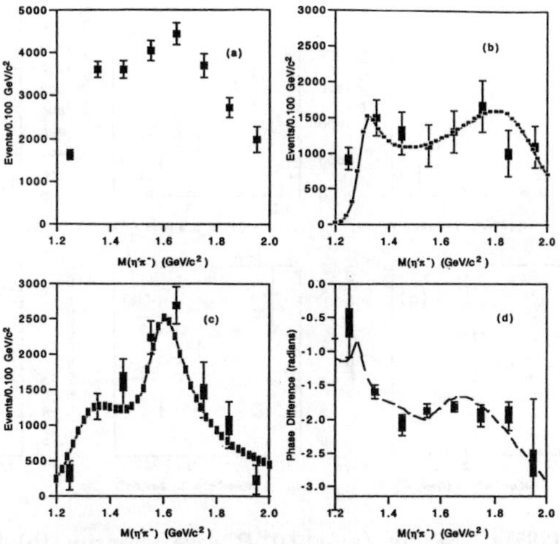

Fig. 8. E852 $\eta'\pi^-$, a) all waves b) D_+ with fit results c) P_+ with fit results d) Phase difference between D_+ and P_+ and fit results.

data to two spin 2 waves, and two exotic spin-1 waves. The D-wave was fixed with a resonant $a_2(1320)$ and a second pole at 1.900 GeV/c^2 with a width of 400 MeV/c^2. The P-wave was parameterized by two poles, fixed at the values determined from the previous $\eta\pi$ and $\rho\pi$ analyses described above. An adequate fit to the $\eta'\pi$ was achieved, and is shown in figure 8.

5. $f_1(1285)\pi$

The exotic picture in the flux-tube preferred decay mode $f_1\pi$ is even less clear. This final state has been studied previously at VES[22] (See figure 2) and at Brookhaven E818,[25] and more recently by E852.[26,27] The three experiments agree on some aspects of the data- all see a large $J^{PC} = 1^{++}$ wave around 1700 MeV/c^2, and each experiment sees a significant contribution of $J^{PC} = 1^{-+}$ in the region 1.5-2.1 GeV/c^2. E818 was very limited in statistics (a total of 747 events) decaying to $K^+K_s\pi^-$. They were only able to perform a rank-1 fit, and the subsequent experiments determined that this was not sufficient. They claimed that nearly half the data (45%) was in the exotic wave $J^{PC} = 1^{-+}$, which they interpreted as being a compound object, based on different branching ratios to $f_1\pi$ and $\eta(1295)\pi$ and the phase motion with respect to the 1^{++} wave. In E852 the phase between the 1^{++} wave and the 1^{-+} wave is relatively flat; if the 1^{++} is in fact an $a_1(1700)$ resonance, this may indicate a resonant exotic state at around the same mass, but results are currently inconclusive. The E852 analysis results are shown in figure 9.

6. Conclusions and Future Prospects

In conclusion, there are credible candidates now emerging that indicates that the meson spectrum is rich, and contains states beyond the naive quark model. E852 has

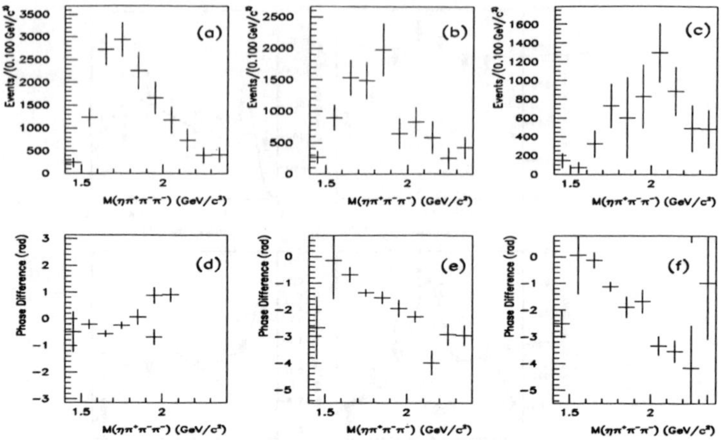

Fig. 9. E852 $f_1(1285)\pi^-$ results. (a) $1^{++}0^+P$ wave intensity. (b) $1^{-+}1^+S$ wave intensity. (c) $2^{-+}0^+D$ wave intensity. (d) Phase motion of $1^{++}0^+P - 1^{-+}1^+S$ (e) Phase motion of $1^{++}0^+P - 2^{-+}0^+D$ (f) Phase motion of $1^{-+}1^+S - 2^{-+}0^+D$.

to date only published data from their first run in 1994- they have roughly three times the statistics from their 1995 run. In addition, K/π separation was achieved via the addition of a Cerenkov counter in their 1997 run, and another run is scheduled for the fall of 1998 with the same configuration. This data should give new information on branching ratios to strange particle decays, which can illuminate the underlying structure of these mesons.

VES is also continuing to process their existing data, and should double their existing statistics. In addition, there are plans to continue the program at Serpukov, including a study of varying nuclear targets, and also studying the energy dependence of resonance production. The Crystal Barrel running program has ended, but there still exists data to be analyzed. In addition, there are new meson spectroscopy programs being developed. For example, there are plans to upgrade the energy of CEBAF in order to study photoproduction of exotic mesons[28,29]

References

1. N. Isgur and J. Paton, Phys. Rev. D **31**, 2910 (1985).
2. C. Bernard *et al.*, Phys. Rev. D **56**, 7039 (1997); P. Lacock *et al.*, Phys. Lett. B **401**, 308 (1997).
3. T. Barnes and F. E. Close, Phys. Lett. B **116**, 365 (1982).
4. I. I. Balitsky, D. I. Dyakonov, and A. V. Yung, Z.Phys. C **33**, 265 (1986); J. I. Latorre, P. Pascual, and S. Narison, Z.Phys. C **34**, 347 (1987); J. Govaerts *et al.*, Nucl. Phys. B **284**, 674 (1987).

5. Y.Uehara *et al.*, Nucl. Phys. A **606**, 357 (1996).
6. S. Ishida, H. Sawazaki, M. Oda, and K. Yamada, Phys. Rev. D **47**, 179 (1992).
7. N. Isgur, R. Kokoski, and J. Paton, Phys. Rev. Lett. **54**, 869 (1985).
8. Frank E. Close and Philip R. Page, Nucl. Phys. B **443**,233-254 (1995).
9. H. J. Lipkin, Phys. Lett. B **219**, 99 (1989).
10. D. Alde *et al.*,Phys. Lett. B **205**, 397 (1988).
11. Y. D. Prokoshkin and S. A. Sadovsky, Yad. Fiz. **58**, 662, [Sov. J. Nucl. Phys. **58**, 606 (1885)].
12. H. Aoyagi *et al.*, Phys. Lett. B **314**, 246 (1993).
13. G. M. Beladidze *et al.*, Phys. Lett. B **313**, 276 (1993).
14. D.R. Thompson *et al.*, Phys. Rev. Lett. **79**, 1630 (1997).
15. D. P. Weygand *et al.*, in *Proceedings of the Seventh International Conference on Hadron Spectroscopy*, edited by S. U. Chung and H. J. Willutzki, 313 (1997).
16. S. U. Chung, Phys. Rev. D **56**, 7299 (1997).
17. S. A. Sadovsky. "On the Ambiguities in the Partial-Wave Analysis of $\pi^- p \rightarrow \eta\pi^0 n$ Reaction", Inst. for High Energy Physics IHEP-91-75, unpublished (1991).
18. W. Dünnweber *et al.*, in *Proceedings of the Seventh International Conference on Hadron Spectroscopy*, edited by S. U. Chung and H. J. Willutzki, 309 (1997).
19. A. Abele *et al.*, Phys. Lett. B **423**, 175 (1998).
20. A. Zaitsev *et al.*, in *Proceedings of the Seventh International Conference on Hadron Spectroscopy*, edited by S. U. Chung and H. J. Willutzki, 461 (1997).
21. G. S. Adams *et al.*, submitted to Phys. Rev. Lett (1994).
22. Yu. P. Gouz *et al.*, in *Proceedings of the XXVI International Conference on High Energy Physics, Dallas, 1992,* edited by J. R. Sanford (American Institute of Physics, 1993), Vol.1, p.572.
23. D. L. Stienike, PhD thesis, University of Notre Dame, (1998).
24. J. H. Lee *et al.*, Phys. Lett. B **323**, 227 (1994).
25. J. H. Lee *et al.*, Phys. Lett. B **323**, 227 (1994).
26. T. Adams *et al.*, in *Proceedings of the Seventh International Conference on Hadron Spectroscopy*, edited by S. U. Chung and H. J. Willutzki, 107 (1997).
27. D. Ryabchikov *et al.*, in *Proceedings of the Seventh International Conference on Hadron Spectroscopy*, edited by S. U. Chung and H. J. Willutzki, 527 (1997).
28. A. Dzierba,*Proceedings of the* 8^+ *workshop* CMU, edited by C. Meyer and D. P. Weygand, 1 (1998).
29. J. Napolitano,*Proceedings of the* 8^+ *workshop* CMU, edited by C. Meyer and D. P. Weygand, 22 (1998).

(HYBRID) BARYONS: QUANTUM NUMBERS AND ADIABATIC POTENTIALS

Philip R. Page*
Theoretical Division, MS B283, Los Alamos National Laboratory, Los Alamos, NM 87545, USA

ABSTRACT

We construct (hybrid) baryons in the flux–tube model of Isgur and Paton. In the limit of adiabatic quark motion, we build proper eigenstates of orbital angular momentum and indicate the flavour, spin, chirality and J^P of (hybrid) baryons. The adiabatic potential is calculated as a function of the quark positions.

1. Introduction

Hybrid baryons are bound states of three quarks with an explicit excitation in the gluon field of QCD. The construction of (hybrid) baryons in a model motivated from and consistent with lattice gauge theory, the non–relativistic flux–tube model of Isgur and Paton, was detailed in ref.[1] There we studied the detailed flux dynamics and built the flux hamiltonian. A minimal amount of quark motion is allowed in response to flux motion, in order to work in the centre of mass frame. Otherwise, we make the so–called "adiabatic" approximation, where the flux motion adjusts itself instantaneously to the motion of the quarks. The main result was that the lowest flux excitation can to a high degree of accuracy (about 5%) be simulated by neglecting all flux–tube motions except the vibration of a junction. The junction acquires an effective mass from the motion of the remainder of the flux–tube and the quarks. The model is then simple: a junction is connected via a linear potential to the three quarks. The ground state of the junction motion corresponds to a conventional baryon and the various excited states to hybrid baryons.

2. Quantum numbers of (hybrid) baryons

The junction can move in three directions, and correspondingly be excited in three ways, giving the hybrid baryons H_1, H_2 and H_3. For each junction excitation, it is found that the junction wave function can be realized to be either totally symmetric, or totally antisymmetric under quark label exchange, indicated by H^S and H^A respectively.

The quantum numbers of the lowest–lying states that can be constructed on the adiabatic surfaces corresponding to each of the six hybrid baryons are indicated in Table 1.

*prp@t5.lanl.gov. Work done in collaboration with Simon Capstick.

Table 1. Quantum numbers of ground state (hybrid) baryons for the adiabatic surfaces B, H_1, H_2 and H_3. Here B denotes the conventional baryon. The mass ordering is $B < H_1 < H_2 < H_3$. In the absense of spin dependent forces all ground states corresponding to a given adiabatic surface are degenerate. The quantum number notation is $(N,\Delta)^{2S+1}J^P$, where N, Δ is the flavour structure of the wave function (i.e. those of the conventional baryons N, Δ respectively) and P the parity.

(Hybrid) Baryon	Chirality	L	$(N,\Delta)^{2S+1}J^P$
B	1	0	$N^2\frac{1}{2}^+$, $\Delta^4\frac{3}{2}^+$
H_1^S	1	1	$N^2\frac{1}{2}^+$, $N^2\frac{3}{2}^+$, $\Delta^4\frac{1}{2}^+$, $\Delta^4\frac{3}{2}^+$, $\Delta^4\frac{5}{2}^+$
H_1^A	1	1	$N^2\frac{1}{2}^+$, $N^2\frac{3}{2}^+$
H_2^S	1	1	$N^2\frac{1}{2}^+$, $N^2\frac{3}{2}^+$, $\Delta^4\frac{1}{2}^+$, $\Delta^4\frac{3}{2}^+$, $\Delta^4\frac{5}{2}^+$
H_2^A	1	1	$N^2\frac{1}{2}^+$, $N^2\frac{3}{2}^+$
H_3^S	-1	0	$N^2\frac{1}{2}^-$, $\Delta^4\frac{3}{2}^-$
H_3^A	-1	0	$N^2\frac{1}{2}^-$

Since quarks are fermions, the wave function should be totally antisymmetric under quark label exchange, called the Pauli principle. The colour structure of hybrid baryons are taken to be identical to those of conventional baryons, i.e. it is totally antisymmetric under label exchange. This imposes constraints on the combination of flavour and non–relativistic spin S of the three quarks that is allowed. The combinations are indicated in Table 1.

"Chirality" gives the behaviour of the junction wave function under reflection in the plane spanned by the quarks, when the positions of the quarks are fixed.

Let L be the orbital angular momentum of the quarks and the junction. It is possible to argue that $L = 1$ for the ground state H_1 and H_2 hybrid baryons, while $L = 0$ for the ground state conventional and H_3 hybrid baryons. The total angular momentum $\mathbf{J} = \mathbf{L} + \mathbf{S}$. Since $L = 0$ for ground state conventional and H_3 hybrid baryons, $J = S$. Since $L = 1$ for ground state H_1, H_2 hybrid baryons, $J = \frac{1}{2}, \frac{3}{2}$ for $S = \frac{1}{2}$, and $J = \frac{1}{2}, \frac{3}{2}, \frac{5}{2}$ for $S = \frac{3}{2}$. These assignments are indicated in Table 1.

3. The potential in which the quarks move

We shall now calculate the junction energy ("adiabatic potential") for the ground and first excited states of the junction as a function of the quark positions.

Define

$$\boldsymbol{\rho} = \frac{\mathbf{r}_1 - \mathbf{r}_2}{\sqrt{2}} \qquad \boldsymbol{\lambda} = \frac{\mathbf{r}_1 + \mathbf{r}_2 - 2\mathbf{r}_3}{\sqrt{6}} \qquad \cos\theta_{\rho\lambda} = \frac{\boldsymbol{\rho}\cdot\boldsymbol{\lambda}}{\rho\lambda} \tag{1}$$

where $\mathbf{r}_i$ denotes the positions of the quarks. The energy is a function of ρ, λ and $\theta_{\rho\lambda}$.

The procedure for evaluating the conventional baryon potential is as follows. We numerically evaluate $V_B(l_1, l_2, l_3)$ by solving the Schrödinger Equation for the junction

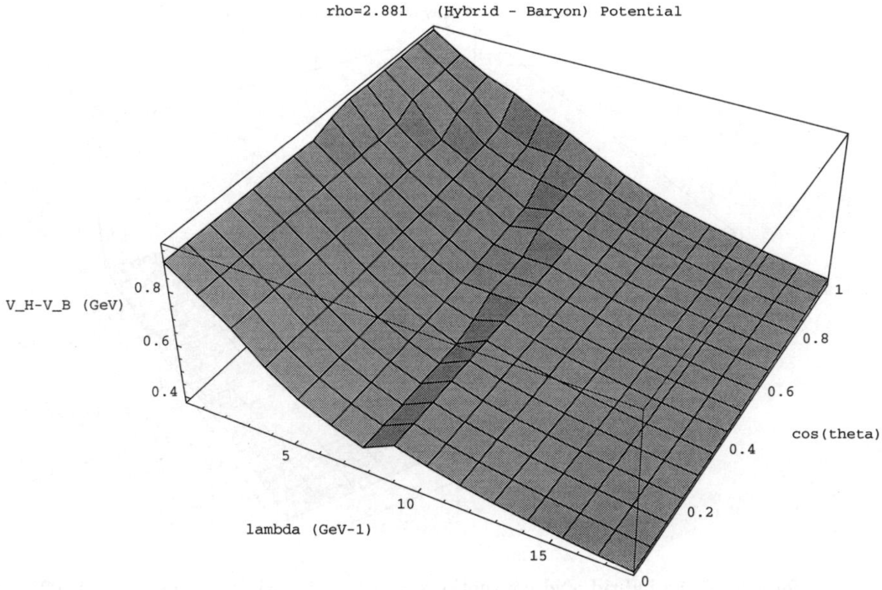

Fig. 1. Difference of the hybrid and conventional baryon potentials (in GeV) for $\rho = 2.881$ GeV^{-1} as a function of λ (in GeV^{-1}) and $\cos\theta_{\rho\lambda}$.

hamiltonian, $(\frac{1}{2}M^{\infty}_{\rm eff}\dot{\mathbf{r}}^2 + V_{\rm J})\Psi_B(\mathbf{r}) = V_B(l_1, l_2, l_3)\Psi_B(\mathbf{r})$, variationally using an ansatz ground state simple harmonic oscillator wave function. $M^{\infty}_{\rm eff}$ is the effective mass of the junction in the limit where the flux–tubes between the junction and quarks are continuous strings. $V_{\rm J}$ is the linear potential between the junction and the three quarks. The hybrid baryon H_1 potential $V_{H_1}(l_1, l_2, l_3)$ is solved using $(\frac{1}{2}M^{\infty}_{\rm eff}\dot{\mathbf{r}}^2 + V_{\rm J})\Psi_{H_1}(\mathbf{r}) = V_{H_1}(l_1, l_2, l_3)\Psi_{H_1}(\mathbf{r})$ with a first excited state simple harmonic oscillator wave function as an ansatz.

The difference between the hybrid and conventional baryon potentials is plotted in Figures 1 - 2. Since the potentials are functions of ρ, λ and $\theta_{\rho\lambda}$, one of the variables is held fixed at a typical value for clarity of presentation.

The (hybrid) baryon potential can be seen to increase when $\rho\lambda$ is small. Numerically, the ratio of the hybrid to baryon potential is found to be $1.44 - 1.6$ for all ρ, λ and $\theta_{\rho\lambda}$.

A preliminary estimate of the ground state H_1 hybrid baryon mass of ~ 2 GeV has also been made by adding the difference between the hybrid and conventional baryon potentials to the phenomenologically successful baryon potential used in ref.[2]

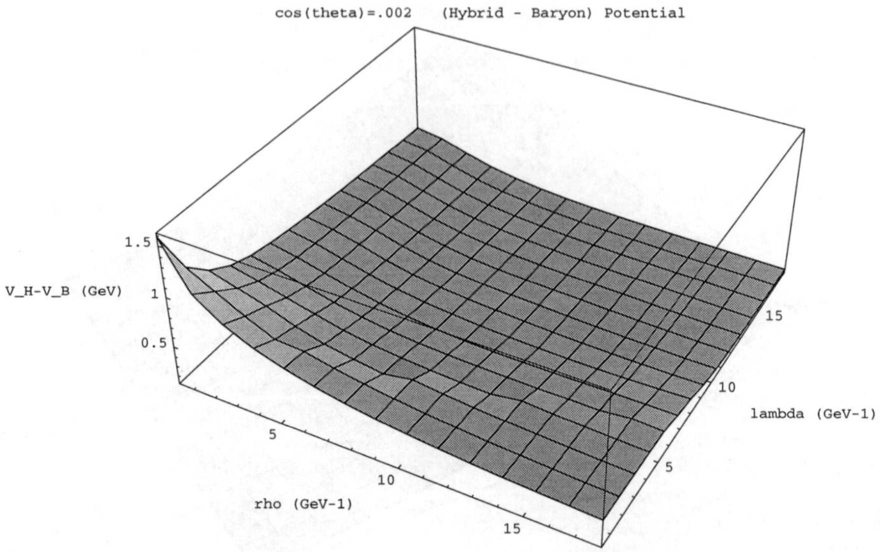

Fig. 2. Difference of the hybrid aNd conventional baryon potentials for $\cos\theta_{\rho\lambda} = 0.002$ as a function of ρ and λ (in GeV^{-1}).

4. Conclusions

The spin and flavour structure of the six hybrid baryons have been specified. Exchange symmetry constrains the spin and flavour of the (hybrid) baryon wave function. The orbital angular momentum of the low–lying hybrid baryon is argued to be unity. The adiabatic potentials have been calculated numerically. The low–lying hybrid baryon mass has been estimated numerically.

References

1. P.R. Page, *Proc. of "Seventh International Conference on Hadron Spectroscopy" (HADRON '97)*, 25–30 August 1997, Upton, N.Y., U.S.A., eds. S.-U. Chung, H.J. Willutzki, p. 553, American Institute of Physics.
2. S. Capstick, N. Isgur, *Phys. Rev.* **D34** (1986) 2809.

EXOTIC MESONS AND SOFT GLUE

A. P. Szczepaniak
Physics Department and Nuclear Theory Center,
Indiana University, Bloomington, IN 47405, USA

ABSTRACT

We summarize the results of recent studies of soft gluonic excitations.

1. Introduction

Understanding of soft QCD in the nonperturbative domain will not be complete until effects of gluonic excitations are accounted for. Interest in theoretical exploration of this aspect of QCD has recently been advanced by BNL, E852 collaboration claims of observation of an exotic state at 1.4 GeV.[1] Even though lattice has become the main tool for studying nonperturbative effects in QCD and it provides many "theoretical observables" otherwise unmeasurable in the real world *e.g* potentials between infinitely heavy $q\bar{q}$ pair, for phenomenological applications it is important to continue with development of models of strong QCD. It is usually stated that an exotic state is a one which does not fit into the simple pattern of valence quark model phenomenology. There is a class of exotic states which are "manifestly exotic" in the sense that their quantum numbers cannot from combining valence quarks alone. One interpretation of the structure of such states is that they are hybrids with both valence quark and gluonic modes being present. Even though the "content" of a specific state is frame, gauge, quantization scheme, *etc* dependent and therefore not very meaningful the success of the nonrelativistic quark model points into a particular direction when building QCD based models of hadron structure. We will therefore use the framework of nonrelativistic QCD in the Coulomb gauge to construct the Fock space and examine various definitions of soft gluonic degrees of freedom in this framework.[2]

The phenomenology of hybrids has been developed in the bag, QCD sum rule, constituent glue, and flux tube models.[3] The nature of the appropriate degrees of freedom could, in principle be established by detailed comparison of model predictions with data. Here we will briefly describe the last two and since the "real" data is still rather controversial we will use lattice to discriminate between the two. In the constituent gluon model, soft gluons are treated on a similar footing as quarks, namely as localized, densities interacting with quarks via static potentials.[4] Color, spin and other intrinsic quantum number assignments of these effective, gluonic degrees of freedom are not well defined and change from model to model. In particular in the Dynamical Quark Models (DQM) build as a low energy approximation to the perturbatively renormalized QCD Coulomb gauge Hamiltonian, gluons behave as massive, transverse, spin-1, color octet objects interacting with the quarks through a static potential.[5] This potential is determined from the static quark limit by comparison with the $Q\overline{Q}$ energies on the lattice. In this approach the main difference between the $Q\overline{Q}$ and Qg potentials comes from the color factors. In particular the model predicts a repulsive Coulomb spike from $Q\overline{Q}$ octet interactions dominating at short distances over the $Q\overline{Q}-g$ attraction. This spike may however be washed out since as the $Q\overline{Q}$ separation decreases at some point it should become energeticaly more favorable for a hybrid to decay into a $Q\overline{Q}$ meson and a glueball. Possible presence of this effect in the lattice data is currently being looked for.[6] Because of the confining interaction, in DQM a single gluon becomes infinitely heavy, but in a colorless bound state it behaves as a massive constituent with a mass around 800 MeV at low momenta. The spectrum of two gluon glueball in this approach has been calculated[7] and more recently the fully renormalized effective Hamiltonian in the gluon sector has been calculated.[8] An alternative approach to soft gluon dynamics is to assume that excited gluon modes around a heavy $Q\overline{Q}$ pair correspond to collective excitations of the chromoelectric flux. The flux tube model of Isgur, Kokoski and Paton,[9] has recently been extended to account for color degrees of freedom and the mapping of the gluon **E** and **B** fields onto the flux tube model space has been established.[10] This in particular allows to calculate, in the flux tube model spin dependent potentials defined in terms of gauge invariant matrix elements of the gluon field. Also, under the assumption that hybrids decay through a local deexcitation of the gluon field, rather then by quark tunneling (which leads to a standard $^3P_0$ decay model), a new approach to hybrid two body hadronic decays has been formulated.[11] These studies seem to favor the flux tube model description of soft gluonic modes. In the following we will summarize the results of the study of hybrid adiabatic potentials in the constituent and flux tube models.

2. Static $Q\overline{Q}$ potentials

Recently the UCSB group measured the $Q\overline{Q}$ energy as a function of quark separation for a number of gluonic field configurations.[12] The configurations are classified according to their transformation properties under symmetries of the static $Q\overline{Q}$ source. These include, rotations around the $Q\overline{Q}$ axis, reflection in the plane containing the $Q\overline{Q}$ axis and combined gluonic parity and charge conjugation. Similarly as in the case of a diatomic molecule, the corresponding $Q\overline{Q}$ adiabatic potential surfaces are labeled by $\Lambda = 0, 1, \cdots$ $Y = \pm$ and $PC = \pm = g, u$ corresponding to the three operations resp. In DQM the static potential, $V(R)$ is given by the spectrum of the constituent gluon in the presence of the $Q(-\mathbf{R}/2)$, $\overline{Q}(\mathbf{R}/2)$ octet source and can be determined from a Shrödinger equation,

$$[K + W]|\mathbf{R}, \Lambda^Y_{PC}\rangle = V(R)|\mathbf{R}, \Lambda^Y_{PC}\rangle \tag{1}$$

with K being the gluon kinetic energy (as described above) and W the residual gluon $Q\overline{Q}$ interaction. The Σ and Π adiabatic surfaces corresponding to $\Lambda = 0, 1$ *resp.* are shown in Fig. 1, together with the flux tube model results.

In Fig. 2. the flux tube model results are compared with the lattice data. The excited flux tube energies are calculated from

$$V(R) = bR + \frac{N\pi}{R}\left(1 - e^{-fb^{\frac{1}{2}}R}\right) \tag{2}$$

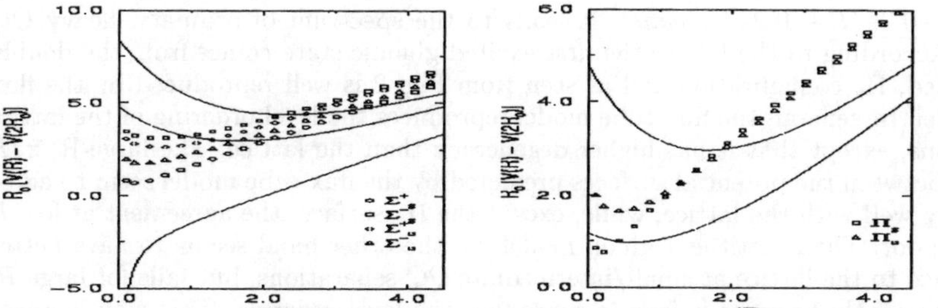

Fig. 1. Σ (left) and Π (right) adiabatic potentials in units of $r_0 \sim 0.5$ fm. DQM (symbols) vs flux tube model (solid lines) predictions.

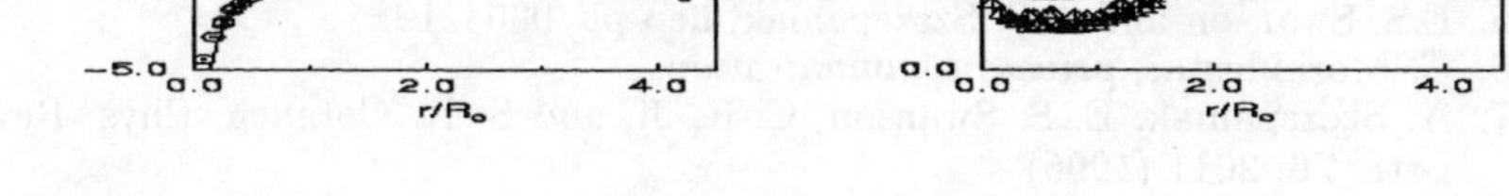

Fig. 2. Same as in Fig. 1 with lattice (symbols) vs flux tube model (solid lines) predictions.

where $f \sim 1$ is a phenomenological constant introduced to smear out the short distance behavior, and $N = \sum_{m=1} m(n_{m+} + n_{m-})$ counts the total momentum of an excited string due to $n_{m\pm}$ left(right)handed modes. The ground state $Q\overline{Q}$ potential corresponds to glue in the $\Lambda^Y_{PC} = \Sigma +_g$ configuration and can be well represented by $V(R) = -0.29/R + 0.24r + const.$ It leads to the spectrum of ordinary, heavy $Q\overline{Q}$ mesons. According to the lattice the first excited gluonic state comes from the doubly degenerated, Π_u configuration and as seen from Fig. 2 is well reproduced by the flux tube model. In general, the flux tube model reproduces the level ordering of the lattice calculations, except that it has higher degeneracy than the lattice. The large-R, π/R splitting between the potential surfaces predicted by the flux tube model seem to agree reasonably well with the lattice, while, except the Π_u surface, the agreement at low R is rather poor. The constituent gluon model, on the other hand seems to have better convergence to the lattice at small/intermediate $Q\overline{Q}$ separations, but fails for large R. More importantly however it fails to reproduce the level ordering. This may however be reinstated by flipping the intrinsic parity of the gluon, *i.e.* making it behave under parity more like the lattice link variable. From Figs. 1,2 is follows that there is clearly need for a large number of degrees of freedom to describe $Q\overline{Q}$ interactions. The same conclusion was reached by studies of spin dependent $Q\overline{Q}$ potentials. In was shown there that a small contribution to the long range part of an effective spin-orbit interaction, which comes from eliminating soft gluon exchange between the $Q\overline{Q}$ pair, can only be reproduced if the gluon is represented as a string-like excitation rather then a constituent particle.[10] Another consistency check comes from decay studies. Constituent gluon models and flux tube model lead to different selection rules in hadronic two body decays of an excited glue in a hybrid state. Just like the S/D ratio in $1^+ \to 1^- 0^-$ is a sensitive probe of the $Q\overline{Q}$ production mechanisms, there are several hybrid decay modes which are sensitive to the decay dynamics.[13] Provided hybrids are narrow,[14] future detailed partial wave analysis of hybrid decay products could provide important constrains on models of soft glue.

References

1. D.R. Thompson et al., E852 Collab., Phys. Rev. Lett. **79**, 1630 (1997).
2. A. Szczepaniak and E. S. Swanson, Phys. Rev. D **55**, 1578 (1997)
3. T. Barnes, F.E. Close, and E.S. Swanson, Phys. Rev. D. **52**, 5242 (1995) and refer. therein.
4. D. Horn and A. Mandula, Phys. Rev. D **17**, 898 (1978).
5. E.S. Swanson and A.P. Szczepaniak, hep-ph/9804219.
6. C. Morningstar, *private communication.*
7. A. Szczepaniak, E. S. Swanson, C.-R. Ji, and S. R. Cotanch, Phys. Rev. Lett. **76**, 2011 (1996).
8. D.G. Robertson, *in these proceedings*; D.G. Robertson, E.S. Swanson, A.P. Szczepaniak, C.-R. Ji, and S.R. Cotanch, *in preparation.*
9. N. Isgur and J. Paton, Phys. Rev. D **31** , 2910 (1985); N. Isgur, R. Kokoski and J. Paton, Phys. Rev. D **31**, 2910 (1985).
10. A.P. Szczepaniak and E.S. Swanson, Phys. Rev. D **55**, 3987 (1997).
11. E.S. Swanson and A.P. Szczepaniak, Phys. Rev. D **56**, 5692 (1997).
12. K. Juge, J. Kuti, and C.J. Morningstar, hep-lat/9709131.

13. P.R. Page, E.S. Swanson and A.P. Szczepaniak, hep-ph/9808346
14. T.D. Cohen, Phys. Lett. B **427**, 348 (1998).

Further Study of 1^{-+} Exotic Meson in the Reaction $\pi^- p \to \eta\pi^- p$ at 18 GeV/c

V. L. Korotkikh for the E852 Collaboration
Nuclear Research Institute of Moscow State University, Moscow 117321, Russia
E-mail:vlk@lav1.npi.msu.su

ABSTRACT

Details of the analysis of the $\eta\pi^-$ system studied[1] in the reaction $\pi^- p \to \eta\pi^- p$ at 18 GeV/c are given. The amplitude analysis indicates the presence of interference between the $a_2(1320)$ and a $J^{PC} = 1^{-+}$ wave between 1.2 and 1.6 GeV/c^2. The phase difference between these waves shows phase motion not attributable solely to the $a_2(1320)$. The data can be fitted by interference between the $a_2(1320)$ and an exotic 1^{-+} resonance with M = $(1370 \pm 16\ ^{+50}_{-30})$ MeV/c^2 and $\Gamma = (385 \pm 40\ ^{+65}_{-105})$ MeV/c^2.

1. Introduction

The $\eta\pi$ system is particularly interesting in searching for exotic (or non-$q\bar{q}$) mesons because the system has spin (J), parity (P), and charge-conjugation (C) in the sequence $J^{PC} = 0^{++}, 1^{-+}, 2^{++}, 3^{-+}...$ for $\ell = 0, 1, 2, 3, ...$. (Here ℓ is the orbital angular momentum of the $\eta\pi$ system.) A $q\bar{q}$ meson with orbital angular momentum L and total spin S must have $P = (-1)^{L+1}$ and the neutral member of its isospin multiplet must have $C = (-1)^{L+S}$. A resonance with a quantum number in the sequence $J^{PC} = 0^{--}, 0^{+-}, 1^{-+}, 2^{+-}, 3^{-+} \cdots$ does not satisfy these conditions and must be exotic. Having isospin I=1, such a resonance could not be a glueball $(2g, 3g, \ldots)$, but it could be a hybrid $(q\bar{q}g)$ or a multiquark $(q\bar{q}q\bar{q})$ state.

Several experiments,[2–5] prior to the publication of the E852 results,[1] had studied the $\eta\pi$ final state, and observed an enhancement in the P wave around 1.4 GeV/c^2. However, they reached conflicting conclusions.

The first credible claim for a 1^{-+} exotic resonance was made by our experiment at BNL.[1] In this paper we present some details of the analysis. We note that since the publication of our letter, an independent confirmation of our results has come from a new measurement by the Crystal Barrel collaboration.[6]

2. Data and PWA

Our data sample was collected in the first data run of E852 at the Alternating Gradient Synchrotron (AGS) at Brookhaven National Laboratory (BNL) with the Multi-Particle Spectrometer (MPS) augmented by additional detectors. The details of the E852 apparatus are given elsewhere.[7]

Of the 47 million triggers, 47,235 events were reconstructed[1] which were consistent with the reaction

$$\pi^- p \to \eta\pi^- p \tag{1}$$

at 18 GeV/c, with the decay mode $\eta \to \gamma\gamma$.

The background is approximately 7% at 1.2 GeV/c^2, and only 1% at 1.3 GeV/c^2.

A partial-wave analysis (PWA)[8-10] based on the extended maximum likelihood method has been used to study the spin-parity structure of the $\eta\pi^-$ system.

The partial waves are parameterized in terms of the quantum numbers J^{PC} as well as m, the absolute value of the angular momentum projection, and the reflectivity ϵ.[11] In our naming convention, a letter indicates the angular momentum of the partial wave in standard spectroscopic notation, while a subscript of 0 means $m = 0$, $\epsilon = -1$, and a subscript of $+(-)$ means $m = 1$, $\epsilon = +1(-1)$. Thus, S_0 denotes the partial wave having $J^{PC}m^\epsilon = 0^{++}0^-$, while P_- signifies $1^{-+}1^-$, D_+ means $2^{++}1^+$, and so on.

The results of the PWA fit for the D_+ and P_+ intensities and the phase difference between these amplitudes, $\Delta\Phi$, are shown in Fig.1.

3. Leakage

The Monte-Carlo study shows that the intensity and phase motion of the P_+ wave do not have the characteristics of a wave which is artificially generated from a pure D_+ wave due to possible incomplete knowledge of the resolution or detection inefficiency. This does not preclude the possibility of some leakage of D_+ wave to P_+ wave being present in the data and distorting the results of the mass dependent fit[1] (MDF). In this section, we describe a test which has been carried out to study the sensitivity of our MDF results to possible residual leakage being present in the data.

The fit has been carried out is a mass dependent partial wave analysis (MDPWA). In such a fit, the PWA is carried out so that all $\eta\pi^-$ mass bins are fit simultaneously and are tied together with a mass-dependent function for each partial wave. That is, the extended maximum likelihood function is generalized to include mass dependence:

$$\ln\mathcal{L} \propto \sum_i^n \ln I(\Omega_i, w_i) - \int d\Omega dw\, \eta(\Omega, w)\, I(\Omega, w). \tag{2}$$

The free parameters in the fit include, in addition to the amplitudes of the partial waves, the Breit-Wigner masses, widths, and intensities as well as mass-independent production amplitude phases.

The results of the MDPWA fit are shown as the smooth curves in Fig.1. Also shown as the points with error bars are the results of the standart PWA.[1] It is clear that the two analyses give consistent results.

4. Conclusion

The results of the MDPWA fit are given in Table 1, where they are compared with those of the combined PWA and mass dependent fit (PWA+MDF). The results are quite compatible when one takes into account the systematic errors. The biggest difference is in the fitted width of the P_+ state which is larger for the MDPWA.

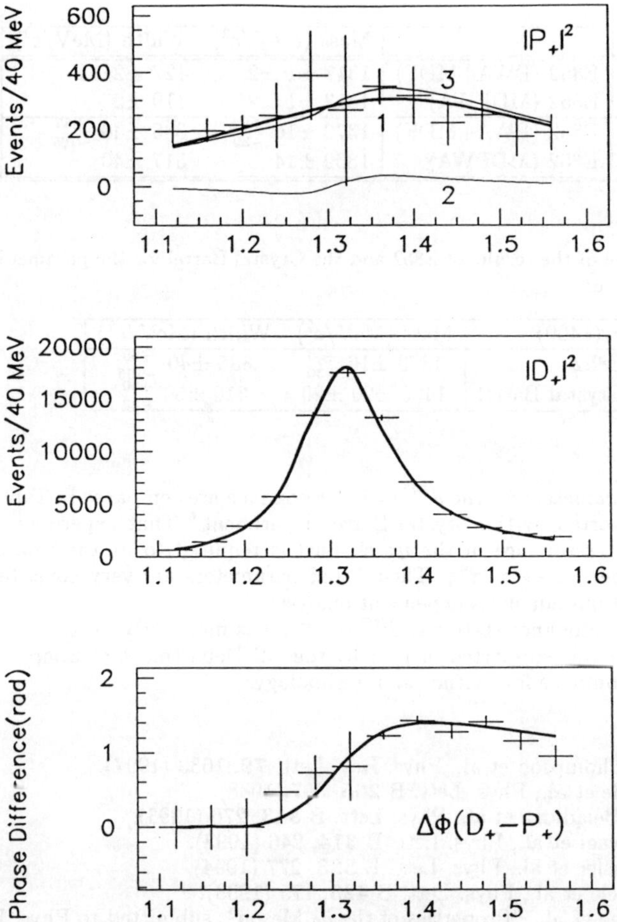

Fig. 1. The fit results of the MDPWA (curves) and the mass independent PWA results (crosses) for the $\eta\pi^-$ system: a) P_+ , b) D_+ intensities and c) their relative phase $\Delta\phi(P_+ - D_+)$. Fig.1a also shows the contributions of the 1^{-+} signal intensity (1), the sum (2) of the leakage and the (signal - leakage) interference term and the complete 1^{-+} wave (3).

Table 1. Comparison of the results of the PWA combined with a separate mass dependent fit (MDF) with those of the MDPWA with leakage.

Meson		Mass (MeV/c^2)	Width (MeV/c^2)
$a_2(1320)$	E852 (PWA+MDF)	1317 ±1 ±2	127 ±2 ±2
	E852 (MDPWA)	1313 ±1	119 ±2
$\pi_1(1400)$	E852 (PWA+MDF)	$1370 \pm 16\ ^{+50}_{-30}$	$385 \pm 40\ ^{+65}_{-105}$
	E852 (MDPWA)	1369 ±14	517 ±40

Table 2. Comparison of the results of E852 and the Crystal Barrel for the parameters of the $J^{PC} = 1^{-+}$ resonance.

$\pi_1(1400)$	Mass (MeV/c^2)	Width (MeV/c^2)
E852	$1370 \pm 16\ ^{+50}_{-30}$	$385 \pm 40\ ^{+65}_{-105}$
Crystal Barrel	1400 ±20 ±20	$310 \pm 50\ ^{+50}_{-30}$

Our fitted parameters for the $J^{PC} = 1^{-+}$ resonance are compared in Table 2 with values recently reported by the Crystal Barrel experiment.[6] That experiment reports that a $J^{PC} = 1^{-+}$ resonance in the $\eta\pi$ channel is required to fit their data in the annihilation channel $\overline{p}n \to \pi^-\pi^0\eta$. Their fitted parameters are very consistent with those determined from our mass-dependent analysis.

So, since the resonance state has $J^{PC} = 1^{-+}$, it is manifestly exotic.

This research was supported in part by the US Department of Energy and the Russion State Committee for Science and Technology.

References

1. D.R. Thompson et al., Phys. Rev. Lett. **79**, 1630 (1997).
2. D. Alde et al., Phys. Lett. B **205**, 397 (1988).
3. G.M. Beladidze et al., Phys. Lett. B **313**, 276 (1993).
4. H. Aoyagi et al., Phys. Lett. B **314**, 246 (1993).
5. C. Amsler et al., Phys. Lett. B **333**, 277 (1994).
6. A. Abele et al., Phys. Lett. B **423**, 175 (1998).
7. S. Teige et al., "Properties of the a_0 Meson", submitted to Phys. Rev.
8. S.U. Chung, "Formulas for Partial-Wave Analysis", Brookhaven BNL-QGS-93-05, unpublished (1993).
9. S.U. Chung, Phys. Rev. D **56**, 7299 (1997).
10. J.P. Cummings and D.W. Weygand, "The New BNL Partial Wave Analysis Program", Brookhaven Report BNL-64637, unpublished (1997).
11. S.U. Chung and T.L. Trueman, Phys. Rev. D **11**, 633 (1975).

ISGUR-WISE FORM FACTOR FOR $\Lambda_b \to \Lambda_c + \ell\nu_\ell$ WITHIN A LIGHT-FRONT CONSTITUENT QUARK MODEL

Fabio Cardarelli and Silvano Simula
INFN, Sezione Roma III, Via della Vasca Navale 84, I-00146 Roma, Italy

ABSTRACT

We present an investigation of the Isgur-Wise form factor relevant for the semileptonic decay $\Lambda_b \to \Lambda_c + \ell\nu_\ell$ performed within a light-front constituent quark model. Adopting different baryon wave functions it is shown that the Isgur-Wise form factor depends sensitively on the baryon structure. Using recent model-independent dispersive bounds on the slope and curvature of the baryonic Isgur-Wise function, our prediction for the semileptonic branching ratio results to be $Br(\Lambda_b \to \Lambda_c + \ell\nu_\ell) = (6.9 \pm 1.5)\ \%\ |V_{bc}/0.040|^2\ \tau(\Lambda_b)/(1.14\ ps)$.

1. Introduction

The theoretical analysis of baryons containing a single heavy quark is greatly simplified by the fact that in the limit $m_Q \to \infty$ the interaction among the heavy quark Q and the light diquark system (qq') becomes flavor- and spin-independent, leading to a spin-flavor symmetry known as the Heavy Quark Symmetry (HQS). For the case of our interest, the HQS relates the $\Lambda_b \to \Lambda_c$ transition amplitudes to the $\Lambda_b \to \Lambda_b$ elastic ones, which in turn can be expressed in terms of a universal function[1], called the Isgur-Wise (IW) form factor $\xi(\omega)$, where $\omega = v \cdot v'$ is the product of the initial (v) and final (v') baryon four-velocities. However, except for the normalization condition $\xi(1) = 1$, the HQS cannot predict the full behavior of the IW form factor, for the complete knowledge of the non-perturbative baryon structure is required. Therefore, in order to obtain quantitative predictions, calculations based on lattice QCD simulations, effective approaches and models are necessary. Here we present the results of an investigation of the Λ-type IW form factor $\xi(\omega)$, based on a relativistic constituent quark model formulated on the light-front[2], as well as our prediction for the semileptonic branching ratio $Br(\Lambda_b \to \Lambda_c + \ell\nu_\ell)$.

2. Light-Front Calculation of the Isgur-Wise Form Factor

In the light-front formalism the Λ_Q wave function can be written as: $\Psi^{1/2,s}_{Qqq'} = \sum_{\sigma'_i} \langle \sigma_i | R_M | \sigma'_i \rangle \Phi^{1/2,s}(\{\sigma'_i\})\ w_{(Qqq')}(\vec{p}, \vec{k})$, where R_M is a (generalized) Melosh rotation, $\Phi^{1/2,s}(\{\sigma'_i\}) = \langle 1/2\sigma'_Q, 00 | 1/2s \rangle\ \langle 1/2\sigma'_q, 1/2\sigma'_{q'} | 00 \rangle$ is the canonical spin wave function and $w_{(Qqq')}(\vec{p}, \vec{k})$ is the S-wave radial wave function, being $\vec{p} = \vec{k}_Q$ and $\vec{k} = (m_{q'}\vec{k}_q - m_q\vec{k}_{q'})/(m_q + m_{q'})$ the Jacobian momenta (with $\vec{k}_Q + \vec{k}_q + \vec{k}_{q'} = 0$). We have considered two functional forms for the radial wave function: harmonic oscillator ($h.o.$) and power-law ($p.l.$) ones, viz.

$$w^{(h.o.)}_{(Qqq')}(\vec{p}, \vec{k}) = N_p e^{-|\vec{p}|^2/2\alpha_p^2} \cdot N_k e^{-|\vec{k}|^2/2\alpha_k^2} \quad (1)$$

$$w^{(p.l.)}_{(Qqq')}(\vec{p}, \vec{k}) = \frac{N_p}{(\alpha_p^2 + p^2)^{n_p}} \cdot \frac{N_k}{(\alpha_k^2 + k^2)^{n_k}} \qquad (n_p, n_k \geq 2) \quad (2)$$

The IW form factor $\xi(\omega)$ has been calculated using the wave functions (1) and (2), following the light-front procedure of ref.[2].

3. The Isgur-Wise Form Factor and the Baryon Structure

The baryon size, $\sqrt{< r^2 >_B}$, depends via Fourier transform on the average values of the squared Jacobian momenta $\langle|\vec{p}|^2\rangle$ and $\langle|\vec{k}|^2\rangle$, which are proportional to the parameters α_p^2 and α_k^2 of eqs. (1) and (2). Thus, introducing the parameter $\beta^2 \equiv \langle|\vec{p}|^2\rangle/\langle|\vec{k}|^2\rangle$, we can express the baryon size through two parameters: β and α_k. In the non-relativistic limit the baryon size governs the behavior of the IW function at low recoil. However, relativistic delocalization effects of the light-quark positions become important when $\langle r^2\rangle_B \leq \lambda_q^2 + \lambda_{q'}^2$, where $\lambda_{q(q')} \equiv 1/m_{q(q')}$ is the constituent quark Compton wavelength. Therefore we expect that the IW form factor will become independent of $\langle r^2\rangle_B$, provided that $\langle r^2\rangle_B \leq \lambda_q^2 + \lambda_{q'}^2$. This expectation is fully confirmed by explicit calculations, which show a clear "saturation" effect for $\alpha_k \geq \alpha_{sat} \simeq 0.4\ GeV/c$ at fixed values of β (see also[2]).

Let us now get an insight on the physical meaning of the parameter β. In coordinate space one gets $\beta^2 \simeq \langle|\vec{r}_q - \vec{r}_{q'}|^2\rangle/\langle|\vec{R}_{Q(qq')}|^2\rangle$, where $\vec{r}_q - \vec{r}_{q'}$ is the distance between the two light spectator quarks and $\vec{R}_{Q(qq')}$ is the distance among the center-of-mass of the spectator pair and the heavy quark (sitting on the baryon center-of-mass). The two limiting cases, $\beta << 1$ and $\beta >> 1$, correspond to a diquark-like and collinear-type configurations, respectively (see[2]). In Fig. 1 we show the IW function calculated at various values of β, in the saturation regime, using the wave functions (1) and (2). It can be noted that within the ω-range accessible in the $\Lambda_b \to \Lambda_c \ell \nu_\ell$ decay ($1 \leq \omega \leq 1.44$) $\xi(\omega)$ is almost totally governed by the value of β and independent of the choice of the functional form of the radial wave function. In Fig. 2 our predictions for $\xi(\omega)$ are compared with other existing results; it can be seen that lattice and QCD sum rule results seem to indicate the dominance of collinear-type configurations in the structure of Qqq' baryons.

4. Dispersive Bounds on the Shape of the Isgur-Wise Form Factor

Recently model-independent dispersive bounds for the slope $\rho^2 \equiv -[d\xi(\omega)/d\omega]_{\omega=1}$ and the curvature $2c \equiv [d^2\xi(\omega)/d\omega^2]_{\omega=1}$ of the baryonic IW form factor at the zero-recoil point have been derived in ref.[8]. The theoretically allowed domain in the $\rho^2 - c$ plane is shown in Fig. 3(a) together with our results obtained for various values of β. The figure clearly shows that the dispersive bounds allow only for $\beta \geq 2$, indicating that diquark-like configurations in Λ-type baryons seem to be unlikely.

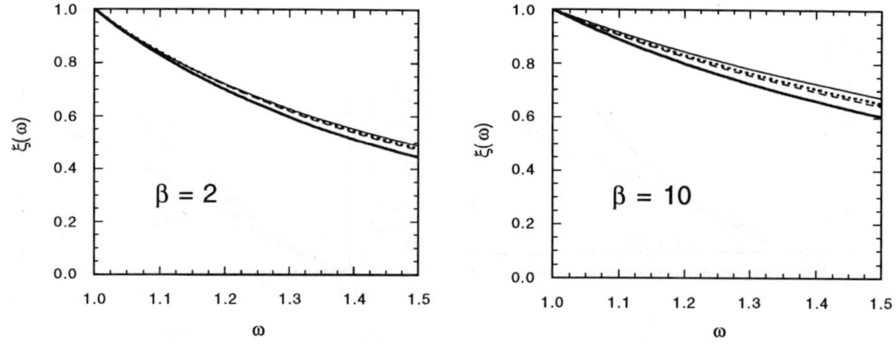

Figure 1. The *IW* form factor calculated at $\beta = 2$ and $\beta = 10$ using different baryon wave functions: *h.o.* (thick solid lines); *p.l.* with $n_p = n_k = 2$, $n_p = n_k = 3$ and $n_p = n_k = 4$ (thin solid, dotted and dashed lines, respectively).

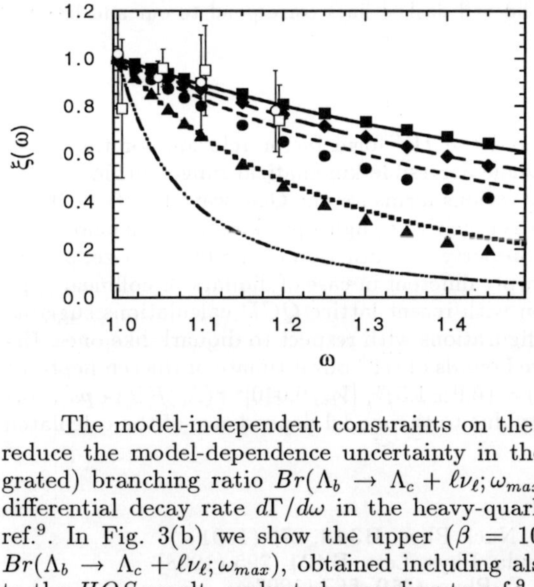

Figure 2. The *IW* function $\xi(\omega)$ calculated for different values of β and compared with other existing results: diquark[3], MIT bag[4] and Skyrme[5] models, QCD sum rule[6] (full triangles, circles, diamonds, squares, respectively). Open dots and squares correspond to the lattice QCD calculations of ref.[7]. The dot-dashed, dotted, dashed, long-dashed and solid lines correspond to $\beta = 0.5, 1, 2, 3$ and 10, respectively. After ref.[2].

The model-independent constraints on the shape of $\xi(\omega)$ can also be used to reduce the model-dependence uncertainty in the evaluation of the (partially integrated) branching ratio $Br(\Lambda_b \to \Lambda_c + \ell\nu_\ell; \omega_{max}) \equiv \tau_{\Lambda_b} \int_1^{\omega_{max}} d\omega d\Gamma/d\omega$, where the differential decay rate $d\Gamma/d\omega$ in the heavy-quark limit can be easily read off from ref.[9]. In Fig. 3(b) we show the upper ($\beta = 100$) and lower ($\beta = 2$) results for $Br(\Lambda_b \to \Lambda_c + \ell\nu_\ell; \omega_{max})$, obtained including also the $1/m_Q$ first-order corrections to the HQS result, evaluated according to ref.[9] and adopting for the heavy baryon binding energy the value $\bar{\Lambda} = 0.75$ GeV from ref.[7]. Our predictions are compared with the allowed range for $Br(\Lambda_b \to \Lambda_c + \ell\nu_\ell; \omega_{max})$ obtained in the recent lattice QCD simulation of ref.[7]. At the physical threshold $\omega_{max} = 1.44$ our estimate is $Br = (6.9 \pm 1.5)$ % $|V_{bc}/0.040|^2\ \tau(\Lambda_b)/(1.14\ ps)$, corresponding to a theoretical uncertainty of only 20 %. It should be noted that if one limits the integration to $\omega_{max} \simeq 1.2$ the uncertainty reduces to $\simeq 10$ %.

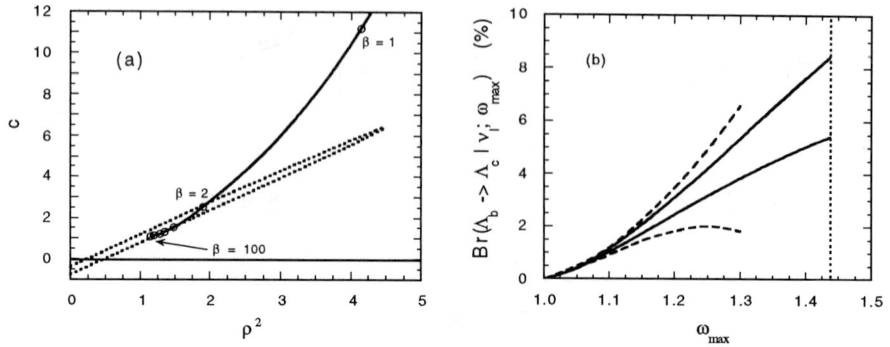

Figure 3. (a) Curvature c vs. slope ρ^2 of the IW function $\xi(\omega)$. The dotted lines identify the model-independent allowed domain determined in ref.[8]. The open dots correspond to our results obtained for various values of β and the solid line is just an interpolation curve. (b) Partially integrated branching ratio $Br(\Lambda_b \to \Lambda_c + \ell\nu_\ell; \omega_{max})$ vs. ω_{max} in units of % $|V_{bc}/0.040|^2\ \tau(\Lambda_b)/(1.14\ ps)$. The solid and dashed lines correspond to our and lattice QCD results[7], respectively (see text).

5. Conclusions

In conclusion, we have calculated the IW form factor relevant for the $\Lambda_b \to \Lambda_c + \ell\nu_\ell$ semileptonic decay in the whole accessible kinematical range within a light-front constituent quark model, using various forms for the Qqq' wave function. It has been shown that the IW form factor is sensitive to light-quark relativistic delocalization effects, leading to a saturation property as a function of the baryon size, as well as to the baryon structure, being sharply different in case of diquark or collinear-type Qqq' configurations. The comparison with recent lattice QCD calculations suggests the dominance of collinear-type configurations with respect to diquark-like ones. Using the model-independent dispersive bounds of ref.[8] our estimate of the semileptonic branching ratio is $Br(\Lambda_b \to \Lambda_c + \ell\nu_\ell) = (6.9 \pm 1.5)\%\ |V_{bc}/0.040|^2\ \tau(\Lambda_b)/(1.14\ ps)$, corresponding to a $\simeq 20\%$ of uncertainty due to the model dependence of the calculated IW form factor.

References

1. N. Isgur and M.B. Wise, Nucl. Phys. **B348**, 276 (1991).
2. F. Cardarelli and S. Simula, Phys. Lett. **B421**, 295 (1998).
3. X.-H. Guo and P. Kroll, Z. Phys. **C59**, 567 (1993).
4. M. Sadzikowski and K. Zalewski, Z. Phys. **C59**, 677 (1993).
5. E. Jenkins, A. Manohar, M.B. Wise, Nucl. Phys. **B396**, 38 (1996).
6. A.G. Grozin and O.I. Yakovlev, Phys. Lett. **B291**, 441 (1992).
7. *UKQCD* collaboration, K.C. Bowler et al., Phys. Rev. **D57**, 6948 (1998).
8. D. Chakraverty, T. De, B. Dutta-Roy and K.S. Gupta, preprint SINP-TNP-98-04, e-print archive hep-ph/9802223.
9. J.G. Korner et al., Prog. Part. Nucl. Phys. **33**, 787 (1994).

Spin Structure of the Nucleon from Semi-Inclusive Deep Inelastic Scattering

Jörg Pretz
Institut für Kernphysik, Universität Mainz, Germany *

on behalf of the Spin Muon Collaboration (SMC)

ABSTRACT

A measurement and analysis of semi-inclusive asymmetries for positively and negatively charged hadrons from deep inelastic scattering of polarised muons on polarised protons and deuterons in the range $0.003 < x < 0.7$ of the Bjorken variable and for a momentum transfer Q^2 >1 GeV^2 is presented. From these asymmetries and the SMC inclusive spin asymmetries we determine the polarised quark distributions of valence quarks, $\Delta u_v(x)$ and $\Delta d_v(x)$, and non-strange sea quarks, $\Delta \bar{q}(x)$, at Q^2=10 GeV^2. Combining all SMC data we find for the first moments $\int_0^1 \Delta u_v(x)\mathrm{d}x = 0.92 \pm 0.11 \pm 0.07$, $\int_0^1 \Delta d_v(x)\mathrm{d}x = -0.53 \pm 0.15 \pm 0.07$ and $\int_0^1 \Delta \bar{q}(x)\mathrm{d}x = -0.02 \pm 0.05 \pm 0.02$, where we assumed $\Delta \bar{u}(x) = \Delta \bar{d}(x) =: \Delta \bar{q}(x)$. We also determine for the first time the second moments of the valence distributions $\int_0^1 x\Delta q_v(x)\mathrm{d}x$ which we find consistent with recent Lattice QCD calculations.

In the non relativistic static quark parton model the contribution of the u and the d quarks to the nucleon spin are $\Delta u = u^\uparrow - u^\downarrow = 4/3$ and $\Delta d = d^\uparrow - d^\downarrow = -1/3$, where $u^\uparrow$ ($u^\downarrow$) is the number of quarks with helicity parallel (anti-parallel) to the nucleon spin. From the axial matrix elements F and D determined in weak baryon decays one gets a first experimental information about the spin structure of the nucleon. Assuming $\Delta s + \Delta \bar{s} = 0$ one finds $\Delta u + \Delta \bar{u} = 0.92 \pm 0.02$ and $\Delta d + \Delta \bar{d} = -0.34 \pm 0.02$ for the first moments. Using in addition information from inclusive polarised muon proton scattering ($\vec{\mu} + \vec{p} \rightarrow \mu' + X$) the European Muon Collaboration (EMC) found the following values: $\Delta\Sigma = \sum_{q=u,d,s}(\Delta q + \Delta \bar{q}) = 0.12 \pm 0.17$ and $\Delta s + \Delta \bar{s} = -0.19 \pm 0.06$.[1] This surprising result, that the helicity contribution of the quarks to the nucleon spin is small, triggered new experiments at CERN, SLAC and DESY. Here the semi-inclusive results of the Spin Muon Collaboration (SMC) at CERN are presented.

To investigate the spin structure of the nucleon, SMC studies deep inelastic scattering of polarized muons off polarised protons and deuterons. In an inclusive measurement, where only the scattered muon is detected ($\vec{\mu} + \vec{N} \rightarrow \mu + X$), the two longitudinal

*Now at Yale University, New Haven, CT

spin asymmetries A_{1p} and A_{1d} are the experimental observables. The observation of a hadron, in addition to the scattered muon ($\vec{\mu} + \vec{N} \rightarrow \mu + h + X$), allows the extraction of further information about the nucleon spin structure from the semi-inclusive spin asymmetry A_1^h, where h is the observed hadron species.

The quark parton model expressions for the inclusive and semi-inclusive asymmetries read:

$$A_1(x) = \frac{\sum_q e_q^2(\Delta q(x) + \Delta\bar{q}(x))}{\sum_q e_q^2(q(x) + \bar{q}(x))}, \quad (1)$$

$$A_1^h(x, z) = \frac{\sum_q e_q^2(\Delta q(x) D_q^h(z) + \Delta\bar{q}(x) D_{\bar{q}}^h(z))}{\sum_q e_q^2(q(x) D_q^h(z) + \bar{q}(x) D_{\bar{q}}^h(z))}, \quad (2)$$

where $\Delta q(x) = q^\uparrow(x) - q^\downarrow(x)$ are the polarised quark distributions and $q(x) = q^\uparrow(x) + q^\downarrow(x)$ are the unpolarised quark distributions. $D_q^h(z)$ are fragmentation functions which represent the probability to produce a hadron of energy fraction $z = E_{had}/\nu$ of the the virtual photon energy ν from a quark q. Here one sees that the inclusive asymmetry $A_1(x)$ provides only information on the sum $(\Delta q(x) + \Delta\bar{q}(x))$, whereas in the expression for the semi-inclusive asymmetry the quarks and anti-quarks have different weights, because in general $D_q^h \neq D_{\bar{q}}^h$. (The probability to produce a π^+ from a u quark for instance is approximately two times larger than the probability to produce a π^+ from a $\bar{u}$ quark in the region $z > 0.2$.) In the SMC spectrometer charged hadrons are detected in addition to the scattered muon. Thus four semi-inclusive asymmetries ($A_{1d}^+, A_{1d}^-, A_{1p}^+, A_{1p}^-$) for positively and negatively charged hadrons on the deuteron and the proton target were measured. These asymmetries are displayed in Fig. 1. To allow an interpretation in the quark parton model a cut on $Q^2 > 1\text{GeV}^2$ and $z > 0.2$ was applied.

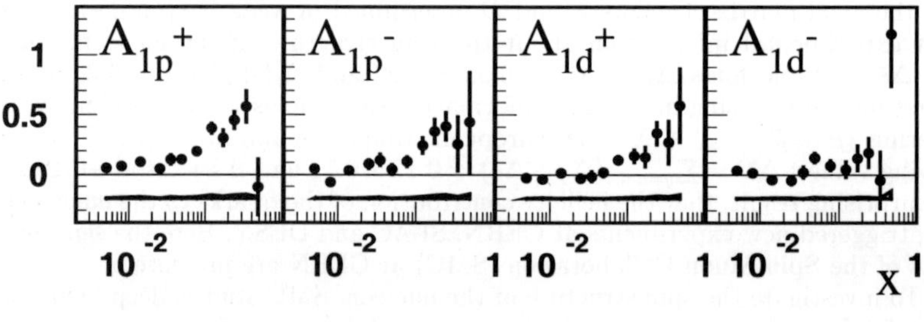

Fig. 1. *Semi-inclusive asymmetries for positive hadrons from the proton, A_{1p}^+, negative hadrons from the proton, A_{1p}^-, positive hadrons from the deuteron, A_{1d}^+, and negative hadrons from the deuteron, A_{1d}^-. The error bars are statistical and the shaded areas represent the systematic uncertainty.*

From these asymmetries and from our inclusive asymmetries we now determine the polarised quark distributions $\Delta q(x)$. The asymmetries in eqs. (1) and (2) are linear combinations of the polarised quark distributions $\Delta q(x)$. Using parametrisations for the unpolarised quark distributions $q(x)$[2] and for the fragmentation functions $D_q^h(z)$[3] at our average $Q^2 = 10$ GeV^2 one can thus extract the polarised quark distributions from the measured asymmetries, which depend only very weakly on Q^2.

We define the polarised valence quark distributions $\Delta u_v(x) = \Delta u(x) - \Delta\bar{u}(x)$ and $\Delta d_v(x) = \Delta d(x) - \Delta\bar{d}(x)$ and assume an $SU_f(3)$ symmetric sea which we denote by $\Delta\bar{q}(x) := \Delta\bar{u}(x) = \Delta\bar{d}(x) = \Delta s(x) = \Delta\bar{s}(x)$. Since we are mainly observing pions the above assumption on $\Delta s(x)$ and $\Delta\bar{s}(x)$ has a negligible influence on the results. The quantity $\Delta\bar{q}(x)$ carries mainly information about the non-strange sea quarks. The results are shown in Fig. 2. The polarised u valence quark distribution, $\Delta u_v(x)$, is positive and the polarisation $\Delta u_v/u_v$ increases with x. The polarised d valence quark distribution, $\Delta d_v(x)$, is negative and the non-strange sea distribution, $\Delta\bar{q}(x)$, is consistent with zero over the measured range of x.

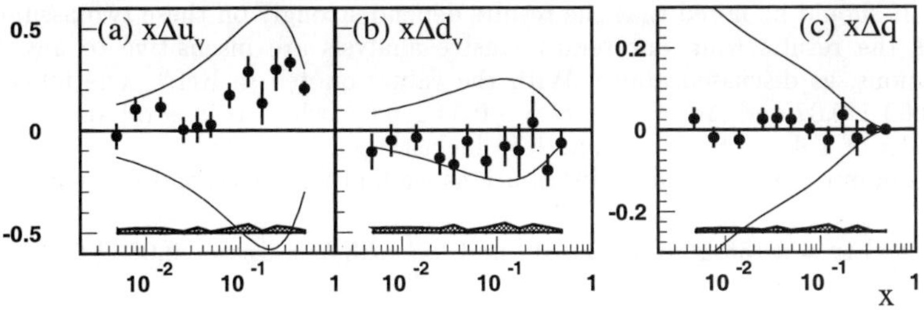

Fig. 2. *The polarised quark distributions $x\Delta u_v(x)$, $x\Delta d_v(x)$ and $x\Delta\bar{q}(x)$ obtained under the assumption $\Delta\bar{u}(x) = \Delta\bar{d}(x)$. The error bars are statistical and the shaded areas represent the systematic uncertainty. The curves correspond to the upper and the lower limits $\pm xq(x)$ from the unpolarised quark distributions[2] evaluated at $Q^2 = 10$ GeV^2. In the last plot the curves are $\pm x(\bar{u}(x) + \bar{d}(x))/2$.*

The results are dominated by the statistical errors. The main contributions to the systematic errors are uncertainties on the assumptions for the unmeasured fragmentation functions and the uncertainties on the target and beam polarisations.

To calculate the first moments $\int_0^1 \Delta q(x)dx$ extrapolations for the unmeasured x regions $0 < x < 0.003$ and $0.7 < x < 1$ are needed. The large x extrapolation is negligible because the upper limit given by the unpolarised quark distribution is small. The low x extrapolation was estimated both from fits to our data and also using parametrisations of the polarised quark distributions from QCD analysis of inclusive data.[4,5] The contributions to the first moments from different x regions are detailed in Tab. 1.

Polarised semi-inclusive deep inelastic scattering is presently a unique tool to measure the polarised valence and sea quark distributions. However, the corresponding

x	$0-0.003$	$0.003-0.7$	$0-1$
Δu_v	0.04 ± 0.04	$0.73 \pm 0.10 \pm 0.07$	$0.77 \pm 0.10 \pm 0.08$
Δd_v	-0.05 ± 0.05	$-0.47 \pm 0.14 \pm 0.08$	$-0.52 \pm 0.14 \pm 0.09$
x	$0-0.003$	$0.003-0.3$	$0-1$
$\Delta\bar{q}$	0.01 ± 0.02	$0.01 \pm 0.04 \pm 0.03$	$0.01 \pm 0.04 \pm 0.03$

Table 1. *The values for the extrapolations to low x (first column), the integrals over the measured range (second column) and the first moments (third column). In the last two columns the first error is statistical, the second systematic.*

first moments can be determined by other methods. With the assumption of a $SU(3)_f$ symmetric sea ($\Delta\bar{u}(x) = \Delta\bar{d}(x) = \Delta s(x) = \Delta\bar{s}(x)$) and $SU(3)_f$ symmetry for the weak decays in the baryon octet ($\Delta\bar{u}_p = \Delta\bar{s}_{\Xi^0}$, ...), the first moments of the valence distributions can be obtained from the axial matrix elements, F and D, of baryon decays. It should be noted that the results depend strongly on these two assumptions, whereas the results from our semi-inclusive analysis are insensitive to any $SU(3)_f$ assumptions, as discussed above. With the values quoted in Ref.[6,7] one finds $\Delta u_v = 2F = 0.92 \pm 0.02$ and $\Delta d_v = F - D = -0.34 \pm 0.02$, which is in good agreement with our results of Tab. 1 obtained from deep inelastic scattering only.

From our polarised quark distributions we find for the second moments:

$$\int_0^1 x\Delta u_v(x)\mathrm{d}x = 0.155 \pm 0.017 \pm 0.010 \quad \text{and} \quad \int_0^1 x\Delta d_v(x)\mathrm{d}x = -0.056 \pm 0.026 \pm 0.011\,, \tag{3}$$

where the first error is statistical and the second is systematic. Due to the additional factor x the low x extrapolations to the unmeasured region give a negligible contributions while the large x contributions remain small. A calculation of the second moments in lattice QCD[8] predicts $\int_0^1 x\Delta u_v(x)\mathrm{d}x = 0.189 \pm 0.008$ and $\int_0^1 x\Delta d_v(x)\mathrm{d}x = -0.0455 \pm 0.0032$, which is in good agreement with our measured values. The SMC results discussed in these proceedings are published in Ref. 9.

References

1. EMC, a. Ashman et al., Phys. Lett. B 206 (1988) 364
2. M. Glück et al., Z. für Physik, C 67 (1995) 433.
3. EMC, A. Arneodo et al., Nucl. Phys. B 321 (1989) 541.
4. M. Glück et al., Phys. Rev. D 53 (1996) 4775.
5. T. Gehrmann, W.J. Stirling, Phys. Rev. D 53 (1996) 6100.
6. R.M.Barnett et al., 'Review of Particle Properties', Phys. Rev. D54 (1996) and off-year update for the 1998 edition (URL: http://pdg.lbl.gov/).
7. F.E. Close and R.G. Roberts, Phys. Lett. B316, (1993). 165
8. M. Göckeler et al. Phys.Lett. B 414 (1997) 340
9. SMC, Adeva et al., Phys. Lett. B 420 (1998) 180

BARYON MASSES AND MOMENTS IN A LOOP EXPANSION WITH A FORM FACTOR

Phuoc Ha and Loyal Durand
Physics Department, University of Wisconsin-Madison
Madison, Wisconsin 53706, USA

ABSTRACT

We study meson loop corrections to the baryon magnetic moments and masses using an approach in which the baryons are treated as composite particles, with the compositeness modeled using a single form factor. Our study of the baryon magnetic moments starts from a QCD-based quark model we derived earlier in a quenched approximation to QCD. For the baryon masses, we start from chiral perturbation theory. Our approach works well in both cases, improving both the agreement between theory and experiment, and the convergence of the perturbative loop expansion.

1. Motivation

We showed recently[1] that the naive quark model (QM) for the baryon moments, and corrections for the binding of the quarks, can be derived from QCD in a Wilson-loop approach that uses well-defined approximations such as the "quenched" approximation and the minimal area law. This extended QM failed to improve the agreement between experiment and theory for the baryon moments. We investigated the problem carefully, and attributed the difficulties to the use of the quenched approximation. A first step in the improving the model is to get beyond that approximation by taking meson loop corrections into account. With this goal, we developed an loop expansion approach for the QCD-based QM in which the composite structure of the baryons is modeled by introducing a form factor, and used this approach to study the baryon moments and masses.[2–4] The results we obtained are remarkably good.

Our approach has implications for heavy-baryon chiral perturbation theory (ChPT). This is generally a useful approach for studying low-momentum processes involving mesons and baryons at the tree level, but encounters problems with the convergence of the perturbation series when loop corrections are considered. In recent work, Donoghue and Holstein[5] and Donoghue, Holstein, and Borasoy[6] proposed to use a momentum-space cutoff to suppress the unreliable short-distance contributions from the loops, and found improved behaviour of ChPT. Our approach using the form factor is similar except that we regard the baryons as physically composite particles. A natural cutoff in momentum space follows from this compositeness.

We have found that the masses in the baryon octet and decuplet are very well described in a chiral loop expansion with form factors.[4] The form factors suppress unphysical short-distance effects associated with the treatment of the hadrons as point particles in standard ChPT, and lead to a controllable perturbation theory. We find that pion loops are as important as kaon or eta loops, rather than being strongly suppressed in standard ChPT. This would be expected when only intermediate- and long-distance contributions are retained. We also find that the contributions of decuplet intermediate states are important in the calculation of the masses and moments, and conclude that those states must be included explicitly in a consistent theory.

2. Wave function effects and the form factor

In our approach, the baryons are composite states and can absorb only limited recoil momentum while remaining in their ground states. This must be taken into account in a dynamical model. In the absence of a detailed theory of the interactions of composite mesons and baryons, we have modeled the compositemess or wave function effects using a form factor at each meson-baryon vertex. In keeping with the heavy baryon picture, we have defined the form factor in the rest frame of the baryon where it can depend only on $\mathbf{k^2}$, the square of the three momentum of the meson.

We have chosen for simplicity to use an one-parameter form factor

$$F(\mathbf{k}) = \frac{\lambda^2}{\lambda^2 + \mathbf{k}^2} \tag{1}$$

normalized at the chiral limit, $\mathbf{k} = 0$. The parameter λ characterizes the natural momentum scale, expected to be much below 1 GeV. With the introduction of this form factor, all the loop diagrams are finite, and no arbitrary subtractions need to be introduced into the theory.

We summarize the results of our form factor approach to the baryon moments and masses here. Details are given elsewhere.[2,4]

3. Applications

3.1. *Baryon moments from a QCD-based QM*

We have derived expressions for the baryon moments from our QCD-based QM with loop corrections[2,3] and used those to fit the experimental octet baryon magnetic moments. We emphasize that all the parameters that appear except for the effective quark moments μ_u, μ_s, and the cutoff λ are known independently. For the loop corrections, we have chosen the coupling constants F, D, $\mathcal{C}$, and $\mathcal{H}$ such that $F + D = 1.25 \approx |\,g_A/g_V\,|$, and the $SU(6)$ relations between the coupling constants $F = 2D/3, \mathcal{C} = -2D$, and $\mathcal{H} = -3D$ are satisfied as expected for $L = 0$ baryons. We have also chosen the decuplet-octet mass difference as $\delta = 250$ MeV, and use $f_\pi = 93$ MeV, $f_K = f_\eta = 1.2 f_\pi$. The fit gives $\lambda = 407$ MeV.

The results of an equal-weight least-squares fit to the seven well-measured octet moments using the three free parameters μ_u, μ_s, and λ are given elsewhere.[2] We find a root-mean-square deviation of the predicted values from the observation of 0.055 μ_N. This is a substantial improvement on the naive quark model, which gives an average deviation of 0.12 μ_N. The transition moment $\mu_{\Sigma^0\Lambda}$ was not included in the fit, but was left as a prediction. The result obtained, $\mu_{\Sigma^0\Lambda} = 1.559\ \mu_N$, differs from the experimental value, $\mu_{\Sigma^0\Lambda} = 1.610 \pm 0.08\ \mu_N$, by 0.05 μ_N, a value within the experimental uncertainty, and one that does not affect the overall fit.

We have also calculated the decuplet baryon magnetic moments using the values of μ_u, μ_s, and λ obtained in fitting the octet moments.[3] Our *prediction* for the Ω^- moment, $\mu_{\Omega^-} = -1.97\ \mu_N$, is in striking agreement with the measured value $-2.02 \pm 0.05\ \mu_N$, and is much better than the prediction of the QM, $\mu_{\Omega^-} = -1.74\ \mu_N$. The theoretical value of $\mu_{\Delta^{++}} = 5.69\mu_N$ falls within the experimental range of 3.7–7.5 μ_N. The other decuplet moments are not known.

A detailed breakdown of the contributions of the loop integrals to the magnetic moments of the octet and decuplet baryons shows that the loop contributions are small relative to the leading QM contributions, suggesting reasonably rapid convergence of the loop expansion. This is in marked contrast to the results obtained in ChPT, where the loop contributions calculated using dimensional regularization are comparable in size to the leading terms in the chiral expansion.[8] We also find that the binding corrections $\Delta\mu_i^{QM}$ found in the QCD-based derivation[1] of the QM are important. It is not possible to obtain as good a fit to the data when these are omitted.

A further interesting point involves the importance of the decuplet intermediate states for the octet moments. We find that that the contributions from the graphs involving decuplet states are substantial. For most of the baryons, those contributions are larger than those from the graphs which involve only the intermediate octet states. This result is insensitive to the value used for the octet-decuplet mass difference. We conclude that the decuplet must be regarded as a set of light baryon states as in the QM, and not as a set of heavy states to be approximated in a chiral momentum expansion.

3.2. Baryon masses in ChPT with form factor

In this problem, we start in the framework of with heavy baryon chiral perturbation theory.[7] The expressions for baryon masses in ChPT with a form factor of the type shown in Eq. (1) are given in our recent paper.[4] We now have total of five parameters to fit the average masses of the eight SU(2) baryon multiplets. Using the same values of the strong coupling constants and the cutoff λ as in the baryon moment problem, we obtain an excellent fit to the masses, with an average deviation from the experimental values of the mean multiplet masses of 0.8 MeV, a deviation significantly smaller than the mass splittings within the multiplets. A refitting of the cutoff parameter gives a best fit with an average deviation of theory from experiment of 0.5 MeV for $\lambda = 443$ MeV.

A more detailed analysis[4] shows that the loop corrections, while substantial, are still small in comparison to the leading contributions to the masses, with the total loop contributions ranging from 23 to 30 percent of the average tree-level masses. This suggests reasonably rapid convergence for the loop expansion. The loop corrections lower the baryon masses, as expected in second order perturbation theory. We find that pion loops, which are negligible in standard ChPT, are quite important as would be expected in a calculation dominated by intermediate- to long-range contributions in the loops. These results are in marked contrast to the results obtained in ChPT for point baryons, where the loop contributions calculated using dimensional regularization are comparable in size to the leading terms, there is little indication that the perturbation series converges, and contributions from pion loops are strongly suppressed.

We again find that the decuplet states must be treated as light on the scale of the octet, and included explicitly. In particular, the contributions of decuplet intermediate states in loops to the octet masses are as important as the contributions of octet intermediate states.

Finally, we emphasize that the excellent overall fits to the baryon masses do not give as sensitive a test of the loop corrections as might be expected. The loop contributions have approximately the same structure as the tree-level masses, and are largely absorbed by readjustment of the tree-level parameters. More sensitive tests are given by the deviations from the usual Gell-Mann–Okubo and decuplet mass relations, which hold identically at tree level. We find that the corrected Gell-Mann–Okubo relation is satisfied within the experimental uncertainties by the measured octet masses, and that the corrected decuplet mass relations are satisfied reasonably well.

4. Conclusions

- It is essential to consider the composite structure of the baryons in calculations of moments, masses, etc., as this structure suppresses the unreliable short-distance contributions in loops obtained when the hadrons are treated as point particles. By modeling the compositeness using a form factor, we obtain a theory for the long-distance physics that appears to incorporate the basic physics and describes low-energy processes involving the baryons very well.
- The loop corrections obtained in this model are small, and the perturbation series appears to be under control.
- In contrast to standard ChPT, we find in our approach that pion loops are very important, with contributions comparable to those from kaon and eta loops.
- The contributions of intermediate decuplet states to masses and moments are very important, of the same order as the contributions from intermediate octet states, and the octet states must be considered explicitly.

References

1. L. Durand and P. Ha, Proceedings of the Como Conference on Quark Confinement and Hadron Spectrum II, Eds. N. Brambilla and G.M. Prosperi (World Scientific, Singapore, 1996); hep-ph/9609495.
2. P. Ha and L. Durand, hep-ph/9804382, Phys. Rev. D **58** (in press).
3. P. Ha, hep-ph/9804383, Phys. Rev. D **58** (in press).
4. P. Ha and L. Durand, hep-ph/9807295, submitted to Phys. Rev. D.
5. J. Donoghue and B.R. Holstein, hep-ph/9803312.
6. J. Donoghue, B.R. Holstein, and B. Borasoy, hep-ph/9804281.
7. H. Georgi, Phys. Lett. **B240**, 447 (1990). E. Jenkins and A.V. Manohar, Phys. Lett **B255**, 558 (1991); *ibid.* **259**, 353 (1991); UCSD/PTH 91-30.
8. L. Durand and P. Ha, hep-ph/9712492, Phys. Rev. D **58**, 013010 (1998).

PROPERTIES OF THE B$_c$ SYSTEM

Lewis P. Fulcher
Department of Physics and Astronomy, Bowling Green State University
Bowling Green, Ohio 43403, USA
email: fulcher@newton.bgsu.edu

ABSTRACT

I summarize a comprehensive account of the energies, splittings and electromagnetic decays of the low-lying states of the bottom charmed meson system. Richardson's potential is used to include running coupling constant effects in the central potential, and the full radiative one-loop expressions derived by Pantaleone, Tye and Ng are used for the spin-dependent potentials. Our predicted result for the ground state energy is 6286^{+15}_{-6} MeV. The ground state lifetime is found to be $\tau_{B_c} = 0.36 \pm 0.05\, ps$. Both of these numbers are consistent with the measured values recently announced by the CDF collaboration at Fermilab.

1. Introduction

The CDF collaboration[1] at the Fermilab Tevatron has recently reported the discovery of B_c mesons. Their announced value for the mass of the ground state is $6400 \pm 390 \pm 130\, MeV$, and their value for the ground state lifetime is $\tau_{B_c} = 0.46^{+0.18}_{-0.16} \pm 0.03\, ps$. This state should be one of a number of relatively stable states below the threshold for the emission of B and D mesons, and thus I want to present some results from a comprehensive calculation[2] of the energies, splittings and electromagnetic decay rates of the low-lying states of the B_c system.

In 1991 Kwong and Rosner[3] predicted the masses of the two lowest states of this system from an empirical mass formula and a logartihmic potential. They obtained a range for the ground state, $6194 \leq M_{B_c} \leq 6292 MeV$. Eichten and Quigg[4] gave a more comprehensive account of the properties of the B_c system. Using several potential models to obtain error estimates, they quoted 6258 ± 20 MeV for the ground state energy. In anticipation of the experimental interest in the B_c system, there have been a number of other potential model calculations.[5]

2. Richardson's potential and the spin-dependent potentials

Richardson's potential[6] contains a confining potential and a short range piece,

$$V(r) = Ar - \frac{8\pi f(\Lambda r)}{(33 - 2n_f)r}\ , \ f(t) = \frac{4}{\pi}\int_0^\infty \frac{\sin tx}{x}\left[\frac{1}{\ln(1+x^2)} - \frac{1}{x^2}\right] dx, \quad (1)$$

where the string constant will be treated as a free parameter instead of a function of Λ in order to achieve a separation of the long and short distance parameters.[7]

To specify the spin-dependent potentials, we will use Pantaleone, Tye and Ng's (PTN) generalization[8] of the Eichten-Feinberg formalism. Thus

$$V_{SD} = \frac{\mathbf{L}\cdot\mathbf{S}_1}{2m_1^2}\left[\frac{1}{r}\frac{d}{dr}(E(r) - Ar + 2V_1) + 2V_5\right] + \frac{\mathbf{L}\cdot\mathbf{S}_2}{2m_2^2}\left[\frac{1}{r}\frac{d}{dr}(E(r) - Ar + 2V_1) - 2V_5\right]$$
$$+\frac{\mathbf{L}\cdot(\mathbf{S}_1+\mathbf{S}_2)}{m_1 m_2}\frac{1}{r}\frac{dV_2}{dr} + \frac{\mathbf{L}\cdot(\mathbf{S}_1-\mathbf{S}_2)}{m_1 m_2}V_5 + \frac{(\mathbf{S}_1\cdot\hat{\mathbf{r}}\,\mathbf{S}_2\cdot\hat{\mathbf{r}} - \frac{1}{3}\mathbf{S}_1\cdot\mathbf{S}_2)}{\mathbf{m_1 m_2}}V_3 + \frac{\mathbf{S}_1\cdot\mathbf{S}_2}{3m_1m_2}V_4,$$

where the component potentials E, V_1, V_2, V_3, V_4 and V_5, include both the tree-level and the full radiative one-loop contributions. These may be expressed in terms of the strong coupling constant $\alpha_{\bar{S}}$, the renormalization scale μ and the constituent masses. As an example, we list the second spin-orbit potential,

$$V_2(r) = -\frac{4\alpha_{\bar{S}}}{3r}\left[1 + \frac{\alpha_{\bar{S}}}{6\pi}\left[(33 - 2n_f)(\ln \mu r + \gamma_e) + \frac{39}{2} - \frac{5n_f}{3} - 9(\ln \sqrt{m_1 m_2} r + \gamma_E)\right]\right], \quad (2)$$

where γ_E is Euler's constant. All of the PTN component potentials are given in the modified minimal subtraction scheme. The central potential parameters are assumed to be flavor independent. These parameters and the constituent masses are determined from the low-lying states of charmonium and the upsilon system. Their values are

$$A = 0.152\,\mathrm{GeV}^2,\ \Lambda = 0.431\,\mathrm{GeV},\ m_b = 4.889\,\mathrm{GeV},\ m_c = 1.476\,\mathrm{GeV}, \quad (3)$$

and n_f is taken to be 3. For the upsilon system the value of the coupling constant $\alpha_{\bar{S}} = 0.30$, and the value of the renormalization scale $\mu = 1.95\,\mathrm{GeV}$. For charmonium these values are $\alpha_{\bar{S}} = 0.486$ and $\mu = 0.80\,\mathrm{GeV}$. In both cases the universal QCD scale $\Lambda_{QCD} = 0.190\,GeV$, a value consistent with other determinations.[9] These parameters give an average deviation of 4.3 MeV for the upsilon states below the continuum threshold and an average deviation of 19.9 MeV for the corresponding charmonium states. The one-loop potentials give a substantial advantage in simultaneously accounting for the fine structure and the hyperfine structure of charmonium and the upsilon system. Thus, it is reasonable to expect they should produce better results for the spin-dependent effects in the B_c system.

3. Results and conclusions

Since the central potential parameters are assumed to be strictly flavor independent and the constituent masses are not allowed to run, the only decision that one has to make to treat the B_c system is to choose a value for $\alpha_{\bar{S}}$. We choose $\alpha_{\bar{S}} = 0.393$, the average of the values for charmonium and the upsilon system. Then preserving the value of the universal QCD scale requires that $\mu = 1.12\,GeV$. Our calculated results for the three lowest levels of the B_c system are shown in Figure 1, where they are compared with earlier potential model calculations[4,5] and recent lattice calculations.[10] The 190 MeV error associated with uncertainty of the overall energy scale of the lattice calculation is not shown. In Figure 1 FUII98 denotes the one-loop results and the FUI98 results are based on tree-level expressions for the spin-dependent potentials, and one can see substantial differences for both the S and the P states. The FUI98 results are very close to the results of Eichten and Quigg, as expected.

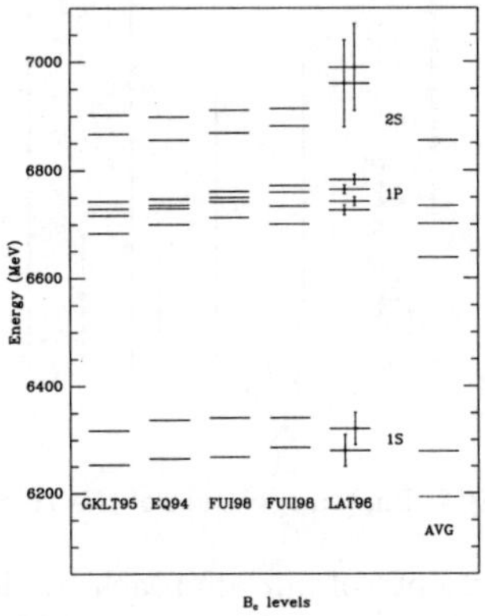

Fig. 1: Low-lying levels of the B_c system. The averages are computed from charmonium and upsilon levels.

An important difference between the spectrum of the B_c system and those of charmonium and the upsilon system arises because the total spin $\mathbf{S} = \mathbf{S}_1+\mathbf{S}_2$ no longer commutes with the Hamiltonian. Thus one must compute a mass-mixing matrix for P states, and it convenient to use the j-j coupling scheme[2,4] for this purpose. After computing the matrix representatives of each of the spin-dependent operators in the spin-dependent potential, one finds that the $J = 2$ states and the $J = 0$ state are the same as in the L-S basis, but that the $J = 1$ states are mixtures. Our result for the lowest $J = 1$ state is

$$\psi_{1m}(1^+) = 0.118\psi_{1m}(3/2, 1/2) + 0.993\psi_{1m}(1/2, 1/2), \tag{4}$$

which is very near the j-j coupling limit, a result consistent with the lattice calculations.[10] In this regard we differ from Eichten and Quigg,[4] whose tree-level results were much closer to the L-S limit. This difference is reflected in the character of the spectrum of photons following the dipole radiative decays of the 1P state. Assuming an equal distribution of initial populations, our results for the photon spectrum of this state contains 6 distinct lines, as shown in Figure 2, instead of the 4 one would expect in the L-S limit.

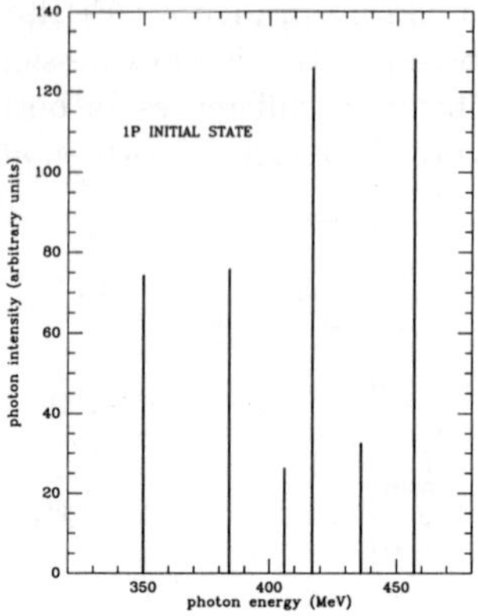

Fig. 2: Photon spectrum of the 1P state.

To estimate the uncertainty of our approach, we have allowed α_S to vary from its charmonium value (0.486) to its upsilon value (0.30). Holding $\Lambda_{QCD} = 0.190\,GeV$ allows an error estimate of the ground state, that is, $M_{B_c} = 6286^{+15}_{-6}\,MeV$. In order to determine the lifetime of the ground state, we have considered $\bar{b}$-quark decay, where the charmed quark is a spectator, and c-quark decay, where the the antibottom quark is a spectator, and the annihilation channels reached by the B_c decay constant. Thus $\Gamma_{tot}(B_c) = 1.85 \pm 0.24 \times 10^{-3}\,eV$, where most of the error arises from the uncertainty[9] in the Kobayashi-Maskawa matrix element V_{bc}.

References

1. CDF collaboration, F. Abe *et al.*, submitted to Phys. Rev. D, available as lanl eprint hep-ex/9804014.
2. submitted to Phys. Rev. D, available as a lanl eprint hep-ph/9806444.
3. W. Kwong and J. Rosner, Phys. Rev. D**44**, 212 (1991).
4. E. Eichten and C. Quigg, Phys. Rev. D**49**, 5845 (1994).
5. S. Gershtein, V. Kiselev, A. Likhoded and A. Tkabladze, Phys. Rev. D **51**, 3613 (1995); Y. Chen and Y. Kuang, Phys. Rev. D **46**, 1165 (1992); V. Galkin, A. Mishurov and R. Faustov, Sov. J. Nucl. Phys. **55**, 1207 (1992).
6. J. Richardson, Phys. Lett. **82B**, 272 (1979).
7. L. Fulcher, Phys. Rev. D **44**, 2079 (1991).
8. J. Pantaleone, S. Tye and Y. Ng, Phys. Rev. D **33**, 777 (1986).
9. Particle Data Group, R. Barnett *et al.*, Phys. Rev. D **54**, 1 (1996).
10. C. Davies, K. Hornbostel, G. Lepage, A. Lidsey, J. Shigemitsu and J. Sloan, Phys. Lett. B **382**, 131 (1996).

PHENOMENOLOGY OF SCALAR MESONS

Leonard Leśniak
H. Niewodniczański Institute of Nuclear Physics,
PL 31-342 Kraków, Poland

ABSTRACT

A separable potential model of three coupled channels $\pi\pi$, $K\overline{K}$ and an effective $2\pi 2\pi$ has been constructed and applied in the analysis of the isoscalar S-wave $\pi\pi$ and $K\overline{K}$ phase shifts and inelasticities. A relatively narrow scalar resonance of mass 1400-1460 MeV has been found. This resonance is consistent with the recent CERN observation of an eventual scalar glueball at 1500 MeV. Other resonances at lower energies like a wide σ meson and a narrow $f_0(980)$ have been confirmed.

Scalar meson spectroscopy is full of open questions. The internal structure of scalars is also very controversial, for example the $f_0(980)$ state is interpreted as $q\bar{q}, q\bar{q}q\bar{q}$ or $K\overline{K}$ quasibound state. Mixing of scalar mesons with scalar glueballs is possible; for example the $f_0(1500)$ resonance, observed by the Crystal Barrel Collaboration in $\bar{p}p$ annihilation at CERN, was interpreted as a candidate for the lowest scalar glueball.[1] The phenomenology of scalar mesons must include studies of interactions between different mesons produced in the decay channels of scalar mesons. In Ref.[2] the scalar meson spectrum was studied in terms of a relativistic $\pi\pi$ and $K\overline{K}$ coupled channel model from the $\pi\pi$ threshold up to 1400 MeV. The phenomenological parameters were constrained by fitting the S-wave data extracted from the experimental cross sections on the $\pi^+\pi^-$ production by π^- scattering on unpolarized hydrogen target[3] and further constraints were imposed by the $K\overline{K}$ phase shift analysis of Ref.[4] .

Recently, authors of Ref.[5] have analysed data obtained on a polarized target by the CERN-Cracow-Munich group for the $\pi^- p_\uparrow \to \pi^+\pi^- n$ reaction. Separation of the π and a_1 exchange amplitudes in this reaction was then possible for the first time. From a set of four solutions for the isoscalar S-wave phase shifts up to 1600 MeV, two of them ("down-flat" and "up-flat") satisfy the unitarity constraint. The "down-flat" solution is in good agreement with the former solution of Ref.[3] up to 1400 MeV. Above 1400 MeV one observes an increase of the phase shifts and larger inelasticities than those seen in [3] . This could be a manifestation of the presence of scalar mesons $f_0(1370)$ or $f_0(1500)$ in that energy range. There a strong four–pion production has been observed in different experiments together with an evidence of clustering into $\sigma\sigma$ or $\rho\rho$ pairs.

In Ref.[6] we have extended the isospin 0 S–wave relativistic 2–channel model of Ref.[2] by adding to its $\pi\pi$ and $K\overline{K}$ channels an effective third coupled channel, here called $\sigma\sigma$. The reaction amplitudes T satisfy a system of the coupled channel Lippmann-Schwinger equations:

$$T = V + VGT, \tag{1}$$

where V, G and T are 3×3 matrices, V is the interaction matrix and G is the diagonal matrix of channel propagators:

$$G_j(E, k_j) = \frac{1}{E - 2E_j(k_j) + i\epsilon}, \quad \epsilon \to 0(+), \quad j = 1, 2, 3. \tag{2}$$

In Eq. (2) E is the total energy, $E_j = \sqrt{m_j^2 + {k_j}^2}$ and m_j is the meson mass in channel j. We consider meson pairs of same mass and momentum k_j in their centre of mass system. In this way 9 meson-meson reactions are simultaneously described by Eq.(1):

$$\begin{pmatrix} \pi\pi \to \pi\pi & \pi\pi \to K\overline{K} & \pi\pi \to \sigma\sigma \\ K\overline{K} \to \pi\pi & K\overline{K} \to K\overline{K} & K\overline{K} \to \sigma\sigma \\ \sigma\sigma \to \pi\pi & \sigma\sigma \to K\overline{K} & \sigma\sigma \to \sigma\sigma \end{pmatrix}. \tag{3}$$

We choose a separable form of the interaction:

$$< p|V_{ij}|q >= \sum_{\alpha=1}^{n} \lambda_{ij,\alpha}\ g_{\alpha,i}(p)\ g_{\alpha,j}(q), \quad i\,, j = 1, 2, 3, \tag{4}$$

where $\lambda_{i\,j,\alpha}$ are coupling constants and

$$g_{\alpha,j}(p) = \sqrt{\frac{4\pi}{m_j}} \frac{1}{p^2 + (\beta_{\alpha,j})^2} \tag{5}$$

are form factors which depend on the relative centre of mass meson momenta p in the final channel or q in the initial channel. In the $\pi\pi$ channel ($j = 1$) we choose a rank-2 separable potential ($n = 2$) and in the other channels a rank-1 potential ($n = 1$). Altogether, this model has 13 parameters: 9 coupling constants $\lambda_{i\,j,\alpha}$ and 4 range parameters $\beta_{\alpha,j}$.

The model is unitary. The diagonal S- matrix elements are parametrized as

$$S_{jj} = \eta_j e^{2i\delta_j}, \quad j = 1, 2, 3, \tag{6}$$

where η_j and δ_j are channel j inelasticities and phase shifts, respectively. Expressions for nondiagonal elements can be found in Ref.[7] . Some of the S-matrix poles in the complex energy plane can be interpreted as resonances. We fit the existing experimental results on the $\pi\pi$ S-wave isoscalar phase shifts together with inelasticity in the $\pi\pi$ channel and with the $K\overline{K}$ phase shifts.

Below 600 MeV we have used data from the K_{e4} decay and above 600 MeV the "down-flat" and "up-flat" solutions of the analysis of Ref.[5] on a polarized target. In addition we have used the results of the analysis of reactions $\pi^- p \to K^+ K^- n$ and $\pi^+ n \to K^- K^+ p$, although targets were unpolarized there[4] . Four fits for $\delta_{\pi\pi}$ and $\eta_{\pi\pi}$

are compared to the experiment in Fig. 1. Above 1400 MeV both "down-flat" and "up-flat" data indicate a decrease towards small values of η ($\eta \approx 0.6$ to 0.7), albeit with large errors. One can notice a better agreement with data of the 3–channel model in comparison with the 2–channel one. The main difference between the 2– and 3–channel fits lies in η above 1400 MeV, where the opening of the $\sigma\sigma$ channel leads to a fast decrease of inelasticity parameters. In order to achieve this behaviour, couplings between the $\pi\pi$ and $\sigma\sigma$ or $K\overline{K}$ channels should be sizable. Fits of similarly good quality were obtained with very different physical parameters in the $K\overline{K}$ and $\sigma\sigma$ channels. Lack of a sufficient number of observables and/or experimental precision, in particular in the effective 4π channel, leads to the existence of several good sets of model parameters with quite different channel and interchannel interactions.

We have studied positions of the S–matrix poles in the complex energy plane ($E = M - i\Gamma/2$). These positions have not been parameters of our fits. In our model we have not included any kind of arbitrary background to fit the data. We have only fitted the meson-meson interaction parameters and then analyzed the analytical structure of the reaction amplitudes. At low energy we find a very broad $f_0(500)$ resonance (also called σ meson) of a width between 510 and 560 MeV. In the 3-channel fits the $f_0(980)$ resonance is seen in a vicinity of the $K\overline{K}$ threshold with a width of about 60 to 70 MeV. The relatively narrow state $f_0(1400)$ appears in the 3–channel fits. Its mass varies from about 1400 MeV to 1460 MeV and the width is about 150 MeV. This width is close to values found by the Crystal Barrel Group (Ref.[1]), however the resonance masses obtained in our fits are smaller than their values of about 1500 MeV.

Finally, let us mention an important difference between the fits presented here for data taken on a polarized target in comparison with those performed with data[3] obtained on a nonpolarized target. If we fit the data of Ref.[3] using our 3–channel model, we obtain a very wide resonance at $M = 1521$ MeV of width 503 MeV. This means that the recent analysis of data[5] supplies new information on the $f_0(1400)$ meson.

In conclusion, we have analysed the isoscalar S–wave $\pi\pi$ and $K\overline{K}$ scattering using the CERN–Cracow–Munich data[5] in the framework of the 2- and the 3–channel models of meson–meson scattering. All fits of the phase shifts analysis[5] indicate presence of a relatively narrow scalar resonance of mass 1400 – 1460 MeV. This resonance is quite compatible with recent observations of a possible scalar glueball at 1500 MeV with a width of 100 MeV.

This work has been performed in the framework of the IN2P3 – Polish Laboratories Convention (project No 93-71).

References

1. A. Abele *et al.*, Phys. Lett. **B385** (1996) 425.
2. R. Kamiński, L. Leśniak and J.-P. Maillet, Phys. Rev. **D50**, 3145 (1994).
3. G. Grayer *et al.*, Nucl. Phys. **B75**, 189 (1974).
4. D. Cohen *et al.*, Phys. Rev. **D22** (1980) 2595.
5. R. Kamiński, L. Leśniak and K. Rybicki, Z. Phys. **C74**, 79 (1997).
6. R. Kamiński, L. Leśniak and B. Loiseau, Phys. Lett. **B413**, 130 (1997).
7. L. Leśniak, Acta Phys. Pol. **B27**, 1835 (1996).

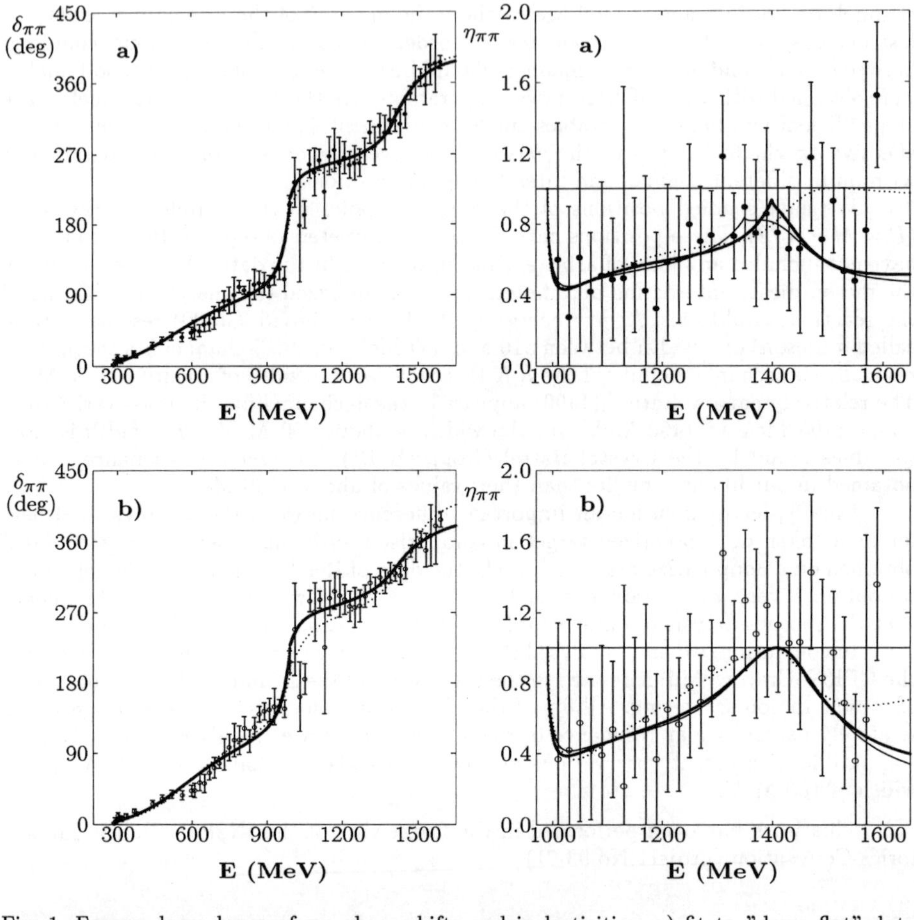

Fig. 1. Energy dependence of $\pi\pi$ phase shifts and inelasticities: a) fit to "down-flat" data of ref.[5] , thick solid line corresponds to fit A, thin solid line to fit B and dotted line to the 2–channel model fit; b) fit to "up-flat" data of ref.[5] , thick solid line corresponds to fit C, thin solid line to fit D and dotted line to the 2–channel model fit.

SUMMARY SESSION PRESENTATIONS

SECTION "A": FEW COMMENTS

G. M. Prosperi
Dip. di Fisica, Università di Milano and
INFN, sez. di Milano, via Celoria 16, I20133 Milano, Italy

ABSTRACT

I present a breef survey of the topics that have been treated in section "A" and a some more specific comments on few of the talks that have been given. Some other papers shall be discussed by E. Shuryak. Only personal interest and no claim of excellence of some papers with respect to others are implied in the choice.

1. General perspective

Tuelve talks have been given and a number of posters presented in the section. All papers have been very interesting for me and I learned a lot from them.

A first group of talks (K.I. Kondo, A. Mueller, H. Suganuma 1 and 2, M. Polyakov) have concerned specifically the mechanism of formation of the flux tube which is believed to be responsable for confinement. Abelian dominance, monopole condensation and developpement of an effective mass for non diagonal gluons in the maximal abelian gauge have been deeply discussed. The problems have been treated from the analytic as well from the numerical point of view and real progresses seem to have been done or more convincing evidences of commonly accepted ideas achieved. Such papers should be considered together with the plenary talks of related subject, like those of Bali and Polycarpov. I shall pay some more attention in the following to the work of Kondo, whyle Shuriak is going to talk more, I think, on that of Suganuma and perhaps of Polyakov.

Of the remaining papers two (T. J. Allen and M. Baldicchi) regard an exploiting in some way of the flux tube piture with a more phenomenological attitude in actual applications. The paper of Allen introduces the consideration of the string vibrations in the "relativistic flux tube model", with the aim of describing, first of all, the hybrid meson spectrum. That presented by Baldicchi concerns the application to the light-light and heavy-light meson spectrum of the three-dimensional reduction of a Bethe-Salpeter equation based on the "area law". On the contrary the works of D. Robertson and A. Weber are in the line of a combination of the hamiltonian and Fock space QCD formalism with more phenomenological assumptions in the gluon as well in the quark sectors. Coulomb gauge, renormalization, the Gribov problem and glueball spectrum are considered in particular. The papers of R. Leed and A. V. Nefediev are devoted to a study of various aspects of confinement in the 't Hooft model of a planar QCD in 1+1 dimensions and specifically to the consideration of the quark-hadron duality and the chiral symmetry breaking.

The last paper, finally, the paper of D. Gromes, concerns an attempt to apply to QCD the so called "optimized δ expansion". This consists essentially in an expansion of the partition function starting from an appropriate zero order lagrangian chosen *a posteriori* by some kind of variational procedure.

2. A more detailled discussion of few papers

In this section I shall make some more detailed remarks on four of the papers that have been presented (Kondo, Allen, Baldicchi, Gromes). As I said some others shall be discussed by Shuriak. I want to state explicitly that my choice has been essentially motivated by personal interest and convenience and no claim of excellence of any paper with respect to others is implied in it.

Let me begin with the paper of Kondo ("A Proof of Quark Confinement in QCD") that requires some special consideration in this conference, at least for its title.

Starting from an idea originally due to Martellini et al. [1], the author tries to reformulate the gauge theory as a "perturbative deformation" of an appropriate "topological field theory". Combining ideas of various people with new ingredients, he claims to be able to prove the area law for a plane rectangular Wilson loop without assumptions in the context of SU(2)-QCD.

Taking in mind also ref. [2], let me try to summarize the basic steps of the paper in the following way.

1) The maximal abelian gauge is adopted by using a gauge fixing term of the form

$$\mathcal{L}_{\rm GF}(x) = i\delta_{\rm B}\bar{\delta}_{\rm B}(\frac{1}{2}A_\mu^a A^{\mu a} + iC^a\bar{C}^a)\,, \tag{1}$$

having written the gauge field as $A_\mu = a_\mu T^3 + A_\mu^a T^a$, with $a = 1,\,2$, and denoted by $\delta_{\rm B}$ and $\bar{\delta}_{\rm B}$ BRST and anti-BRST trasformations involving (beside the gauge and the quark fields) the gost fields $C(x)$ and $\bar{C}(x)$ and the lagrangian multiplier $\Phi(x)$.

2) The positions are made

$$\begin{aligned} \psi(x) &= U(x)\Psi(x) \qquad (2)\\ A_\mu(x) &= U(x)V_\mu(x)U^+(x) + \frac{i}{g}U(x)\partial_\mu U^+(x) \end{aligned}$$

and $U(x)$, $V_\mu(x)$ and $\Psi(x)$ are treated as independent variables in the functional integral, the correct number of degrees of freedom being restored by a second gauge fixing constraint. This second costraint involves an additonal lagrange multipier $\beta(x)$ and new gost fields $\gamma(x)$ and $\bar{\gamma}(x)$. The result is the following expression for the partition function of the theory

$$\begin{aligned} Z_{\rm QCD}[J] &= \int [dU][dCd\bar{C}][d\Phi]\exp i\Big\{S_{\rm TFT}[U,C,\bar{C},\Phi] + \\ &+ \int d^4x(J^\mu\Omega_\mu + J_C C + J_{\bar{C}}\bar{C} + J_\Phi\Phi) + W[U;J,\eta,\bar{\eta}]\Big\}\,, \qquad (3) \end{aligned}$$

where

$$S_{\rm TFT}[U,C,\bar{C},\Phi] = \int d^4x i\delta_{\rm B}\bar{\delta}_{\rm B}(\frac{1}{2}\Omega^a_\mu\Omega^{\mu a} + iC^a\bar{C}^a) \tag{4}$$

and

$$\begin{aligned} e^{iW[U;J,\eta,\bar{\eta}]} &= \int [dV][d\gamma d\bar{\gamma}][d\beta][d\Psi d\bar{\Psi}] \exp i\left\{S_{\rm pQCD}[V,\Psi,\bar{\Psi},\gamma,\bar{\gamma}] + \right. \\ &+ \left. \int d^4x(\mathcal{I}^A_\mu(U,J)V^A_\mu + \frac{i}{2}\mathcal{K}^{AB}(U)V^A_\mu V^{\mu A} + {\rm tr}(\bar{\eta}U\Psi + \eta\bar{\Psi}U^+))\right\}. \end{aligned} \tag{5}$$

In such equations $A,B=1,2,3$, $\Omega_\mu = \frac{i}{g}U\partial_\mu U^+$, $S_{\rm pQCD}$ denotes an expression formally identical to the pure QCD action (with the second gauge fixing), the various J, η, $\bar{\eta}$ denote sources and $\mathcal{I}^A_\mu$ and $\mathcal{K}^{AB}$ are certain functions of their arguments. Notice that $S_{\rm TFT}$ defines a topologocal field theory. Then the essential idea is that the singular configurations of the gauge field should be completely taken into account through Ω^a_μ and so the integrations over U, C, $\bar{C}$ and Φ performed exactly, those on the remaining variables treated perturbatively.

3) By using the Diakonov-Petrov version of the non abelian Stokes theorem the Wilson correlator for the loop C is written as

$$\langle W[C]\rangle_{\rm QCD} = \langle e^{\frac{i}{2g}\int_S d^2x\epsilon^{\mu\nu}{\bf n}\cdot\partial_\mu{\bf n}\times\partial_\nu{\bf n}}\langle e^{\frac{i}{2}\oint_C dx^\rho n^A V^A_\rho}\rangle_{\rm pQCD}\rangle_{\rm TFT}\,, \tag{6}$$

where S is a surface delimited by C and

$${\bf n}(x) = (\sin\theta\cos\phi,\ \sin\theta\sin\phi,\ \cos\theta)\,, \tag{7}$$

with

$$U(x) = e^{i\phi(x)T_3}e^{i\theta(x)T_2}e^{i\chi(x)T_3}\,. \tag{8}$$

If the term inside the internal brackets is replaced by 1, the theory becomes a pure thopological field theory. Then, for a rectangular Wilson loop of sides R and T, the theorem of dimensional reduction of Parisi and Sourlas can be applied and $W[C]$ can be put identical to the corroponding expression for the non linear sigma-model in two dimension which satisfies the area law,

$$\langle e^{\frac{i}{2g}\int_S d2x\,\epsilon^{\mu\nu}{\bf n}\cdot\partial_\mu{\bf n}\times\partial\nu{\bf n}}\rangle_{\rm TFT} = \langle\ldots\rangle_{\rm NLSM} = e^{-\sigma RT}\,. \tag{9}$$

Furthermore, if the internal term in (8) is replaced by its expansion in g, corrections can be evaluated and in [2] the author obtains

$$\log\langle W[C]\rangle_{\rm QCD} = -\sigma RT - Cg^2(R+T) + \frac{g^2}{4\pi}\frac{T}{R} + \ldots\,, \tag{10}$$

or, in terms of static potential,

$$V(R) = -\lim_{T\to\infty} = \sigma R + Cg^2 - \frac{g^2}{4\pi} + \ldots\,. \tag{11}$$

As a general comment on the paper of Kondo let me say that the discussion given by the author is very interesting, in my opininon, particularly for the new perspective under which the problem is considered. However, the claim made in the title can not be shared. As matter of fact, in spite of the internal consistency implied by eq. (10), the idea that all the non perturbative configurations can be properly taken into account by the topological term remains a very problematic assumption. Indeed, like the authors of ref. [1], Kondo does not succeed to connect explicitly the string tensione σ with any natural scale in QCD, like Λ_{QCD}.

A second paper which is appealing to me is that of Allen et al.

Some years ago M. G. Olsson et al. succeeded to quantize numerically the "relativistic flux tube model" consisting in considering a quark and antiquark drawing a rigid chromoeletric flux tube [3]. The tube was supposed to bring a fixed ammount of energy per unity lenght and was allowed only of transversal momentum, whyle the quarks could have transversal as well longitudinal momentum. Such model achieved considerable success in providing a physical picture of the dynamical confinement of meson states and in particular in reproducing correct slopes for Regge trajectories and good Isgur-Wise functions for semi-leptonic decay.

In the paper presented at the conference the rigid tube is replaced by a dynamical string and the action of the model is assumed to be

$$S = S_{\mathrm{quark}} + S_{\mathrm{string}}\,, \tag{12}$$

with

$$S_{\mathrm{quark}} = \frac{1}{2} \sum_{i=1,2} \int d\tau (e_i^{-1} \dot{x}_i^\mu \dot{x}_{i\mu} - e_i m_i^2) \tag{13}$$

and

$$S_{\mathrm{string}} = S_{\mathrm{Nambu-Goto}} = -\frac{a}{2} \int d\tau \int_0^\pi d\sigma \eta^{ab} \partial_a X^\mu \partial_b X_\mu\,. \tag{14}$$

The string world sheet $X^\mu(\sigma,\tau)$ and the quark and the antiquark world lines $x_{1,2}(\tau)$ are supposed related by the boundary condition

$$X(0,\tau) = x_1(\tau), \qquad X(\pi,\tau) = x_2(\tau)\,. \tag{15}$$

First the case of fixed quarks is considered. Then the usual machinery for string quantization can be used and the energy eigenvalues can be written

$$P^0 = \sqrt{(ar)^2 + 2\pi a(N+c)}\,, \tag{16}$$

where

$$N = \sum_{n=1}^{\infty} n(N_n^+ + N_n^-)\,, \tag{17}$$

N_n^+ and N_n^- being the number of phonons in the mode n with helicities 1 and -1 respectively and c a constant. In a similar way the component of the angular momentum along the quark-antiquark axis turns out to be

$$\Lambda = \hat{\mathbf{r}} \cdot \mathbf{L} = \sum_{n=1}^{\infty} (N_n^+ - N_n^-)\,. \tag{18}$$

The string excited states can be classified in a molecular spectroscopy fashion and denoted by the usual symbols $\Sigma^{\pm}_{g,u}$, $\Pi^{\pm}_{g,u}$, $\Delta^{\pm}_{g,u}$, ... Expression (16) is interpreted as $Q\bar{Q}$ static potential for that particular state. The results have been compared with the the corresponding potential obtained by C. Morningstar et al. by lattice numerical simulations [4] and found in reasonable agreement, with the exception (for some reason) of the states Σ^-.

Furthermore, if the two quarks are suposed to rotate with angular velocity Ω, the total energy and angular momentun can be written (in the equal mass case)

$$E = P^0 + p_1^0 + p_2^0 = \left[\left(\frac{a\,r\,\pi\Omega/2}{\sin(\pi\Omega/2)}\right)^2 + 2\pi a(N+c)\right]^{1/2} + 2\left[\mathbf{p}^2 + m^2 + \frac{2a}{\pi N}\right]^{1/2} \tag{19}$$

and

$$L^2 = \left[\frac{ar^2}{8}\left(\frac{\pi\Omega - \sin(\pi\Omega)}{\sin^2(\pi\Omega/2)}\right) + L_Q\right]^2 + \Lambda^2 . \tag{20}$$

The above equations can be quantized numerically by a method analoguous to that used for the original relativistic flux model. Work is in progress for the heavy-heavy quark case and results on hybrid meson spectrum are expected soon. Also the intermediate masses case is under consideration.

To the relativistc flux tube model, in some way, even the paper of Baldicchi (in which I too am personally involved) is related.

More precisely, this paper is in the framework of a Bethe-Salpeter formalism which has been developped in Milano, starting from a Feynman-Schwinger type path integral representation of the "second order" gauge invariant Green functions in QCD [5,6]. Such representation permits to express the 4-point and 2-point functions simply in terms of Wilson loop correlators. Using only the assumption that the logaritm of such a correlator can be written as the sum of its perturbative expression and an area term,

$$i\ln\langle W\rangle = i\ln\langle W\rangle_{\rm pert} + \sigma S_{\rm min}\,, \tag{21}$$

(and the so called straight line approximation for the second term) then, one succeeds to derive a "second order" quark-antiquark Bethe-Salpeter equation and a quark Schwinger-Dyson equation in which the kernel is expressed in principle as an expansion in the strong coupling constant α_s and in the string tension σ (in pratics only the lowest order approximation has been evaluated). Then, if one were able to solve in some reasonable approximation first the SD equation and then the BS equation (using in this second equation the full quark propagator ontained in the first step), one should be able to handle the entire heavy-heavy, heavy-light and light-light meson spectrum, apart from corrections due to pair creation effects. Unfortunatly this is a very difficult task. However, if one excludes the light pseudoscalar mesons (for the delicate interplay existing between chiral symmetry breaking and the existence of a zero mass solution of the BS equation in the vanishing quark mass limit), sensible results can be expected even from a three dimensional reduction of the BS equation. Actually replacing the full quark propagator in the BS equation by a free propagator and performing an instantaneous approximation on the kernel, one obtains the eigenvalue equation (reduced

Salpeter equation) for a relativistic hamiltonian of the form

$$H = \sqrt{m_1^2 + k^2} + \sqrt{m_2^2 + k^2} + V\,, \tag{22}$$

where V is a very complicated quantity involving spins and momenta which has however various significant limits. Precisely, in the heavy quark limit V reduces to the semirelativistic potential which has been obtained (starting from similar assumption on W) in previous works and successfully applied to heavy quarkonium spectrum. Instead, if the spin dependent terms are neglected, V becomes identical (apart from a question of ordering) to the potential obtained from the relativistic flux tube model. Furthermore, in the static limit one find the local potential

$$V = -\frac{4}{3}\frac{\alpha_s}{r} + \sigma r\,, \tag{23}$$

which has been used as such in (22) in some attempts to reproduce the entire spin averaged spectrum. Notice, however, that for a reasonable overwhole fit eq. (23) requires the ad hoc addition of flavor dependent constants and a value $\sigma \sim 0.22/, \mathrm{Gev}^2$ for the string tension which would give for the Regge trajectories the slope

$$\alpha' = \frac{1}{8\sigma} = 0.57\,\mathrm{Gev}^{-2}\,, \tag{24}$$

rather far from the experimental slope $\alpha' = 0.88\,\mathrm{Gev}^{-2}$. On the contrary the potential appearing in (22) has a second local limit for high angular momentum

$$V = -\frac{4}{3}\frac{\alpha_s}{r} + \frac{\pi}{4}\sigma r\,, \tag{25}$$

which gives

$$\alpha' = \frac{1}{8\frac{\pi}{4}\sigma} = \frac{1}{2\pi\sigma} \tag{26}$$

and reproduces the correct result for $\sigma \sim 0.18\,\mathrm{GeV}^2$, the value required by the best fit for heavy quarkonium.

The paper presented by Baldicchi concerns a direct evaluation of the spectrum using the full V expression apart from the spin orbit like terms (that have been omitted in this first application due to their complication). The procedure consists in solving first the eigenvalue equation for the static potential (23) by the Rayleigh-Ritz method and then in evaluating the quantities $\langle H \rangle$ for the eigenfunctions obtained in the first step.

By using the parameters

$$\begin{aligned} \alpha_s &= 0.363 & \sigma &= 0.175\,\mathrm{GeV}^2 \\ m_b &= 4.81\,\mathrm{GeV} & m_c &= 1.40\,\mathrm{GeV} \\ m_s &= 200\,\mathrm{MeV} & m_u &= m_d = 10\,\mathrm{MeV} \end{aligned}$$

one succeeds to reproduce reasonably well the bottonium and the charmonium spectra, the Regge trajectories (with correct slope and intercepts) for the ground triplet states of the $u\bar{u}$, $u\bar{s}$, $s\bar{s}$ systems, the known triplet and singlet S-states and the spin averaged P-states for the light-heavy systems. Notice that no ad hoc constant is added and that the first four parameters are pratically determined by the heavy quarkonium fit, whyle the light quark masses are fixed *a priori* on typical current values. In particular the heavy-light sector is completely parameter free.

Let me finally consider Gromes paper. As I mentioned, Gromes tries to aply to QCD the so called optimized δ-expansion. The method, which has been already applied with some success in field theory, can be effectively illustrated on a toy model.

Let us consider the quantity

$$Z(g) = \frac{1}{\sqrt{2\pi}} \int_{-\infty}^{\infty} dx\, e^{-(\frac{x^2}{2}+gx^4)}\,, \tag{27}$$

which can be considered as the partition function of a 0-dimensional theory. Ordinary perturbation theory would correspond to simple expanding in g inside the integral. However, the expansion for $Z(g)$ obtained in this way has zero radius of convergence and even asymptotically gives reasonable results only for extremly small g. On the contrary one can consider the alternative quantity

$$Z'(g,\lambda,\delta) = \frac{1}{\sqrt{2\pi}} \int_{-\infty}^{\infty} dx e^{-\lambda\frac{x^2}{2}} e^{\delta[(\lambda-1)\frac{x^2}{2}-gx^4}\,, \tag{28}$$

which includes two new parameters and becomes identical to the preceding one for $\delta = 1$ irrespective of the value of λ. Then one can expand Z' in δ, truncate the expansion at a given order n, put eventually $\delta = 1$. Obviously the truncated expression does depend on the artificial parameter λ, whyle the exact expression does not. This suggests to choose λ as function of g according to some principle of "minimun sensitivity" $[\frac{\partial Z_n(g,\lambda)}{\partial\lambda} = 0]$ or of "fastest apparent convergence"$[Z_n(g,\lambda) = (Z_{n-1}(g,\lambda)]$. If one proceeds in this way the first order approximation already reproduces the exact result up to a few per cent in the entire range $(0,\infty)$ of values of g.

In extending the method to field theory one can introduce an arbitrary function in place of the single parameter λ. In such an order of ideas in his attempt to apply it to QCD Gromes replaces the ordinary action for pure QCD,

$$S = -\frac{1}{4}\int d^4x F^a_{\mu\nu}(x) F^{\mu\nu a}(x)\,, \tag{29}$$

by the expression

$$S' = -\frac{1}{4}\int d^4x d^4y [(1-\delta^2)K(x-y) \;+\; \delta^2\delta(x-y)] \times\; F^a_{\mu\nu}(x;\delta) U^{ab}(x,y;\delta) F^{\mu\nu b}(x;\delta)\,, \tag{30}$$

in which $F^a_{\mu\nu}(x;\delta)$ and $U^{ab}(x,y;\delta)$ denote the field and the Schwinger string respectively after the substitution $g_0 \to \delta g_0$ (again S' reduces to S for $\delta = 1$) and $K(x-y)$ is an arbitrary function. Notice that in the δ expansion the quantity

$$D^{ab}_{\mu\nu}(q) = \delta^{ab}\frac{1}{q^2\tilde{K}(q)}(g_{\mu\nu} - \xi\frac{q_\mu q_\nu}{q^2}) \tag{31}$$

plays the role of free gluon propagator. The idea would be to parametrize $\tilde{K}(q)$ appropriately and choose such parameters according to one of the mentioned principles.

So far Gromes has evaluated the gluon vacuum polarization only up to second order in δ. In order to be able to calculate interesting quantities, like the β function, even the third order terms should be included. There are complications and renormalization difficulties. The author believes to be just at the start in "climbing mount QCD". I find, however, his approach very suggestive.

3. Conclusions

In conclusion I found all the talks and even the posters presented in section "A" very stimulating and instructive. I fiel to thank warmly all participants for the interest and the level of the entire discussion.

References

1. F. Fucito, M. Martellini and M. Zeni, Nucl. Phys. **B 496**, 259 (1997)
2. K. I. Kondo, preprint, hept-th/9805153 (May 1998).
3. C. Olson, M. G. Olsson and K. Williams, Phys. Rev. **D 45**, 4307 (1992).
4. This conference.
5. N. Brambilla, E. Montaldi and G. M. Prosperi, Phys. Rev. **D 54**, 3506 (1996); G. M. Prosperi, hep-th/9709046 (1997).
6. As general reference see also e. g.: N. Brambilla and G. M. Prosperi ed.s "Quark Confinement and the Hadron Spectrum", World Scientic, 1994; "Quark Confinement and the Hadron Spectrum II", World Scientic, 1996.

(PARTIAL) SUMMARY OF SECTION A: CONFINEMENT MECHANISM, FLUX TUBES AND STRINGS

Edward Shuryak
Department of Physics and Astronomy
State University of New York, Stony Brook Ny 11794, USA

ABSTRACT

An overview of some issues discussed is given, including the following ones: (i) the mechanism of confinement, (ii) the structure of the string, (iii) non-perturbative glue in vacuum and hadrons, and (v) quark-hadron duality. In doing so I try to indicate issues important for JLab. Finally I discuss another issue debated during this conference, namely (vi) whether one can use some common effective potential for description of various hadrons, using heavy quarkonia as an example.

1. Overview

My first remark is about a general situation in non-perturbative QCD/ hadronic physics. It is striking that it has several sub-communities, with rather weak interaction between them. Roughly speaking, there are three different schools which think about those issues in completely different terms. (i) The first one (naturally, the dominant one at this conference), concentrate on *confinement* phenomenon and strings. Hadrons are viewed basically as bare quarks connected by strings, and hybrids are string excitations. The vacuum is "empty" of any non-perturbative glue and the glueballs are closed strings. The stronghold of this point of view is physics of charmonium/bottonium and Regge trajectories. There are of course extended claims for all other hadrons as well. (These proceedings is probably the best possible review of development in these directions.)

(ii) The second one (to which I belong) concentrate at hadrons made of light quark. The basic phenomenon underlying it is *breaking of chiral symmetry* , both spontaneous ($SU(N_f)_A$) and explicit by the anomaly ($U(1)_A$). This school view light quarks as collective excitations propagating on top of the quark condensate $< \bar{q}q >$, and thus obtaining "constituent quark" mass, being the main part of masses of all light hadrons. The residual interaction binds those into hadrons, in some cases even without a confining potential (which is treated as small effect). Originally this school was using Nambu-Jona-Lasinio-type models. Recent QCD-based development views the vacuum (as well as glue inside glueballs and hadrons) as a superposition of *instantons*, which naturally provide the (non-local) interaction between quarks needed for chiral breaking. The obvious stronghold of this ideology is chiral dynamics of pions and its relatives. There is also a claim that all other lowest mass hadrons made of light quarks (and heavy-light ones) are explained by it. For recent review I propose that by T.Schaefer and myself.[1]

(iii) The third school tries to *re-sum perturbative diagrams* using well known QED tools like Schwinger-Dyson or Bethe-Salpeter equations. They hope that when the coupling constant is not too small, both chiral symmetry breaking and confinement will appear, e.g. from modified propagators. An example is the $G \sim 1/q^4$ behavior of the gluon propagator at small q. Similarly, the one-gluon exchange, if strong enough, may also lead to chiral symmetry breaking, quark constituent masses etc. The first point I want to make is that these three pictures of hadrons are indeed so different, that they cannot be simultaneously right. And available lattice/phenomenological data should be used to pick the right one. One example (which was several times mentioned at this conference): although the $G \sim 1/q^4$ gluon propagator may provide confinement, it definitely does *not* correspond to a string-like configuration of the field. Thus it is already ruled out by lattice studies, which do see a rather narrow string between static charges. (Other similar examples are to follow below.)

My next general point provides an argument *against* the perturbative views (iii). In fact we do not have *one* QCD vacuum / hadronic spectrum but *many*, numbered by the famous θ angle related with the phase of the instanton amplitude. For example, at $\theta = 0$ (the physical case) and $\theta = \pi$ there is no CP violation, but the instanton-induced amplitude change sign. The energy difference between these two vacua (in YM theory without quarks) can be estimated from lattice data on topological susceptibility, and gives a number of the order of 1 GeV/fm^3 (which is large compared to those associated with chiral breaking or confinement). Strong arguments exist showing that there should be drastic differences between these vacua, and as θ changes different vacua cross should each other and cause some phase transitions. The perturbation theory (no matter how resummed, with or without renormalons) simply does not know *anything* about the θ angle! What it means is that there cannot be any unique way of defining QCD starting from perturbative series, unless it somehow includes this parameter. The same question applies to "abelian projected" version of QCD (much advocated at this conference). It does not seem to have enough topology to care about the θ angle as well. How these approaches are going to reproduce the θ dependence of the vacuum and hadrons?

2. The Physical Mechanism of Confinement

Impressive lattice studies reported (see especially talks by Bali and Michael) has strengthened the case made by earlier works on the field distributions in static configurations. Properties of glue excitations with different orbital momenta they found leave little doubt that a (rather thin) string is indeed formed in such configurations. This by itself rules out some models of confinement. I already mentioned the model with the gluon propagator $G \sim 1/q^4$. Another one which I think belongs to the same class is a suggestion[2] that confinement is due to the tail of the instanton distribution over sizes $dn/d\rho$ at large ρ: I will return to it at the end of this section. *What is the internal structure of a string?* It seems that lattice data are consistent with the dual superconductor picture, see talks by Bali, Polikarpov, Ichie and Suganuma. In addition, as M.Baker told us, the dual picture reproduces the spin-dependent potentials well. The interesting discussion here was initiated by the observation that abelian projected field shows string radius to be only about .15 fm, while the complete SU(2) fields indicate about twice larger radius. The point (emphasized by Suganuma, but also by others) is that the non-abelian fields have to be located outside the abelian one, since the monopole current should wind around abelian field, and the monopoles are made of it. This convincing argument however seem somewhat to contradict another set of ideas, presented here by Kondo and Suganuma. They proposed that the non-abelian fields can be to first approximation ignored, and then taken into account perturbatively. The reason for that, they argued, is because its phase is a simple "random variable". However it seems to me that this view point cannot be quite correct, since it would destroy the monopoles and evantuallu the outer region of the string. *How the string interact?* Recall that attraction means dual superconductor of the 1-st kind, and repulsion means the 2-nd. Bali has shown that there are clear signals of weak attraction. C.Michael has also studied the issue, and his conclusion is the octet string tension is nearly exactly twice the triplet one*: thus he sees no significant interaction. It also suggests that QCD strings are close to the border line. *The theoretical status of the dual superconductor picture* is however still controversial. Its proponents say that recent Seiberg-Witten solution of N=2 Supersymmetric QCD has provided support to their point of view. Well, not quite. In this theory the Higgs VEV does break the gauge group, and both monopoles and dual photons are indeed real excitations of the theory. In QCD we definitely have no "dual gluons" or Higgses with masses $\sim 1GeV$ †. The fact that those do appear in the abelian dual Landau-Ginzburg Lagrangian basically invalidate it, no matter how good is its description of the string. It is not the right effective theory of QCD at low energies: we have to look for another one, without the color group being broken. *Is confinement generated by monopoles?* It was repeatedly mentioned that several abelian projections were shown to preserve the full string tension. Another strong contender (discussed here by Polykarpov) is further reduction of the abelian fields to only Z_N parts of them. As shown in[3] it also reproduces the string tension, reviving an old viewpoint that *vortices* rather then monopoles (or the 2-d objects rather than the 3-d ones) are responsible for confinement. A debate which of the two are more physical objects is still going on, and extensive lattice work is needed to resolve the issue. Let me finish this section with a discussion of two (instanton-related) confinement mechanisms which (I think) do *not* work. The first is based on the correct observations that (i) an abelian projection of an instanton produces a monopole loop (of size ρ) and (ii) in a sufficiently dense ensemble of instantons those loops merge into a long loops, in a manner similar to a percolation transition. Nevertheless, the potential V(r) (extracted

*The famous "Casimir scaling" for the non-triplet strings was debated in literature for ages: it seem not to work. I never understood what it is based on anyway.

†The data for shown by Suganuma indicate that the abelian part has the propagator which is consistent with *zero* mass. Has it any specific physics consequences?

by calculating the Wilson loops with the ensemble of instantons) show smooth (and approximately linear) dependence on the instanton density, ignoring the percolation point. The second is the suggestion[2] that large-ρ tail of the instanton distribution may be responsible for confinement. Indeed, if $dn/d\rho \sim \rho^{-3}$ one can get it. However, if the origin of confinement are kind of large domains of weak field, we would not get thin strings. Furthermore, if it is due to some tail of the distribution, confinement should be "intermittent" on configuration-by-configuration basis: many lattices should have no confinement and some have very strong one. It is completely contrary to lattice observations, in which confinement fluctuates little.

With a tail $dn/d\rho \sim \rho^{-5}$ which fits well (still rough) lattice data, one gets approximately linear potential $V(r) = Kr$ up to r$\sim$ 3 fm. There was a claim by Suganuma et al[4] that the tension K has the right magnitude, but it must be mistaken since our calculation[5] as well as by others[6,7] get in this case the tension $K \sim 100 - 200 MeV/fm$, few times less than the empirical/lattice value $K \sim 1 GeV/fm$.

3. The Non-perturbative Glue

3.1. How the Non-perturbative Glue in Vacuum actually look like?

Some people reject this question[‡], but I think it remains to be the most important issue, a key to the whole hadronic physics. Let me make the question more specific, emphasizing dimensionality of the objects: *Is it mostly made of (i) closed string (2-d membranes), (ii) instantons (4-d objects), (iii) monopoles (3-d objects) or (iv) vortices(2-d objects)?* Obviously our lattice friends can answer this question simply by making lattice fields smooth (in order to get rid of "quantum noise" of perturbative nature) and then see what they look like. What they see are 4-d "bumps". Many of them have nonzero topological charge and are identified as instantons or anti-instantons. Some have zero topological charge and smaller action, they look like close instanton-anti-instanton pairs (and disappears under further "cooling", oraction minimization). Something of the order of 90 percent of the action is saturated by those objects. For those who think that a "smoothening" of gauge fields is a questionable procedure, there is another way of looking at the problem, this time from the point of view of light quarks. One can look how the wave function of the quark condensate (which all approaches mentioned above claim to generate). In order to do so, one should take $|\psi_\lambda(x)|^2$ for the lowest eigenmodes λ of the Dirac operator and study how the corresponding eigenfunction is distributed in space. What one sees on the lattice are again isolated bumps correlated to instantons. I never head that anybody was able to identify 3-d (or 2-d) structures either in action distribution or in quark condenate: for me it put large question mark for claims that monopoles (vortices) are physical. The "instanton dominance" is clealry established, as far as gauge field action of chiral symmetry breaking is concerned. However it is not so if we for confinement (see the previous section). In fact, a "slightly smoothened" vacuum preserves confinement with the original string tension (see e.g.[8]), and further studies should be able to find out what it is due to.

---

‡Paton commented for example that in his model the vacuum has no non-perturbative glue at all.

3.2. *Glue in Glueballs and Hybrids*

At this conference this subject was discussed from a string picture point of view. It does not work for recently claimed hybrids, as those turned out to be lighter than expected. Let me present an argument that it cannot be true at least for the scalar glueball as well, based on its size and shape. The *size* of the scalar glueball is surprisingly small. This was first seen on the lattice from the magnitude of finite size effects[9] and then confirmed directly from measurements of the wave functions.[10,11] The size of the scalar glueball (defined through the exponential decay of the wave function) is $r_{0^{++}} \simeq 0.2$ fm, while $r_{2^{++}} \simeq 0.8$ fm[11 §]. It indicates that spin-dependent forces between gluons are much stronger than between quarks, at least in the scalar channel. Instanton-based calculation done by T.Schaefer and myself[12] have obtained both the mass and size values of the scalar glueball in agreement with those obtained on the lattice. The parameter enhancing gluon interaction with instantons is nothing else but large classical amplitude of the instanton field at the center $G^2(0) = 192/(g^2\rho^4)$ (basically, a parameter of the same nature as the one enhancing induced photon emission inside lasers). The wave function of the scalar glueball (defined by a split-point correlator) has a strong peak at the origin. This is in direct contradiction with the closed string picture, suggesting a hole there. Closing this subsection, let me say that all these arguments do not of course invalidate estimates based on the string picture: it is possible (and even quite probable) that resonant excitations of many different types can coexist, as they do in countless examples from condense matter/nuclear physics.

3.3. *How to look for Glue inside Ordinary Hadrons?*

One method, is to extract the *next twist matrix elements* for nucleons from power corrections to deep inelastic scattering. The theory of these corrections based on general Operator Product Expansion (OPE) was worked out in early 80's.[13] As emphasized in these works, while the leading twist operators are related to *probabilities* to find quarks or gluons with certain momenta inside the nucleon, the next twist ones describe *correlations* between partons. In particularly, mixed quark-gluon operators of the structure $< N|\bar{q}...G_{\mu\nu}...q|N >$ tell us what is the gluon field at the same (scattering) point where the quark is. Needless to say, it would be very important to have such information, and JLab experiments are well suited to fill this gap.

Meanwhile, one can estimate such matrix elements from various models of the nucleon. Such estimates based on the instanton liquid model[14] were discussed here by M.Polyakov. His results show that in fact these matrix elements are smaller than a naive estimates based on the magnitude of the fields inside the instantons: estimates have some small factors like diluteness of the instanton vacuum. The reason for that is that the quark operators which appear as corrections to DIS all are vector or axial vectors, so they are not chirality changing and cannot thus be represented by instanton zero modes. It is a part of a general tendency of the instanton effects to be "hidden" in vector and axial channels. Much larger gluonic effects are expected in scalar/pseudoscalar channels or isosinglet axial channel related with the anomaly.

---

§For comparison, a similar measurement for the π and ρ mesons gives 0.32 fm and 0.45 fm.

Very intriguing tool in this respect is *virtual/real charm annihilation* processes. Let me conclude this section with three examples of this kind. (i) The process $\eta_c \to \bar{N}N$ is forbidden by chirality conservation rules of perturbation theory: nevertheless it is observed and has a surprisingly large branching. Its relation to instantons was discussed in.[15] (ii)Recently, CLEO collaboration has reported measurements of inclusive and exclusive production of the η' in B-decays :

$$Br(B \to \eta' + X \,; 2.2\, GeV < E_{\eta'} < 2.7\, GeV) = (7.5 \pm 1.5 \pm 1.1) \cdot 10^{-4} \,, \tag{1}$$

$$Br(B \to \eta' + K) = (7.8^{+2.7}_{-2.2} \pm 1.0) \cdot 10^{-5} \,. \tag{2}$$

Simple estimates show that these data are in severe contradiction with the standard b-quark decay into light quarks: Cabbibo suppression V_{ub} leads to decay rates two orders of magnitude smaller than the data (both inclusive and exclusive ones). Alternative mechanism, suggested by Halperin and Zhitnitsky,[16] is based on the Cabbibo favored $b \to c\bar{c}s$ process, followed by a transition of virtual $\bar{c}c$ into the η'. The latter transition implies large "intrinsic charm" component of the η'. Its quantitative measure can be expressed through the matrix element

$$\langle 0|\bar{c}\gamma_\mu\gamma_5 c|\eta'(q)\rangle \equiv i f^{(c)}_{\eta'} q_\mu \,. \tag{3}$$

and one needs $f^{(c)}_{\eta'} \approx 140\, MeV$ in order to explain the CLEO data, see.[16] This value is surprisingly large, being only a few times smaller than the analogously normalized residue $\langle 0|\bar{c}\gamma_\mu\gamma_5 c|\eta_c(q)\rangle = i f_{\eta_c} q_\mu$ with $f_{\eta_c} \simeq 400\, MeV$ known experimentally from the $\eta_c \to \gamma\gamma$ decay. By expansion in inverse powers of charm quark mass, the problem can be reduced to the matrix element of a particular dimension-6 pseudo-scalar gluonic operator: $\langle 0|g^3 f^{abc} G^a_{\mu\nu} \tilde{G}^b_{\nu\alpha} G^c_{\alpha\mu}|\eta' >$. It was estimated in[17] and found to be indeed large, enhanced by very strong gluonic fields of small-size instantons. Quantitative description of CLEO data still remains to be work out, however. (iii) The last example (strongly related with the previous one) is the contribution to the spin of the nucleon of the polarized charm "sea" (to be measured at next generation DIS experiments like CERN COMPASS). It is also related with the matrix element of the current $\bar{c}\gamma_\mu\gamma_5 c$, but this time averaged over the nucleon. As explained in recent paper by Blotz and myself,[18] using the same gluon operator one can do it in the instanton model. The resulting charm contribution of about -0.03.

4. Quark-Hadron Duality

The issue in early 70's was a check of QCD itself: and we did found that, for example, the total cross section of $e^+e^- \to hadrons$ is close to total cross section of annihilation to all species of quarks. Now the topic was revived, and the issue is the *nature and magnitude of the oscillating component*, the difference between the two. One useful direction of recent work is to study what happens in exactly solvable models. At this conference Lebed has presented results[19] for heavy quark decay in 2-d QCD at large N, known as 't Hooft model. The issue was also studied by Shifman et al.[20] Two groups have nicely demonstrated that the duality itself works. However they strongly disagree about the magnitude of oscillating component. Lebed claimed that it is O(1/M), based on numerical fit, while Shifman et al find much smaller effect O(1/M^3) (M is the mass of the heavy quark). What is the general reason why the oscillating part exist at all? Shifman et al[20] simply argued that since the exponentially decaying contribution to correlators for space-like (Euclidean) momenta $\Pi(Q) \sim O(exp(-Q * const))$ does exist , such contributions become oscillating O(sin(Q*const)) for the time-like (Minkowski) momenta. What I find exciting about this argument is that oscillation period is related with singularities of amplitudes in x space away of x=0 (like the radius of the instanton in expressions like $1/(x^2+\rho^2)$). Let me add two important comments on that. First, since the derivative of the scattering amplitude over energy determines the *lifetime of the intermediate system*, the oscillating component should exist because otherwise we would not be able to get from the amplitude the duration of the final sate interaction. (In the canonical case of total cross section of $e^+e^- \rightarrow hadrons$, this is the time at which the string is broken and final hadrons appear, or widths of corresponding resonances.) The second: one can ask what happens when resonances overlap¶. My general expectation is that oscillating function is changed to a *random* one. However its correlators like

$$K(\epsilon, \delta\epsilon) =< \sigma(\epsilon+\delta\epsilon)\sigma(\epsilon) > \tag{4}$$

should be regular functions which still display the characteristic correlation scale in energy of the order of the inverse time of final state interaction $\delta\epsilon \sim \Gamma(\epsilon)$. Similar phenomenon known in nuclear physics under the name of Erickson fluctuations. Summary: the physical nature and accuracy of parton-hadron duality is not yet clarified. Clearly much more theoretical and experimental work is needed. A good place for the latter is Jefferson Lab. which can check on accuracy of/deviations from the so called Bloom-Gilman duality (a difference between sum over hadronic excitations and smooth partonic structure function).

5. Can one use a common Effective Potential for various hadrons?

This issue was not part of section A I have to review, but it was discussed in a very interesting lecture by P.Lepage and, since it is vital to the field, I am tempting to try to clarify the issue somewhat. Let me briefly repeat what he said in his talk‖. His basic idea is that lattice simulation allows one to get wave functions of quarkonia states like J/ψ and Υ, which can in turn be used to get an effective potentials (by putting those into Schroedinger equation). Among the questions Lepage has addressed are: (i) whether these potentials are universal and (ii) how close they are to the one extracted for static quarks. His results show that (i) for J/ψ and Υ the potentials are very close, but (ii) both are about 20 percent different from the static one. Let me start with the issue of the time scales relevant for the problem. We discuss heavy

¶For example, in Regge trajectories $M_n \sim \sqrt(n)$ and spacing between resonances is $O(1/\sqrt(n)$ while the width $\Gamma_n \sim M_n$.

‖The reader of the proceedings can of course read the Lepage's contribution, while now I only have recollections of his talk. In case my points are also made there, I apologize.

quarks Q interacting with a non-perturbative glue, so let us imagine that those have characteristic time scales τ_Q and τ_g respectively. The static potential would be correct if $\tau_Q >> \tau_g$. It is however easy to see that the heavy quark limit ($M \to \infty$) in fact leads to the *opposite* relation. It is true that the velocity of heavy quarks decreases with M, but sizes of quarkonia decrease even more and $\tau_Q \sim 1/(M\alpha_s^2)$. From old papers of Voloshin and Leutwyler we know that one can use in this limit dipole expansion, leading to oscillatory $V \sim r^2$ potential with the state-dependent coefficient. This explains why Lepage's upsilon potential deviates from the static one. But why it is the same for Υ as for J/ψ? My answer to that is that (if we still use the dipole approximation to second order) the effective potential is proportional to the following vacuum average over the gluoelectric fields

$$V \sim r^2 \int d\tau < \vec{E}(\tau) U \vec{E}(0) U^+ > exp(-\tau\omega_{PS})$$

where U are color trasnport matrices and ω_{PS} is the energy difference between the original S-wave state (J/ψ or Υ) and that of P-wave states the dipole excites**. Although the charmonium states are mostly sensitive to the linear potential while bottonium are more due to the Coulomb part, by some play of numbers the essential frequency ω_{PS} is about the same for both. It means that the frequency of rotation of quarks in both states are the same, and we probe the *same* correlator of the vacuum fields. (Incidentally, it is about coincides with the correlation time of the vacuum field, see e.g. the talk by Dosch in these proceedings.). Of course, if we go outside the dipole approximation in the second order, there are differences between two cases (e.g. the relativistic corrections with higher power of the velocity). Lepage's results then imply that those are sufficiently small. In summary: neither very heavy quarkonia (nor real $J/\psi, \Upsilon$) should be described by static potentials. One may construct those using lattice data for wave functions (and/or Gauge field correlators), but those are going to be state-dependent.

References

1. T. Schafer and E. Shuryak. Rev.Mod.Phys.70:323-426,1998, hep-ph/9610451.
2. M. Fukushima, A. Tanaka, S. Sasaki, H. Suganuma, H. Toki and D. Diakonov Presented at Lattice 96: Nucl.Phys.Proc.Suppl.53:494-496,1997 Also in Lattice 96:494-496 (QCD161:I715:1996), hep-lat/9610003. Phys.Lett.B399:141-147,1997.
3. M. Faber, J. Greensite and S. Olejnik, hep-lat/9807008.
4. F. Araki, H. Suganuma, H. Toki. Talk given at The 17th RCNP International Symposium on Innovative Computational Methods in Nuclear Many-Body Systems (INNOCOM 97), Osaka, Japan, 10-15 Nov 1997. e-Print Archive: hep-ph/9804292.
5. D.Chen, J.Negele and E.Shuryak. Instanton-induced static potential in QCD, revisited. CTP-MIT-2719. NTG-98-42.

**For simplicity I ignore their splittings.

6. T. DeGrand, Anna Hasenfratz, Tamas Kovacs (Colorado U.). COLO-HEP-396, Talk given at Yukawa International Seminar on Non-Perturbative QCD: Structure of the QCD Vacuum (YKIS 97), Kyoto, Japan, 2-12 Dec 1997. hep-lat/9801037;
7. D.Diakonov, private communications.
8. M. Feurstein, E.M. Ilgenfritz , H. Markum, M. Muller-Preussker, S. Thurner. Nucl.Phys.Proc.Suppl.63:480-485,1998. hep-lat/9709140.
9. P. Van Baal and A. S. Kronfeld, Nucl. Phys. B (Proc. Suppl.) 9, 227 (1989).
10. K. Ishikawa, G. Schierholz, H. Schneider, and M. Teper, Nucl. Phys. B227, 221 (1983); T. A. DeGrand, Phys. Rev. D36, 176 (1987); R. Gupta, A. Patel, C. F. Baillie, G. W. Kilcup, and S. R. Sharpe, Phys. Rev. D43, 2301 (1991).
11. F. de Forcrand and K.-F. Liu, Phys. Rev. Lett. 69, 245 (1992).
12. T. Schafer and E. Shuryak, Phys.Rev.Lett.75:1707-1710,1995, hep-ph/9410372.
13. E.V. Shuryak, A.I. Vainshtein, Nucl.Phys. B199:451,1982, Nucl.Phys. B201:141,1982. R.L. Jaffe, M. Soldate, Phys.Rev.D26:49-68,1982.
14. M.V. Polyakov, C. Weiss, Phys.Rev. D57 (1998) 4471-4474; Acta Phys.Polon. B28 (1997) 2751-2764 and these proceedings.
15. M.Anselmino and S. Forte, Phys.Lett.B323:71-77,1994, hep-ph/9311365.
16. I.Halperin, A.Zhitnitsky, Phys.Rev.Lett.80:438-441,1998. hep-ph/9705251.
17. E.V. Shuryak and A.R. Zhitnitsky, Phys.Rev.D57:2001-2004,1998, hep-ph/9706316.
18. A.Blotz and E. Shuryak, INSTANTON INDUCED CHARM CONTRIBUTION TO POLARIZED DEEP INELASTIC SCATTERING. hep-ph/9710544, Phys.LettB, in press.
19. B.Grinstein and R.F.Lebed QUARK HADRON DUALITY IN THE 'T HOOFT MODEL FOR MESON WEAK DECAYS: DIFFERENT QUARK DIAGRAM TOPOLOGIES. hep-ph/9805404
20. B. Blok , M. Shifman and Da-Xin Zhang, Phys.Rev.D57:2691-2700,1998; hep-ph/9709333. I. Bigi, M. Shifman, N. Uraltsev and A. Vainshtein. hep-ph/9805241

NEWS FROM THE LATTICE

C. Morningstar
University of California at San Diego, La Jolla, California 92093-0319

ABSTRACT

A summary of some recent results from lattice simulations is presented. These include first calculations of the strangeness magnetic moment of the nucleon, three new studies of the gluon propagator, flux tube attraction in $U(1)$ gauge theory, and the static-quark potential and its gluonic excitations.

1. Introduction

Lattice simulations afford a means of studying confinement from first principles. Faster computers and more efficient simulation techniques (duality transformations, improved actions, anisotropic lattices) are allowing access to physics in previously unexplored territories: the strangeness content of the nucleon, the infrared region of the Landau-gauge gluon propagator, the long flux in $U(1)$, the rich glueball spectrum, and the static-quark potential and its gluon excitations for quark-antiquark separations as large as 4 fm. In this talk, the new lattice results presented at this conference are summarized.

2. Strangeness magnetic moment of the nucleon

The strangeness content of the nucleon is currently of much interest. A significant $\bar{s}s$ content in the nucleon can resolve the long-standing discrepancy between the pion-nucleon sigma term extracted from low-energy pion-nucleon scattering and that from the octet baryon masses. A large and negative strange-quark polarization has emerged from spin structure function studies at CERN and SLAC, combined with neutron and hyperon β decays. Substantial contributions from strange quarks in the axial-vector and scalar current matrix elements would seem to imply similar such contributions to the vector current, but this would jeopardize the $SU(6)$ prediction of the neutron to proton magnetic moment ratio, which lends credence to the valence quark picture. The SAMPLE collaboration has recently measured the neutral weak magnetic form factor in elastic parity-violating electron scattering at backward angles, and obtained the strangeness magnetic form factor by subtracting out the nucleon magnetic form factor. Elastic ep and $e\,^4He$ parity-violation experiments are currently planned at TJNAF to measure the asymmetry at forward angles to extract the strangeness electric mean-square radius.

Williams presented results[1] from the first lattice calculation of the strangeness magnetic $G^s_M(q^2)$ and electric $G^s_E(q^2)$ form factors, where q is the four-momentum

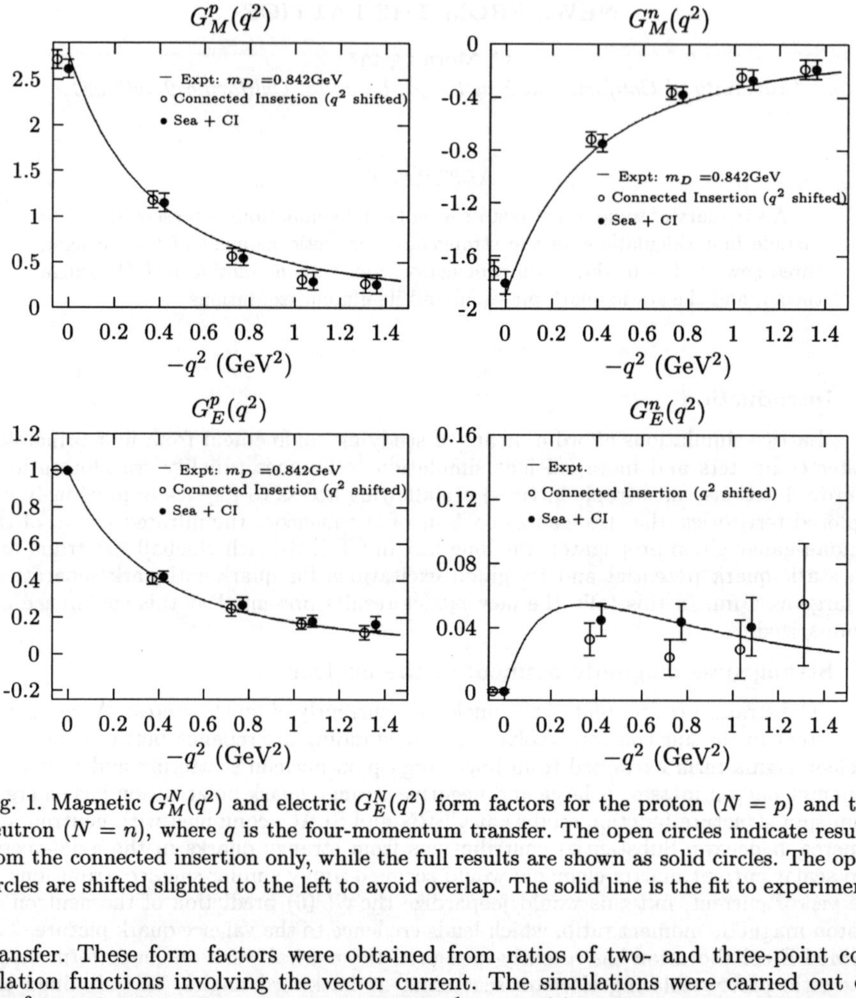

Fig. 1. Magnetic $G_M^N(q^2)$ and electric $G_E^N(q^2)$ form factors for the proton ($N = p$) and the neutron ($N = n$), where q is the four-momentum transfer. The open circles indicate results from the connected insertion only, while the full results are shown as solid circles. The open circles are shifted slighted to the left to avoid overlap. The solid line is the fit to experiment.

transfer. These form factors were obtained from ratios of two- and three-point correlation functions involving the vector current. The simulations were carried out on a $16^3 \times 24$ lattice with an inverse spacing $a^{-1} = 1.72(4)$ GeV and were possible only because of recent developments in Z_2 noise techniques for quark propagators. The calculations were done in the quenched approximation; contributions due to sea quarks were explicitly included at one fermion-loop order (the disconnected insertion diagram). The connected insertion diagram yields the contributions from the valence quarks and their Z graphs. Three values for the Wilson hopping parameter were used, corresponding to quark masses of 120, 200, and 360 MeV.

Results for $G_M^s(q^2)$ and $G_E^s(q^2)$ were obtained at four nonzero values of q^2. Extrapolating to the limit $q^2 \to 0$ using a single pole form yielded $G_M^s(0) = -0.36 \pm 0.20$ and a value for $G_E^s(0)$ consistent with zero. The electric mean-square radius was $\langle r_s^2 \rangle_E = -0.061 \pm 0.003\,\mathrm{fm}^2$. For the u and d quarks in the disconnected insertion diagram, $G_{M,\mathrm{dis}}^{u/d}(0) = -0.65 \pm 0.30$ was found. Adding all of the sea quark contributions from the disconnected insertion diagram gives $\mu_{dis} = (2/3G_{M,\mathrm{dis}}^u(0) - 1/3G_{M,\mathrm{dis}}^d(0) - 1/3G_M^s(0))\mu_N = -0.097 \pm 0.037\mu_N$, which, when combined with the connected insertion diagram, brings the ratio of neutron to proton magnetic moments $\mu_n/\mu_p = -0.68 \pm 0.04$ into agreement with the experimental value of 0.685. The accidental cancellation between the disconnected insertion diagram and the rest-frame Z-graph component of the connected insertion diagram leaves the nonrelativistic valence contribution dominant, which explains why the $SU(6)$ valence quark picture works well for the μ_n/μ_p ratio; $SU(6)$ fails in the axial-vector and scalar cases because this cancellation does not occur. The results for the form factors are shown in Fig. 1. Contributions from the strange quarks are not particularly small, but they are largely cancelled by the u and d quark contributions, resulting in a small overall effect to the form factors (the effect is greatest in the nucleon electric form factor). Issues related to the quenched approximation, chiral extrapolations, finite lattice-spacing artifacts, and finite volume errors must still be explored.

3. Gluon propagator

The connection between confinement and the gauge-dependent gluon propagator is unclear. However, knowledge of the infrared behaviour of the gluon propagator is important for various approaches to modelling confinement. Some studies based on Dyson–Schwinger equations claim that an infrared enhanced (such as $1/q^4$) propagator is required for confinement; others assume that a dynamically generated gluon mass leads to an infrared finite behaviour. Lattice simulations can tell us the true infrared behaviour of the gluon propagator from first principles. However, large finite-volume effects have prevented previous lattice studies from accessing the momentum region where the relevant nonperturbative behaviour is expected. At this conference, three groups reported on new lattice studies of the gluon propagator; two worked in the Landau gauge and one used the maximally-abelian gauge in $SU(2)$ with residual $U(1)$ Landau gauge fixing.

Williams presented results,[2] shown in Fig. 2(a), revealing new structure in the gluon propagator for momenta as small as 0.4 GeV. Three separate simulations were done in order to control systematic errors from finite lattice spacing and finite volume. They obtained results in the quenched approximation on a large ($32^3 \times 64$) lattice for a lattice spacing near 0.1 fm, which enabled them to probe momenta as small as 400 MeV. Gribov copies were ignored. They also verified the tensor structure of the propagator, given in the continuum by

$$D_{\mu\nu}^{ab}(q) = \delta^{ab}\left(\delta_{\mu\nu} - \frac{q_\mu q_\nu}{q^2}\right) D(q^2), \tag{1}$$

where q is the four-momentum of the gluon, a and b are colour indices, and μ and ν are the space-time indices. One of their main findings was to rule out a $1/q^4$ infrared

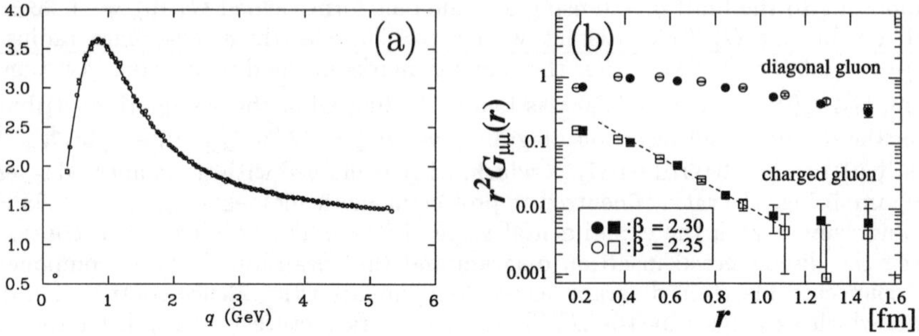

Fig. 2. (a) The Landau-gauge gluon propagator in momentum space multiplied by q^2, with nearby points averaged.[2] The line is a best fit using the form given in Eq. (2). The scale is set using the string tension. (b) The $SU(2)$ gluon propagator[3] in the maximally-abelian gauge with residual $U(1)$ Landau gauge fixing. The colour diagonal and off-diagonal components are shown.

enhancement (for the case when quark-antiquark pair creation is neglected). They fit several phenomenological functions to their data, but most resulted in very large χ^2 values. In particular, the Gribov, Stingl, Marenzoni, and Cornwall forms were all shown to fail. One of the few functions to provide a satisfactory description of the data over a wide range of momenta was found empirically to be

$$D(q^2) = Z\left(\frac{A}{(q^2)^{1+\alpha} + (M_{\rm IR}^2)^{1+\alpha}} + \frac{1}{q^2 + M_{\rm UV}^2}\left(\frac{1}{2}\ln\frac{q^2 + M_{\rm UV}^2}{M_{\rm UV}^2}\right)^{-d_D}\right), \qquad (2)$$

where $Z = 1.78^{+45}_{-20}$, $A = 0.49^{+17}_{-6}$, $aM_{\rm UV} = 0.20^{+37}_{-19}$, $aM_{\rm IR} = 0.43^{+5}_{-1}$, and $\alpha = 0.95^{+7}_{-5}$ were the fit parameters, and $a^{-1} = 1.885$ GeV and $d_D = 13/44$. This function does incorporate some ultraviolet information from perturbation theory. In the future, these authors intend to use an improved-gauge field action and increase the lattice volume in order to probe deeper into the infrared.

Results for the $SU(2)$ gluon propagator in the maximally-abelian gauge (with residual $U(1)$ Landau gauge fixing) were presented by Suganuma[3] and are shown in Fig. 2(b). Two simulations on a $12^3 \times 24$ lattice were done, but the lattice spacings were too similar to allow any serious study of systematic errors from finite spacing and finite volume. Their focus was to demonstrate that the colour off-diagonal component has a Yukawa fall-off corresponding to a large mass so that long-range physics ($r > 0.35$ fm) must be dominated by the colour diagonal component. This, they claimed, is the origin of abelian dominance in nonperturbative QCD. The dependence of the off-diagonal gluon's effective mass on the residual $U(1)$ gauge fixing was not studied. It should be noted here that abelian dominance of nonperturbative QCD has not yet been convincingly established, and that confinement is much more than the area law of the Wilson loop.

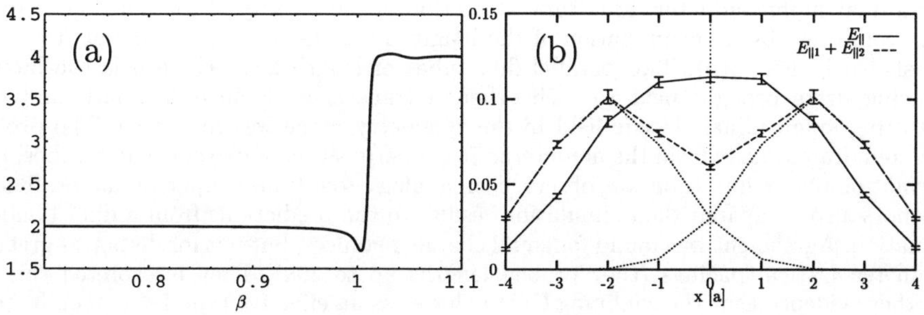

Fig. 3. Results from dual simulations of flux tubes in $U(1)$. (a) Ratio of the string tension of a doubly-charged toroidal flux tube over that of a singly-charged flux tube with respect to the coupling β. (b) Longitudinal electric field profile of two interacting flux tubes in the symmetry plane ($E_{\|}$, solid line) for $\beta = 0.96$. The equal charges are separated by $4a$, where a is the lattice spacing. Dotted lines are field profiles $E_{\|1}$ and $E_{\|2}$ for single flux tubes at $x = -2a$ and $x = +2a$. The noninteracting superposition $E_{\|1} + E_{\|2}$ is shown as a dashed line.

Nakajima and Furui[4] presented preliminary results from not only quenched simulations, but also from Langevin simulations including dynamical fermions (for three values of the Wilson hopping parameter $\kappa = 0.1$, 0.15, and 0.2). Smeared gauge fixing was used to resolve Gribov ambiguities. Their lattices were very small ($4^3 \times 8$) and much more work is needed to control systematic errors before any conclusions can be drawn on the role of quark-antiquark pair creation in the gluon propagator.

4. Dual lattice simulations of $U(1)$ flux tubes

The study of confinement in the strong-coupling phase of compact $U(1)$ in four dimensions can be viewed as a stepping stone on the way to understanding confinement in QCD. Confinement in the $U(1)$ theory is driven by magnetic monopoles whose currents squeeze the electric field into flux tubes. A dual transformation of the $U(1)$ path integral exists so that simulations can be carried out much more efficiently in the dual theory, which corresponds to a certain limit of a dual Higgs model. Results obtained in the dually-transformed theory were presented[5] at this conference, along with some first computations in the dual Abelian Higgs model. This study focussed on the question of whether the vacuum corresponds to a type-I or a type-II superconductor, and the role of quantum fluctuations of the dual degrees of freedom were investigated.

First, it was shown that the string tension scales proportionally to the charge rather than the square of the charge in the confinement phase. This is shown in Fig. 3(a) in which the string tension of a closed double flux tube is compared to that of a single closed flux tube as a function of the coupling β. These results were obtained on a $8^3 \times 16$ lattice. In the Coulomb phase (large β values), the ratio of the string tension of the double flux tube to that of the single flux tube is 4 in agreement with the expected result for the energy of a homogeneous field of double strength; below the transition in the confinement phase, the ratio is 2, in agreement with the expectations for a dual superconductor. Flux tube attraction in the vicinity of the phase transition was confirmed by an examination of the longitudinal electric field profile at $\beta = 0.96$ as shown in Fig. 3(b). Two parallel flux tubes of length $22a$, where a is the lattice spacing, were brought next to each other, a transverse distance $4a$ apart, and the resulting longitudinal electric field in the symmetry plane was measured. The profile differed dramatically from the noninteracting superposition of two single flux tubes; the formation of one flux tube was observed, in analogy to a type-I superconductor. These authors also compared their simulation results to the predictions from a dual London equation. Agreement was found for small charge distances, but not for distances greater than ten lattice spacings (they probed lengths up to $20a$). They interpreted this as further evidence that the confining $U(1)$ behaves as an effective type-I superconductor.

5. Static-quark potentials

Three groups reported on calculations of static quark-antiquark potentials; four-quark energies in $SU(2)$ lattice gauge theory were also presented.

The static-quark potential and its gluon excitations are very useful probes of confinement. It is generally believed that at large quark-antiquark separation r, the linearly-growing ground-state energy of the glue is the manifestation of the confining flux whose fluctuations can be described in terms of an effective string theory. The lowest-lying excitations are then the Goldstone modes associated with spontaneously-broken transverse translational symmetry. Expectations are less clear for small r. The gluon excitations of the static-quark potential are also useful for studying hybrid heavy-quark mesons using a Born-Oppenheimer expansion.

The first comprehensive determination of the low-lying spectrum of gluonic excitations in the presence of a static quark-antiquark pair was presented at this conference.[6] The glue energies for r from 0.1 to 4 fm were extracted from Monte Carlo estimates of generalized Wilson loops in eight simulations using an improved gauge-field action. The use of anisotropic lattices in which the temporal lattice spacing a_t was much smaller than the spatial spacing a_s was crucial for resolving the glue spectrum, particularly for large r. Finite volume errors were shown to be negligible. Particular attention was paid to the volumes for very large r: not only were checks carried out using additional simulations, but also by studying volume effects for a naive Nambu-Goto string. Agreement of energies obtained using different quark-antiquark orientations on the lattice was used to check the smallness of finite-spacing errors and to help identify the continuum rotational quantum numbers corresponding to each level (there are only discrete symmetries on the lattice). Results for lattice spacings ranging from 0.12 to 0.29 fm were obtained to facilitate extrapolation to the continuum limit. The hadronic scale parameter $r_0 \approx 0.5$ fm was used to set the scale.

The continuum-limit extrapolations are shown in Fig. 4. The energies are labelled by the magnitude Λ of the projection of the total angular momentum of the gluons onto the molecular axis, and the symmetry η_{CP} under spatial inversion with charge conjugation. The capital Greek letters $\Sigma, \Pi, \Delta, \Phi, \ldots$ are used to indicate states with $\Lambda = 0, 1, 2, 3, \ldots$, respectively. States with $\eta_{CP} = 1(-1)$ are denoted by the subscripts g (u). The Σ states are additionally labelled by a superscript $+$ ($-$) depending on whether they are even (odd) under a reflection in a plane containing the molecular axis. The

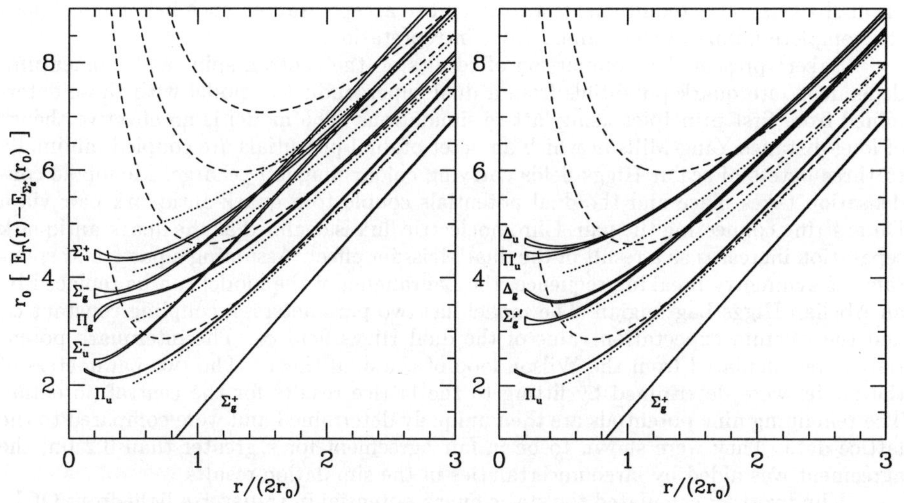

Fig. 4. Gluon excitations of the static-quark potential. The dashed lines indicate the locations of the $m\pi/r$ gaps above the Σ_g^+ curve for $m = 1$, 2, 3, and 4. The dotted curves are the naive Nambu-Goto energies in four-dimensions. Note that a Φ_u interpretation for the curve labelled Π'_u cannot be ruled out.

ground-state Σ_g^+ is the familiar static-quark potential. The lowest-lying excitation is the Π_u. There is definite evidence of a band structure at large r: the Σ'_g, Π_g, and Δ_g form the first band above the Π_u; the Σ_u^+, Σ_u^-, Π'_u/Φ_u, and Δ_u form another band; and the Σ_g^- is the highest level. The level orderings and approximate degeneracies of the gluon energies at large r match, without exception, those expected of the Goldstone excitations which are a universal feature of any effective string theory description of the long confining flux. However, the precise $m\pi/r$ gap behaviour is not observed. For separations less than 2 fm, the gluon energies lie well below the Goldstone energies and the Goldstone degeneracies are no longer observed. The two Σ^- states are in violent disagreement with expectations from a fluctuating string. These somewhat surprising results cast serious doubts on the validity of treating glue in terms of a fluctuating string for quark-antiquark separations less than 2 fm.

Some authors[7,8] claim that the glue can be described merely as a mathematical string using the Nambu-Goto action (turning a blind eye to the fact that the Nambu-Goto string cannot be consistently quantized in four dimensions). These results clearly contradict this claim. For r greater than 1 fm, the Π_u and Π_g energies disagree the least with the naive Nambu-Goto energies. In all other cases, the discrepancies are significant. The Σ_u^- and Σ_g^- levels look nothing like the predictions from the naive Nambu-Goto string. The departures of these levels from the Nambu-Goto energies are so severe that explanations in terms of mixings with other string modes or consequences of small-r symmetry requirements are difficult to accept. Rather, such behaviour signals the complete failure of the Nambu-Goto intrepretation.

Baker[9] presented a comparison of results on the central, spin- and momentum-dependent interquark potentials from a dual superconductor model with those determined from first principles using lattice simulations. The model is an effective theory of long-distance Yang-Mills in which an octet of dual potentials are coupled minimally to three octets of scalar Higgs fields carrying colour magnetic charge. Monopole condensation takes place and the dual potentials couple to a quark-antiquark pair via a Dirac string connecting the pair. Chromoelectric flux is confined as the quark-antiquark separation increases as a result of the dual Meissner effect. Assuming a particular spontaneous symmetry breaking sequence, the Lagrangian of the model can be replaced by an Abelian Higgs Lagrangian. The model has two parameters, a coupling constant α_s and the vacuum expectation value of the dual Higgs field ϕ_0. The interquark potentials were calculated from the Wilson loop of the dual theory. The two parameters of the model were determined by fitting to the lattice results for the central potential. The remaining nine potentials are then uniquely determined and were compared to the lattice data. They were shown to be in fair agreement for r greater than 0.2 fm; the agreement was aided by large uncertainties in the simulation results.

Klindworth[10] calculated the static quark potential in transverse light-front QCD. In this approach, the two light-cone coordinates $x^{\pm} = (x^0 \pm x^3)/\sqrt{2}$ are continuous and their associated degrees of freedom are non-compact, while the other two transverse directions $x_\perp$ are discretized and the gauge fields associated with these directions are expressed in terms of compact, non-unitary link variables. The calculation was carried out in the large N_c limit, where N_c is the number of colours. In so doing, quark-antiquark pair creation could be ignored. The static quark potential was then calculated using Lanczos diagonalization techniques which necessitated a Fock-space truncation. A linearly confining potential was found, and an excited potential was also determined. The calculation served mainly as an initial test of this new regularization and computational scheme.

Furui[11] presented results for four-quark energies from quenched $SU(2)$ simulations. The aim of the analysis was to obtain a compact expression for the four-quark interactions for an arbitrary spatial configuration of the quarks. They have examined square, rectangular, tilted rectangular, linear, quadrilateral, non-planar (at least one link on axis), and tetrahedral (two links off axis) geometries on a $16^3 \times 32$ lattice at two different lattice spacings. Their results suffer from significant lattice artifacts.

6. Conclusion

A plethora of new results from lattice simulations are shedding light in some of the previously dark corridors of the labyrinth surrounding confinement. Computer simulations of gluons and quarks remain an important tool for navigating this maze.

References

1. S.J. Dong, K.F. Liu, and A.G. Williams, hep-ph/9712483.

2. D.B. Leinweber, C. Parrinello, J.I. Skullerud, and A.G. Williams, Phys. Rev. D **58**, 031501 (1998); hep-lat/9803015.
3. H. Suganuma and K. Amemiya, hep-lat/9712028.
4. S. Furui and H. Nakajima, contribution to these proceedings.
5. M. Zach, M. Faber and P. Skala, hep-lat/9709017.
6. K.J. Juge, J. Kuti, and C. Morningstar, hep-lat/9809015.
7. C. Michael, hep-ph/9809211.
8. T. Allen, M. Olsson, and S. Veseli, hep-ph/9804452.
9. M. Baker, contribution to these proceedings.
10. R. Klindworth and M. Burkardt, contribution to these proceedings.
11. S. Furui and B. Masud, contribution to these proceedings.

AFTER DINNER PRESENTATION

CONFINEMENT III: AFTER DINNER TALK*

Howard Georgi
Lyman Laboratory of Physics, Harvard University, Cambridge, MA 02138, USA

When Nathan asked me to give a ten minute talk at the banquet at this conference, I hesitated. Since I am largely an outsider to the communities represented here, I wasn't sure that I would have anything useful to say. But then the "ten minute" part of the invitation triggered a thought. So I will intersperse a few serious observations about the conference and the subject with:

The top 10 reasons to study confinement and attend the confinement conference —

10 — You are tired of SUSY.

Actually, I was surprised not to see many SUSY people at the conference. Frankly, I didn't think that there were 150 theorists left in the world who were not studying SUSY. But also, I think that there may be interestng insights there. One obvious question that would be fun to discuss with SUSY people is whether the monopoles of Seiberg-Witten have anything in common with monopoles that has been discussed so much at this conference.

9 — You would really like to understand why the quark model works.

This is a semi-serious one, just because I have an old interest in the quark model. In particular, I have been interested in the recent claims of people like Glozman that pion exchange does a much better job of describing the spin splittings in the $\ell = 1$ baryons that color gluon exchange. With a student, Hael Collins, I realized that it is possible, purely at the level of the mathematics, without worrying whether you are getting the physics right, to make a completely fair comparison of the two, with a one-to-one mapping of the parameters describing the baryon masses. What I expected to happen was that neither would work very well, and that both would be required. Somewhat to my surprise, I found that pion exchange really works.[?] I don't know what this means. Nathan tells me that it doesn't mean anything.

8 — You want to play with a really fast computer and you need an excuse .

7 — You are not a control freak, so you don't like controlled approximations.

The difference between the nuclear physics and particle physics traditions always amazes me here. Trained as a particle physicist, I tend to like to control my

*Research supported in part by the National Science Foundation under grant number NSF-PHY/98-02709. Harvard preprint HUTP-98/A063.

approximations even if it means distorting the physics. A lot of the the speakers here clearly didn't care. I don't know whether this is good or bad.

6 — Reality is over-rated. You prefer "Reality".

One of the really interesting things to me about the conference was the extent to which the modelers had adopted lattice results as a source of "data". This struck me as a very healthy development, and I would like to see even more of it. It seems to me that models need to be flexible enough to describe alternative "realities", in addition to the one we are stuck with. This flexibility ought to regarded as an important check on the usefulness of the models, and it is terrific that lattice simulations are getting to the point where they can provide the virtual realities required.

5 — You want your experimental collaboration to fit on one page of the journal.

4 — You have trouble visualizing 11 dimensions, so you like your string theory in 4.

3 — Nora won't allow you to do anything else.

2 — If you understand confinement well enough, maybe you can work backwards to understand what Maldacena is doing.

1 — Where else can you mix Virginia wine with Tritium beer?

References

1. H. Collins and H. Georgi, Harvard preprint HUTP-98/A064 in preparation.